389

Dorothy K Boberg.

Evolution and Reason
— Beyond Darwin

Dorothy Kurth Boberg

Clarion Pacific Publishers

First Edition

Summary: New theory — **Complementary Systems Evolution.** Review of reasoning patterns and introduction of Boberg's new logics of paradox. Survey of the 3.8-million-year fossil record and recent anthropological discoveries. Explanation of complementary DNA processes resulting in **salmutations** (new genera from symbiotic recombinations of DNA) mediated by microorganisms through eons, mediated recently by human genetic engineering.

Composition and Typography by To-The-Point Press.
Printed in the United States of America by Delta Lithograph.
Bound by Stauffer Edition Binding Co. Inc.

Library of Congress Cataloging in Publication Data

Boberg, Dorothy Kurth 1930-

Evolution and Reason — Beyond Darwin

Bibliography, Glossary, Taxonomy, Illustrations.

1. Evolution — Philosophy 2. Biology 3. Paleontology
4. Anthropology 5. Geology 6. Physics 7. Chemistry 8. Logic
I Title

B818.B663 1993 92-75583

ISBN # 1-881804-07-0

DNA drawing on cover by Warren Schepp Graphics, based on axiel view of tenfold B form DNA by Dr. Robert Langridge, Computer Graphics Lab., University of California, San Francisco.

...symbiogenesis as a source of evolutionary innovation is not limited to the microcosm.

— **Lynn Margulis**
Symbiosis in Cell Evolution

...the reshuffling and subsequent evolution of DNA have not simply recycled the original products; they have produced wonders.

— **Stephen Jay Gould**
The Panda's Thumb

It is in these far reaches of our search for understanding of the world that we encounter deep paradoxes.

— **Gunther S. Stent**
Introduction to *Mind from Matter?* by Max Delbruck

It is enough for me to comtemplate the mystery of conscious life perpetuating itself through all eternity, which we dimly perceive, and to try humbly to comprehend even an infintesimal part of the intelligence manifested in nature.

— **Albert Einstein**
Living Philosophies

Evolution is systematic. Change and survival of organisms require such complex interactions of complementary systems that life as we know it almost defies probability.

— **Dorothy Kurth Boberg**
Evolution and Reason — Beyond Darwin

Publisher's Preface

Evolution theory is speculative, controversial, rapidly changing and fraught with dilemma and paradox. Author Dorothy Kurth Boberg's *Evolution and Reason* is a major work in evolution philosophy. It organizes and clarifies concepts. It explains apparent "gaps" in evolution sequences, resolves some conflicts between theorists and provides processes for resolution of paradox.

Natural philosopher Boberg's epochal contribution to philosophy of science and reasoning, **complementary systems evolution,** is founded on logics — new reasoning processes. It posits microbial mechanisms of evolution and introduces a new taxonomic Kingdom, "Virusa."

Evolution and Reason is three books in one volume: a perceptive survey of evolution theory, a sequence of the evolution of reason and an intriguing account of the fossil and anthropological record of the 3.8-billion-year development of lifeforms, with magnificent illustrations. Through these three, the innovations of Boberg's theories are substantially validated in her documentation.

In challenging the Aristotelian logic of science, Boberg moves **beyond Aristotle** to elucidate new reasoning patterns. In delineating practical application of Einstein's theory of relativity, she goes **beyond Einstein** in time/space concept; and her masterwork establishes a perspective of modern evolution philosophy that takes her **beyond Darwin** to **complementarity dynamics** resolution of much evolution controversy.

Scientists and professors were enlisted in the editing of *Evolution and Reason.* Each reviewed the sections pertaining to his/her own scientific and philosophic expertise, but we found not one who could assess the entire manuscript because of the magnitude of its scope. Those who were bound too tightly in the strictures of Darwinian thought were outspokenly defensive; those innovative spirits who could break through those strictures were enthusiastic in their praise. The following are excerpts from the latter:

Pre-Publication Reviews

"A brilliant mind! Perceptive analysis!"

Norman Cousins, Author/Philosopher,
Former Editor of *Saturday Review of Literature*

"*Evolution and Reason* is a mind-boggling interdisciplinary achievement. It shakes the either/or logic foundations of classical science. The focus on complementarity will expand students' perceptions and reasoning processes. Boberg's philosophy will have significant impact on education and social sciences."

Dr. Estelle Lit, Assistant Professor of Education,
California State University

"... an enjoyable and valuable resource for professors of the natural and physical sciences and their students."

Dr. John Minch, Professor of Paleontology and Geology;
California Environmental Resources Consultant

"... innovative, important overview of modern evolution theory. Boberg's extensive documentation from diverse fields, and her 165 unique illustrations, will enhance understanding. The text clarifies obscurities of scientific language, and moves the study of evolution from the narrow confines of fossils, bones and teeth to a holistic concept including man's ability to express and use intelligence."

Dr. Carol Smallenburg, Professor Emeritus,
School of Education,
California State University, Los Angeles

Clarion Pacific is proud of its role in the production of this text introducing Boberg's new theory, **complementary systems evolution.**

Preface

It is audacious and unorthodox for an interdisciplinarian to propose a new theory of evolution. Nothing is ever fully original, and yet different perspectives give rise to "new" truths.

My method is to synthesize concepts of many scientists and philosophers with my own, in the search for our long evolutionary heritage. My goal is to share kaleidoscopic perspectives in development of new reasoning patterns.

I want to displace "survival of the fittest" with the theory of ***Complementary Systems Evolution***. In the process I hope to humanize some classical scientific and philosophic theories with simplification, with enthusiasm, with a touch of poetic license and some speculative challenges to the status quo.

Dorothy Kurth Boberg

Evolution and Reason
— Beyond Darwin

Table of Contents

1

New Reasoning Power in Evolution

Entrenched Aristotelian reasoning and Darwinian theory have blocked accurate assessment of ever-changing and evolving lifeforms. Future evolution of human life on planet Earth requires transcendence of old-guard science and heed of Albert Einstein's plea for "new modes of thinking." Climb with me on thought paths beyond Aristotle, beyond Darwin and beyond Einstein.

Evolution from simple organisms to human beings is an intriguing scientific concept. Charles Darwin's principle of natural selection has become a basic premise of western biological science. Although both popular and scientific perceptions of evolution remain associated with his theory, generations of scientists have introduced modifications, alterations and expansions of evolution ideas. Darwin wrote *The Origin of Species* more than 130 years ago; it has become outdated. It is outmoded by discoveries of the 20th Century.

Evolution theory is under siege!

Darwinism Challenged

I have conceived new concepts of evolution which challenge classical theory. They do not refute the basic premise that evolution has occurred, but rather, direct new light on its *processes*. These concepts are based not only on new data developed in the 20th Century, but also on *new reasoning patterns* — new methods to assess the ideas of Darwin and others — a virtual evolution of reason.

Darwin, having studied both living and fossil animals, proposed that organisms vary over time, but he did not know the cause of this variation. In an early study of genetic variation published several years after Darwin's masterwork, Gregor Mendel revealed a hereditary gene mechanism in garden peas. Mendel's insights were generally unknown to the scientific community until their rediscovery during the early part of the 20th Century.

Years passed before several scientists combined the knowledge of Darwin and the insights of Mendelian genetics into new theory. From the 1930s to 1960s, giants in the biological sciences — Julian Huxley, Theodosius Dobzhansky, George Gaylord Simpson, J. B. S. Haldane and Ernst Mayr — produced a "modern synthesis" of evolution theory called neo-Darwinism. However, by the 1970s, serious questions about the adequacy of this synthesis as well as that of Darwin's original theory began to emerge.

In 1970, Lynn Margulis proposed a new theory of evolution: Microorganisms of the early Precambrian Period joined together in symbiotic relationships. Bacteria, chemically complex but small organisms fused and evolved into morphologically more complex, nucleated-celled lifeforms. Her concepts led to the realization that neither Darwinian nor neo-Darwinian theory is adequate to explain evolution.

Other work in genetics directly challenged the adequacy of classical theory. By 1953, the discovery of the explicit code structure of DNA (deoxyribonucleic acid) by James Watson, Francis Crick, Rosalind Franklin and Maurice Wilkins had invalidated the assumption that evolution occurs only by random variations of individuals. Their discoveries confirmed that the genetic code of life is structured by four macromolecules, nucleotides known as adenine, thymine, cytosine and guanine. Various combinations of these nucleotides are the code for each individual and species of plant and animal. This deterministic code requires reevaluation of evolution theory. It is no longer possible to posit only random phenotype variations. Variations in both the genotype (gene configuration) and the phenotype (body characteristics) are specifically coded in DNA.

In recent years, scientists have learned that viruses can reorganize the DNA of their hosts. This fact also requires new thinking about evolutionary processes. By the early 1980s, a few scientists had begun to reconcile knowledge of DNA, RNA (ribonucleic acid) and viruses with classical evolution theory. However, the new discoveries cannot readily be understood by the pattern of reasoning of Darwin nor by the logic of the scientists of the modern synthesis. The symbioses of single-celled organisms, the purposeful movements of DNA, and the complementary interactions of cells and viruses require new modes of thinking.

Adventures in Evolutionary Thought

Biological scientists are not the only ones who have been fascinated by the idea of evolution. Students of many scientific disciplines have tried through the ages to puzzle out the same questions: How did living things arise? How have they changed? What are the relationships among living organisms, and between them and their environment? We are challenged to seek fuller answers to these questions, but to do so we need to employ methods of reasoning which go beyond classical logic. The study of evolution then becomes an adventure in expanded vision.

Most scientists who have contemplated the mysteries of evolution have specialized in the study of one or several lifeforms. Few have tried to blaze a pathway through the many missing links (saltations) between bacterial organisms and human beings. Let us assume that it is possible to find such a pathway. Perhaps the search for this hypothetical trail could only inspire adventurous philosophers, generalists not chained to the rigors of a single discipline.

Just as a physical structure has a physical foundation, so evolution theory is grounded in logic. As a prerequisite to our adventure, an analysis of the logic inherent in Darwin's theory and a survey of reasoning patterns employed by other scientists and philosophers is needed. This is not an attempt to flesh out each system of logic in all its beautiful symmetry; rather, it is a presentation of essential concepts in simplified form. The purpose here is not to discuss philosophy for its own sake, but to use philosophies of logic to understand evolution of organisms and evolution of reason.

Monistic logic, the foundation of Charles Darwin's work, is the classical logic of science. This inductive and deductive approach, elucidated by Aristotle, was developed in its scientific application by Rene Descartes (although he is primarily known for dualistic reasoning). Monistic logic is a reasoning pattern based upon one subject and its parts. It is sometimes considered *mechanistic* because the parts function together as a whole like a machine. The adventurer must recognized the limitations of monistic logic before he can fully appreciate other systems of logic.

New Definition of Logic

Scientists and philosophers of evolution have used both the logic of Aristotle and systems of reasoning which do not fit the traditional definition of logic. Reasoning in perspectives of dualism, general systems theory, quantum mechanics, and relativity add new dimensions of understanding. These reasoning patterns do not supplant the basic logic of science but expand it to reach broader perspectives. It is therefore necessary to provide a new definition of logic, one which encompasses the concepts implicit in the unfolding literature of science:

> A system of logic is a reasoning pattern which expands insight into complex and-or paradoxical questions whether or not it yields proofs.

On this trajectory, new ways of thinking about evolution occur and new theories arise. Major systems of logic following monistic logic are considered in the following categories:

Dualistic logic of Georg Hegel, Henri Bergson and Alfred North Whitehead is a *dialectic,* formulated from two perspectives, a thesis and antithesis, which are then considered in their interactions and lead to a synthesis. This synthesis carries meaning beyond the concepts inherent in the two original perspectives.

Ludwig von Bertalanffy is the father of *systems logic,* reasoning patterns concerned with *interactions* of many parts of a whole. The progression from one principle or subject to the perspective of two principles, to analysis of systems with many interactions, illustrates a natural advance in complexity.

The *relativity logic* of Albert Einstein is a logic of time. Consideration of whether two events are simultaneous or consecutive is important to questions of interactions, whether of celestial bodies, atoms, or speciation events.

Quantum logic of Max Planck and Werner Heisenberg and *complementarity logic* of Niels Bohr ushered in the atomic age. Reality was no longer considered by them to be fixed and immutable. Reality has a statistical nature and involves the thinker as both subject and object on the stage of life. Relativity, quantum and complementarity logics are logics of paradox. They underlie consideration of paradoxical questions of time and space, of matter and energy. The logic of Einstein and Bohr inspired my development of new logics to explain paradoxical questions in biology and other fields.

I initiate, and present here two logics of paradox: *reciprocal time logic,* and *complementarity dynamics.* I will also define and elucidate a third, *complementary systems logic,* which has been inherent in the development of modern space technology and molecular biology. To understand the many paradoxes of evolutionm these logics are needed:

1. **Reciprocal time logic** is a corollary of Einstein's logic of time. It posits that perception may oscillate within a parameter I call a *reciprocal moment.* This reasoning process

enables us to envision interactional events which cannot be understood to occur in strictly sequential time patterns. This concept is essential to understand interactions among cells and DNA-RNA interactions within the germ cells of speciating organisms.

2. **Complementarity dynamics** is a logic of the dynamic equilibrium between two opposites or complements. It is initiated in order to visualize that opposites are not necessarily mutually exclusive. Two antithetical entities may function in complementary patterns which may not be sequentially related or divided into separate spheres of operation. The unity of complements or opposites is a concept which expands the either/or mind set of monistic thinking. It adds the flavor of continuous interaction to relationships of opposite entities and/or opposite concepts — an ever changing kaleidoscope. This synthesis of reasoning patterns is as logical as the inductive and deductive logic of classical science and leads to expanded vision of evolutionary processes.

3. **Complementary systems logic** is a reasoning pattern of the space age — the logic of interacting systems, both hierarchical and complementary. This logic illuminates the dynamics of interacting systems of Sun, Moon, Earth, tectonic motion, climate, time and evolving life.

Ambivalence and Change

Monistic logic underlying Charles Darwin's theory of natural selection restricted perception of the main means of evolution to *competitive* processes and led to misinterpretation of other processes. From the beginning, Darwin had his critics. Even scientists who wrote that the main processes of evolution are *not competitive* as proposed by Darwin, but rather are *cooperative*, symbiotic processes, have generally used monistic logic. Judging that *cooperative* processes are the main means of evolution, they made an either/or choice between the two processes. By contrast, complementarity dynamics synthesizes the dynamic equilibrium interactions between competitive *and* cooperative processes.

Monistic logic was the foundation of the biological sciences when the study of ecology emerged. New thinking patterns were evoked in reaction to the rigid requirements of this logic. A systems approach was needed by scientists who studied the multitude of interactions occurring in biological communities. Systems logic developed in ecology, as well as in other disciplines, without scientists' clear recognition of the changes occurring in their thought patterns. Consequently, studies in ecology have sometimes seemed confusing to those steeped in monistic logic.

Ambivalent uses of logic abound in the scientific literature. The logic underlying biological studies has varied, both with the subject matter and with the timing of considered events. Using monistic logic, evolutionists have evaluated the changes in animal lines, searching for patterns of descent that occur over eons of time. On the other hand, ecologists using systems logic have evaluated interactions of lifeforms during short time spans in limited locations. In the two disciplines different logics underlie the different interpretations of the *relationships* between organisms.

Principle of Responsive Salmutation

The new systems of logic presented in more detail in following chapters culminate in origination of the *principle of responsive salmutation*. I have coined the word "salmutation" from the words "saltation" and "mutation." A saltation is a jump (so-called missing link) between organisms in the fossil record. A mutation is a reorganization of DNA that can be recognized in the genotype and often in the phenotype of an organism, which is slightly different from that of its progenitor(s).

A *salmutation* is a macromutational and-or megamutational process of reorganization of DNA, sometimes mediated by viral DNA/RNA, in which a new *taxon line* is formed, in anticipation of and in response to environmental change. The concept of salmutation bridges gaps between known taxa in the fossil record and clarifies relationships among lifeforms known in the modern natural world.

The principle of responsive salmutation arises from the paradoxical fact that DNA reorganization both anticipates and reacts to environmental change. Prerequisite to understanding responsive salmutation is comprehension of the reciprocal moment, a percept presented in Chapter 6. With this perspective it becomes possible to envision the processes by which new organism lines, which can live in the varied environments of earth, are created. The idea of DNA reorganization in relation to environmental change may also require recognition of intelligence functioning through natural processes.

Initiation of New Genera

Although mutations may pattern new species of a genus by means of recombinations of DNA at the gene level, truly different organisms, new taxa lines, new genera may be patterned by recombinations of DNA at either the gene or chromosome level, and viral DNA/RNA may be an essential component of this recombination.

Whether organisms will, in the future, be classified as a different variety (race) or species, based upon a mutation, or in a different genus based upon a salmutation event, depends upon the degree of DNA change and other criteria used by systematists.

Scientists associated with the neo-Darwinian theory conceived that *gradual* changes from one taxon line to another occurred by means of gene mutations. The concept that major changes of organism *lines* may occur at the genus level, or above, has not been generally recognized, although this had been anticipated by Richard Goldschmidt in the 1960s and by a contemporary and friend of the systematist Linnaeus as early as the 1700s.

Peter Artedi, Linnaeus' friend and collaborator, organized his system of animal classification with an emphasis on genus rather than on species. Many systematists currently concentrate their efforts on changes evident at the genus level of fossils. However, the connection between reorganization or insertion of genes on different chromosomes (or change in chromosome configuration or number) and origination of new taxa lines has not been fully appreciated.

This emphasis on genus, especially in vertebrates, and coordination of the genus classification with gene and chromosomal recombination events requires a new theory of evolution, one based on the principle of responsive salmutation. This is a principle essential for progress in the fields of both genetics and evolution. This and other principles will be described and illustrated in *Evolution and Reason* by survey of selected microorganisms, plants and animals of the time period from 3.8 billion years before present (3.8 BBP) to modern day. The new concepts and perceptions spark questions and challenges for evolution studies into the 21st Century.

2

The Logic of Science — Monistic Systems of Logic

Logic is a system of reasoning leading to a conclusion, a synthesis, or expanded insight. It is not necessarily a system of proof, as complex questions are often not subject to one final proof. Rather, logic is here considered to be the study of patterns of reasoning which enlighten the mind, whether or not a specific conclusion is reached. Different systems of logic help to elucidate diverse areas of human understanding and none should be neglected.

The evolution of reason, from Plato to present, began with monistic logic — the foundation logic of science. The Western World generally credits the Greek culture of Plato and Aristotle with the transition from superstition and primitive thought to systematic reasoning. The logic called *inductive* did not flow full-blown from the pen of Aristotle. Indeed, Democritus, Anaxagoras and many others contributed to its development. When Aristotle gathered together the wisdom of his age, new modes of reasoning arose. Sometimes, as a practical matter, it is necessary to credit a concept to the individual who made it well known, rather than trace all its antecedents and contemporaries, and this work follows that pattern.

Monistic Logic of Aristotle (384-322 BC)

Inductive Logic

Monistic logic entails one subject and the parts which compose it. One Aristotelian logic is a system of *induction* in which observations are made about the known parts of a whole. This logic was later called *mechanistic* because, like a machine, all the parts fit together to form the whole, and if the parts are known the whole is known. In the following example the whole is the classification "birds;" the parts are the various kinds of birds.

To begin inductive reasoning, facts are gathered. A collection of birds may be made; perhaps robins, blackbirds, doves, storks and falcons. Other information about birds is gathered. The facts are then analyzed and an hypothesis is proposed. What is a definitive characteristic of all birds? Do they all have wings? If so, a hypothesis may be stated: "All birds have wings."

This hypothesis is the basis for further investigation, and if no birds can be discovered which do not have wings, the hypothesis is considered to be valid. However, the essence of inductive logic is that there is always the possibility that a wingless bird may be found. Therefore, the search for the wingless bird (the null hypothesis) continues. The original hypothesis is not considered to be proven. It remains a theory. This logic has been the basis for much of science from the time of Aristotle to the time of Einstein and remains the logic of much scientific work today.

Implicit in the concept of induction is the requirement that choices be made. It is necessary to determine whether an individual bird specimen has or does not have wings. In the above example, either wings are present or they are not present; either they exist or they do not exist. It must be determined as an either/or choice.

Aristotle's "ultimate doctrine" and "most certain of all principles" is that the same thing cannot *be* and *not be* at the same time. He said that contraries cannot at the same time belong to the same subject. If the contrary of an opinion is the negation of that opinion the same person cannot believe the object *to be* and *not to be.* On the surface, this seems reasonable and few would dispute it, but when Aristotle's corollary of his most certain of all principles is presented, the logic of induction and either/or choices is seen in a different light.

The corollary of Aristotle's ultimate doctrine is the law of the excluded middle: "...there cannot be anything between two contradictories..." An idea "must either be asserted or denied." Therefore, according to Aristotle, if anyone says that anything *is* or that it *is not,* that person either speaks *truly* or speaks *falsely* (Aristotle, 1966: 58, 70).

This logic, by allowing no middle ground between a subject and its contrary or contradictory, does not allow (in the above example) a concept of partial wing, incipient wing, atrophied wing, or variously expressible (wing) genes. There is no gray area between black and white. When induction is presented as a scientific method, it is usually recognized that it is a format for hypotheses and theories. However, the fact that the logic requires either/or choices without possibility for a spectrum of choices, is not always recognized. Inductive logic is a system for gathering and analyzing data. It is not a system which leads to a firm and indisputable conclusion.

Deductive Logic

Aristotle recognized that the conclusions of inductive logic are tentative. He, therefore, described a logical system in which the conclusion necessarily follows from the premises. The conclusion is true if the major premise is true and all-inclusive, and the subject of the minor premise is a part of the subject of the major premise. This is *deductive logic.* The classical example is the syllogism:

Major premise: All birds have wings.

Minor premise: A robin is a bird.

Conclusion: Therefore, a robin has wings.

The problem with the logic is: How is a major premise developed which is both true and all-inclusive? Aristotle said such a major premise could be developed using *inductive logic*. The concept that birds have wings, which is substantiated by extensive evidence, and which has not been refuted, can be stated as a major premise for a deductive syllogism. There is still not certainty. The question remains whether the major premise will always be found to be true, but until it is proven to be false, it can be used as a major premise in deductive reasoning.

Thus, inductive and deductive systems may be considered together to improve the quality of a conclusion. Their separate deficiencies are not always recognized. Induction does not lead to an indisputable conclusion, and deduction leads to a true conclusion only when the major premise is true and all inclusive, and only when the subject of the minor premise is a part of the subject of the major premise. Therefore, when the logics are combined in one process, the conclusion remains subject to continuing uncertainty and requires additional verification.

Another essential element of Aristotle's logic is that it is impossible to demonstrate a truth if one does not start from the major premise, or the intuitive foundation and link one's reasoning in a chain until one reaches the conclusion. Aristotle believed that each science has its fundamental premises which "must be assumed as a condition of the science and cannot be proved by the science" (Evans, 1977: 34). Thus, although the conclusion from a study using inductive logic may be used as the major premise for deductive logic, other unprovable fundamental premises may be needed as foundation for a science. These premises may have a source apart from the logical system.

The history of science has shown that these fundamental premises have become so much a *given* of each scientific discipline, that they often remain unchallenged for generations. In this work, I will identify some of these fundamental premises of evolution theory and evaluate them.

Monistic Logic of Rene Descartes (1596-1650)

Rene Descartes, a French citizen, introduced modern philosophy and science of thought at the age of 24. He is considered a *dualist* by some because he introduced the concept of the dual nature of mind and matter; that these two concepts encompass the whole of reality. His philosophy effectively separates consideration of the functioning of the mind and the body or matter, into two separate realms of reality. Therefore, I believe his logic of science, which in essence considers only the realm of matter, should be considered to be *monistic*.

His major contribution to science is a method of inquiry which breaks up a problem into its constituent parts and arranges them in order. This order consists of stating a first principle which requires no other principle, and then arranging all other concepts so that their proof depends solely on an ordered sequence. This is known as *linear inference* and its essence is unilateral and deductive (Roth, 1937: 71-78).

Linear inference is a cause and effect line of reasoning. From a few self evident principles consequences are deduced which are the propositions of science (Whitehead 1934: 78). Descartes believed that unilateral deduction from first principles is applicable to biology as well as to physics.

An example of this reasoning is:

> The sun (cause) supplies energy (effect) for plants. This energy (cause) enables plants to develop chlorophyll and the process of photosynthesis (effect). Photosynthesis (cause) powers the processing of carbon dioxide and water into oxygen (effect). This oxygen (cause) provides a major ingredient of the air (effect) in which animals may survive.

Linear inference is a valuable tool for reasoning, but it can be seen from the above that the concept of *causality* is inadequate to understand the totality of complex scientific questions. The sun's energy cannot be considered to be the cause of photosynthesis, nor the cause for development of an aerobic environment, although it is an essential factor.

Descartes acknowledged that there is a gap between general laws deduced by linear inference and the particular facts of each case, but he also maintained that as science progresses, facts will be developed in ever greater numbers. He believed that, as more facts become known, the weight of evidence will prove or disprove the basic principles of science. As with the logic of Aristotle, Cartesian logic, despite its major contribution to scientific thought, does not offer indisputable proofs.

Descartes believed, as Aristotle did, that the refusal to accept a self-contradictory proposition is innate. The human mind requires that a choice be made between contradictory alternatives. Nothing can be both black and white, dead and alive, present and absent. He believed that human reason insists on an either/or choice between contradictory alternatives (Sahakian, 1966).

Charles Darwin made many monistic either/or choices when he wrote *The Origin of Species*. Although Hegel had lived and written prior to Darwin's publication, there is little evidence that Darwin considered Hegel's dualistic logic when writing his own manuscripts. Many of the other philosophers and scientists, whose logic is considered here, had not been born when Darwin first published his theory of evolution. Dualistic logic of Hegel and others, the first breaks from the strictures of the monistic either/or mind set, gives an expanded dimension to concepts of evolution.

3

Dualistic Systems of Logic

In contrast to monistic, *dualistic logic* begins by positing two opposing statements or principles. Their relationships are then considered and evaluated. This reasoning leads to a synthesis, that is, some combination and-or expansion of the two poles and the interaction between them. Georg Hegel, Henri Bergson and Alfred North Whitehead were three philosophers who developed systems in this tradition.

Georg Hegel was a German whose form of logic was not inherent in the English and American foundation patterns of science. However, it is becoming evident that the quantum logic of the international atomic scientists is an expansion of dualistic logic, with a debt owed to Hegel, Bergson and Whitehead. The logic of complementarity dynamics also has roots in these earlier dualistic systems.

Dualistic Logic of Georg Hegel (1770-1831)

Hegel's system of logic uses a conversational method. It is a dynamic reasoning process called a *dialectic*. The Hegelian dialectic starts with a thesis, which is a statement of fact. It is opposed by an equally assertable but contradictory statement, the antithesis. The two are reconciled in a synthesis.

Hegel's dialectic can be applied where the thesis and antithesis are in the form of opposites which appear, on the surface, to be irreconcilable. Thesis: this animal is alive because it is breathing; antithesis: this animal is dead because it is not breathing; synthesis: this animal is hovering between life and death because its breathing is irregular and discontinuous.

The thesis, antithesis, synthesis succession provides a means for reconciling opposites. It implies that objects or ideas which may appear to be contrary to each other, in fact may be two poles of a larger truth. This larger truth shares aspects of both poles but is new and different from each pole. The whole is more significant than the sum of its parts. The systematized whole is *Truth* (Hegel, 1975).

This system of logic is *dualistic* because truth stems from two sources, often opposite sources. Aristotle and Descartes require a choice between opposites, and construct a causal chain of reasoning to reach a conclusion. Hegel acknowledges opposites as components of the interrelatedness of all things. Even if ideas or objects cannot be related as parts of a whole, they are still related by synthesis of opposites.

Dualistic Logic of Henri Bergson (1859-1941)

The dominant logic in evolutionary thought in France during the latter half of the nineteenth century was Darwin's mechanistic, or inductive approach. Henri Bergson was steeped in Darwinism in his early years but later found the logic inadequate to explain evolution. When he wrote *Creative Evolution* in 1911, he turned to a dialectic approach.

Dialectic had its roots in the thought of early Greeks, as did inductive and deductive logic. Various reasoning patterns arose early and have reappeared in different forms through time. Bergson did not use dialectic in the Hegelian tradition, with rigidity of a synthesis emanating from a thesis and antithesis. Rather, he conceived a *fluid dialectic* in which he compared and contrasted two concepts, then reached a conclusion which might be a synthesis or might be beyond synthesis, evolving from the original two concepts.

Unlike Darwin, who embraced the idea of *chance*, and rejected the concepts of purpose, finality, or plan in evolution, Bergson believed that neither chance nor purpose was a singularly adequate concept. He conceived that life processes demonstrate a "creative impulse," which encompasses both chance and purpose, with an element of freedom.

He saw life as movement, and movement as creative (Gallagher, 1970). This concept put him in conflict with the accepted mechanistic theory of evolution, which considers life to be material, subject only to *chance* interactions with other life, with the possibility of chance mutations, beneficial or detrimental.

Bergson also contrasted concepts of time and duration and applied them to evolution theory. He considered the idea of time accepted by scientists of his day to be too rigid. This concept of time required the use of intelligence to mark off discrete moments which could be described as static elements of a dynamic reality. Bergson believed that humans have an intuition of duration that evolved from instinct, rather than originating from intelligence. He believed that intuition of duration is necessary to understand the continuous creation of unforeseeable novelty, which evolution demonstrates.

Bergson expressed necessity for change in the classical concepts of time, perhaps not realizing that in Switzerland Albert Einstein had already published new concepts of time in the field of physics. Bergson also posited a philosophical concept of *indeterminacy* before Werner Heisenberg conceived the idea in relation to the movement of electrons in physics. Bergson said there is an element of indeterminacy or fluidity in consideration of time and duration, and thus in the reality of events.

He also posited the idea of *complementarity*, probably predating Niels Bohr's development of the concept. Complementarity is similar to synthesis in that it unites two opposites into one whole. However, the opposites under consideration by Bohr were usually very complex problems of physics. Bergson considered that immobility and flux are complementary concepts. He said that matter and mind are complementary and should not be considered mechanistically. Matter and mind are complementary reality, both appearing as perpetual becoming.

Although Bergson created great enthusiasm for his insights among intellectuals of his day, the scientific community clung to mechanism as the logic of science and considered the ideas of Bergson to be metaphysical, which he, indeed, agreed was the case. Metaphysics, after all, is the study of the theoretical principles underlying any particular science. Bergson believed there was need to revise the theoretical principles underlying evolution theory.

Dualistic Logic of Alfred North Whitehead (1861-1947)

Like Bergson, Whitehead conceived the relationship of two opposing concepts as *interactive,* anticipating the complementarity concepts of Niels Bohr. He said that for some philosophers, mind is the sole reality and matter is appearance; that for others, mind is ephemeral and matter is the sole reality. He recognized the mind-body relationship as one of the oldest and most persistent problems of philosophy.

Whitehead (1934) believed that the viewpoint of Descartes on this dualism was a disaster for the philosophy of science. Descartes considered that there was matter with spatial extension, and mind which was external to matter or the body, that neither required the other for its essence, and that the concept of interrelationships was unnecessary to consider their separate existences.

Descartes' view that space and matter are the fundamental subjects of science dominated the field for four centuries. Whitehead believed that this view had finally given way to the concept of *process.* Whitehead conceived process as a complex activity which changes the emphasis from structure to function. Thus, he argued, the laws of nature are all-pervading patterns of behavior, though many of the discontinuities and shifts remain a mystery. Physical nature is henseforth to be considered as a complex of *interrelationships.*

Whitehead (1934: 63, 76) identified the dual character of sense perception as "... partly irrelevant to the body and partly referent to the body." He visualized the person as a unity of perception and reality. He saw the human individual as "one fact, mind and body" and considered that the individual also formed a unity with the environment. He believed that two great failures of 19th Century science were: disregard of the relationship of each organism to its environment, and disregard of the intrinsic worth of the environment.

The contribution of Whitehead (1925: 198) to the logic of science is his insistence that opposing viewpoints be reconciled by looking for consideration of *both* "comprehensive truths and finer perspectives." He stated, "Wisdom is the fruit of a balanced development."

Understanding evolution requires dualistic logic as well as monistic logic. Other logics are also needed.

4

Systems Logic

Systems logic deals with multiple strands of related information as interacting wholes. It facilitates understanding of the complexities of various life processes such as those evident in embryological development. It improves understanding of interactions of lifeforms in colonies and communities, as exemplified in the study of ecology. It may be used to consider evolutionary questions. Many of the techniques of systems reasoning were not available to Darwin. His conclusions, therefore, lacked the comprehensiveness of this approach.

Systems Logic of Ludwig von Bertalanffy (1901-1972)

Ludwig von Bertalanffy, an Austrian biologist and philosopher, is known as the father of *General Systems Theory* (GST). Although he called his work "general systems teaching," and others called it "a new world outlook," I consider it to constitute several systems of logic.

The *basic* logic of Bertalanffy (1968) consists of systematic organization of principles and methods of reasoning. He did not delineate one formal logic, but rather elucidated principles which consider the whole, not merely as the sum of its parts, but rather as complexes of elements functioning in interactions. He discussed the hierarchical nature of things, from atoms to the universe. It seems inevitable that his principles and methods will eventually be organized into more formal systems of logic.

In 1925, Bertalanffy said that the prevalent mechanistic approach in biology appears to neglect or deny the *essence* of the phenomena of life. He advocated organismic biology which considers lifeforms as interacting, organized systems. As he experimented with metabolism and growth, he advanced the concept that organisms are *open systems*. To study an open system, one in which outside resources are incorporated and expelled, it is necessary to devise models which address the problems of order, organization and wholeness. This evolved into systems logic.

Classical science uses analytical procedures which require that the organism be divided, literally or figuratively, into its constituent parts which can then be studied as causal trains of individual units. For example, the body of an animal may be dissected to determine the cause of disease in an organ or cell mass. The diseased part is excised, studied, and perhaps inserted into another animal to test whether it causes disease there. The effect is duly noted in the scientific literature. There is no doubt that this method has been effective in biology and medicine. However, Bertalanffy believed that such analysis is inadequate for study of many problems in biology and other disciplines.

Systems concepts, developed in other fields during many years, were brought together in Bertalanffy's *General Systems Theory* (1968). He identified approaches such as calculus, computerization and simulation, cybernetics, information theory, theory of automata and game theory. He thought these methods are important, but he viewed his own work as a synthesis beyond them.

Bertalanffy's Principles of Systems Logic

General systems logic can be characterized by the following principles (Bertalanffy, 1968; Davidson, 1983):

1. Comprehensive Natural Laws

There are natural laws of organization governing systems on all levels, and these laws can be organized into a comprehensive philosophy. These are laws of interactive relationships within and between wholes. Bertalenffy's aim is to comprehend *wholeness*.

According to Mark Davidson (1983: 23) Bertalanffy felt that the complexity of biological expression "requires more than mechanistic analysis, more than cause and effect analysis, more than the concept that understanding the parts will lead to understanding the whole."

Bertalanffy recognized that the laws of organization he elucidated are not all-inclusive, nor can they be applied in every case, but he wanted to begin the process of defining them. He believed that the fundamental task of biology is to discover the laws of biological systems: the laws of organization of an organism's parts and processes which govern survival of the whole. He advocated that the search for such laws be interdisciplinary, because laws that apply to the study of biology are applicable to many other disciplines.

2. Hierarchy

Reality is hierarchical. Bertalanffy saw the universe as a tremendous hierarchy — from elementary particles to atoms, to molecules, to microscopic structures of cells, to complete cells, to organisms, to organizations of individuals, to organization of the planet, to the solar system and to the universe. A similar hierarchy applies to both structures and functions. In the physical world, energy transforms into matter, leading to growth and development of biological structures that express a flow of ever more complex processes.

He found hierarchies evident in every subject he approached, and considered it necessary to visualize these relationships in subsystems, systems and superordinate systems. The human body may be recognized as a hierarchy of a superordinate nervous system governing such systems as digestion, respiration and musculature, all composed of organ and cell subsystems.

3. Open Systems

Living organisms are open systems, as contrasted to the closed systems of conventional physics. A closed system is one which is isolated from its environment. It is characterized by the chemical equilibrium that occurs in a closed container following the interaction of the contained elements. It is governed by the second law of thermodynamics; the concept: in a closed system the process moves from a condition of available energy to unavailable energy (as when coal burns) — a state of chemical and thermal equilibrium. The second law is a statement about *entropy*: a closed system tends toward a state of most probable distribution — disorder or disorganization (i.e. the probable distribution of blue, green, and red marbles dropped randomly into a container is disorganized).

Bertalanffy conceived that every living organism — as an open system — maintains itself in a dynamic chemical and thermodynamic equilibrium, while experiencing continuing inflows and outflows, building up and breaking down components. Although an organism may look relatively the same over time, its cells are continuously being replaced. For example, the life span of a human respiratory or digestive tract cell is only a few days. Such rapid turnover is a state of high statistical improbability. The essence of life is metabolism, a dynamic equilibrium process; the essence of evolution is the tendency toward increasing order and complexity. Living organisms are examples of negative entropy or "negentropy."

In closed systems the final state is determined by the initial conditions e.g. the positions of the planets at a certain time can be determined from the positions at a previous time, or, in a chemical reaction, the final concentrations of the various reactants depend on their initial concentrations. The final state will be changed if the initial conditions or processes are changed.

In an open system, on the other hand, the same final state may be reached in different ways from different initial conditions. Bertalanffy called this *equifinality*. Examples abound in the study of embryology. A normal animal of some species can develop from one fertilized ovum, from each of the first two cells of a divided ovum, or from fusion of two whole ova. In addition, each half of a mature individual such as a planarian worm, can regenerate a lost half and become a complete individual again. An example of equifinality in botany is R. H. Whittacker's (1953) description of very different initial vegetations developing into very similar climax forests.

The transition toward higher order, organization and originality is characteristic of open systems in embryology, botany and evolution. Bertalanffy believed that the open systems model presents principles and methods for understanding not only biological concepts such as embryology and the climax formation of biomes, but also such diverse subjects as neurology, transactional philosophy and evolution.

4. Integration

Reality is integrative. The concept of *wholeness* permeates Bertalanffy's logic — not just the wholeness of one system but the wholeness implicit in the integration of ever extending wholes. He stated that, in any realm of biological phenomena, whether embryology, metabolism, growth, or consideration of life in a biome, the behavior of an element is not the same in isolation as within the system.

A cell on a petri dish cannot be understood in the same manner as a cell operating within the tissue. The relations between the system (the cell) the subordinate systems (the molecules) and the superordinate systems (the organs and the body) have to be taken into account to understand the behavior of the cell.

He believed that an organism, whether one-celled or a human being of great complexity, manifests an important wholistic dimension of existence. This can be lost in analysis and identified in synthesis. The organization of the parts of an organism and its processes is as essential as the parts and processes themselves. His unifying principle is that organization occurs at all levels.

5. Interactions

Reality is interactional and dynamic. The parts and processes of an organism must be exactly harmonized to maintain dynamic equilibrium. Interactions occur throughout nature and are harmonized at ever higher levels of complexity. Bertalanffy identified a principle of centralization, expressed in biology as the emergence of a leading part, such as the brain. Progressive centralization occurs with progressive individualization. Thus, organisms evolve from simple to complex, becoming more integrated but also more specialized, with a brain actively dominating the interactional system.

Bertalanffy saw the interactional hierarchy extending from cells to organs, organs to organisms, and organisms to organizations beyond individuals, with each level harmonized in dynamic equilibrium. He criticized modern science for stressing research on isolated units acting in one-way causality, and recognized the appearance in all fields of science of the concepts: wholism, organismic models and gestalts, which demonstrate the necessity to think in terms of systems in mutual interaction.

6. Relativity

Reality is relative; there is no absolute time or space. Bertalanffy understood that the concepts of open systems, integration, and interaction could not be constrained in causal chains of linear reasoning. He also recognized the problems inherent in systems reasoning — not only the complexity of dealing with large numbers of variables, but also the problems of considering interactions without becoming immersed in identifying sequences. He conceived that wholes are composed *instantaneously*. He was critical of the feedback loop characteristic of cybernetics, because he considered it a circular causal chain and thus limited in its application to general systems.

Bertalanffy perceived that interaction in a system reaching equilibrium means that interactions may be a certain time or distance from a final state. The system, then, seems to aim at that future state or may be considered to depend on the idea of a future final state.

The idea of *finality* or purposefulness is a distinguishing characteristic of religion and metaphysics but an enigma to scientists of many generations. This may be an unacceptable prospect for some, although there are many scientists who have held the idea of finality or purposefulness, in logic-tight compartments of their minds for a lifetime.

Bertalanffy accepted the implications of finality because he reasoned that open systems are accessible to scientific interpretation and open systems *require* the concept of purposefulness. He believed that science must take account of finality since it represents an essential feature of reality.

He perceived, however, that the concept requires new understanding of the relationship of time and space. He believed that finality can be linked to Einstein's theory of relativity. He did not develop this completely, but said, "The biologist finds that there is no absolute space or time but...they depend on the organization of the perceiving organism" (Bertalanffy, 1968: 229).

Methods of Systems Logic

General systems logic is characterized by a number of different constructs or methods. These are models representing certain perspectives of reality. Bertalanffy said that there is no unique and all-embracing world system. The various systems approaches are models of different aspects of reality and they are not mutually exclusive. They are often applied in combinations and further syntheses are desirable.

W. R. Ashby (1958) said that systems studies generally can begin either with consideration of the regularities of observed systems or with consideration of the set of all conceivable systems, and then reduction of the set to a functional size. In 1960, Bertalanffy outlined an intuitive series of principles and envisioned an interplay of intuitive, experiential and deductive procedures. He warned of allowing theoretical models to become closed and definitive too early, as systems logicians still grope for foundations.

It is beyond the scope of this work to evaluate the methods of General Systems Theory. The following is a summary of some of the methods discussed by Bertalanffy (Davidson, 1983):

1. **Calculus** — simultaneous differential equations and allometric analysis (ratios between variables).

2. **Computerization and simulation** — facilitates calculations with a large number of variables, models reality in mathematical formulas.

3. **Graphs** — represent relational structures in topological space.

4. **Cybernetics** — uses block and flow diagrams of regulatory mechanisms.

5. **Information modeling** — demonstrates negative entropy.

6. **Automata modeling** — uses algorithmic machine for input, output, trial and error learning.

7. **Game modeling** — uses system of antagonistic forces.

8. **Decision modeling** — uses choices among alternatives.

Bertalanffy envisioned the use of mathematical models as fundamental. He found that an astonishingly broad range of phenomena fits the pattern of the allometric equation which represents a ratio between variables. For example, in a large number of animals the ratio between basal metabolic rate and unit of body surface is 2/3. Bertalanffy called this a "highly simplified approximate formula" but thought that this and similar formulas are essential modeling tools.

He also said that the value of purely qualitative models should not be underestimated. He recalled the impact of Darwin's theory, which changed the general frame of reference of science, and the Copernican revolution, which allowed ancient problems to be seen in a new light. Bertalanffy originated a number of new qualitative models based on his principles.

An understanding of the quantitative and qualitative modeling techniques is not necessary for appreciation of the underlying principles of systems logic and its relation to other thought systems. Systems logic, although still in its infancy, is a harbinger of future logical development. Its use is helpful to understand evolutionary events.

5

Relativity Logic

Henri Bergson and Ludwig von Bertalanffy both identified the need to consider the *relativity of time,* to understand the interactions of events in biology and evolution. Although it has been said that few understand Albert Einstein's concepts, Einstein wrote a book for the general reader, free of mathematical formulas, which is accessible to those who desire to learn the meaning of the relativity of time (Einstein, 1961).

Relativity is considered here to be a *system of logic* because it develops a pattern of reasoning — leading to valuable insights about the relationships of events — which is inaccessible via other reasoning patterns.

Charles Darwin's concept of time was *linear,* meaning that events are recognized to occur in order from past, to present, to future. He considered that time is *absolute* — the same for all observers of phenomena — i.e. an event happened at two o'clock on June 10th, with no concern about a time difference in different locations. The understanding of time which has emerged from the work of Einstein raises questions of which Darwin was unaware, about the timing of speciation events.

Relativity Logic of Albert Einstein (1879-1955)

The logic of relativity is a logic of time and space relationship. Like Darwin, most scientists prior to Einstein considered time to be linear and absolute. Time was measured by movement of sunlight around the Earth's surface (or after Copernicus, by the movement of the Earth around the Sun).

The problem that all observers shared the same time designation, was partially solved by the Greenwich Convention. This Convention set time zones to encapsulate an average time for individual areas between Earth's meridians so that an hour of time was measured between each zone. As it focused on the hourly difference between zones, this Convention tended to obscure the fact that Sun time was slightly different for areas closely adjacent.

The difference between time zones caused few problems to those who stayed at home (and off the long distance phone lines). Time seemed the same for them and all their neighbors. But travelers experienced time problems, especially those traveling at jet speeds. To gain a day in traveling half-way round the Earth from east to west, or to lose a day traveling from west to east must be experienced to understand the disorientation. Jets had not been invented when Einstein considered the problems of time measurement. But he anticipated through thought problems that time was very different for travelers than it was for a stationary observer.

Prior to Einstein, the logic which expressed the relationship of an observer to the fixed time of an event was monistic, the logic of the one-way causal chain of events. The world was viewed from the position of the only observer who mattered: "I," "me." The Cartesian "I think, therefore I am" expanded to: "I am here, therefore all time is measured in relationship to my position."

Einstein, on the other hand, was able to think of *two* observers, one fixed and one moving, and how they would perceive time. He did not conceive of jet aircraft, but mentally compared an observer traveling in a train — moving close to the speed of light — to an observer standing in a fixed position next to the tracks. To mentally view position and time from *two* perspectives — rather than one — is rewarding. Overcoming a long history of viewing motion and time from one's own single perspective is not easy, but it is exhilarating.

If one seeks to understand evolution, one must comprehend the Einsteinian logic of time. The key to understanding is the idea of the *relativity of simultaneity*: that two events which appear to be simultaneous (happening at the same time) to one observer, may not appear to be simultaneous to another. The second observer may observe the events to occur in succession. Therefore, the time of the events depends upon the location of the observer.

To illustrate the concept of the relativity of simultaneity, Einstein (1961: 25-26) conceived: an observer which I shall represent as O, standing on an embankment at point O, next to a train track. O is standing there and a train is traveling at a velocity near the speed of light (v velocity) moving toward the east. In the center of that speeding train sits E in a position to see all around him, perhaps in a bubble car. This person, E, will observe all events from the perspective of his train. Observer O will observe all events from the perspective of his position on the embankment (Figure 5.1).

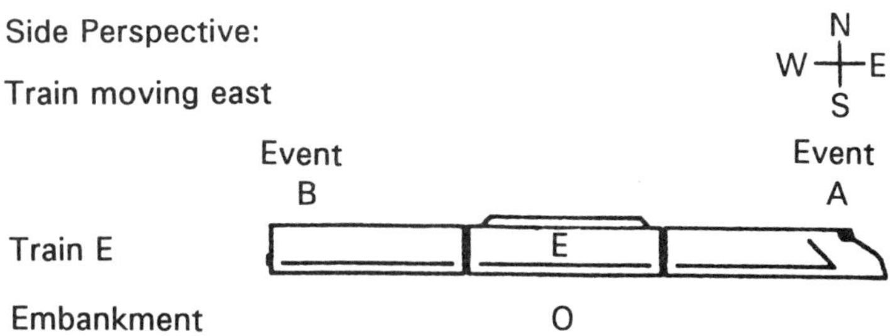

Figure 5.1 Adaptation of Albert Einstein's concept of the relativity of simultaneity. (After Einstein).

The question then arises: Are two events (e.g. two strokes of lightning A and B) above the train — seen as simultaneous by the observer O on the embankment — also simultaneous relative to E on the train moving speedily toward the east? It would appear that the answer is no.

When it is said that lightning strikes A and B are simultaneous as observed by O on the ground, this means: the rays of light from points A and B, respectively, toward the observer at point O meet each other at O at the same time, having traveled the same distance. This assumes that light moves at a constant speed, as Einstein believed.

The events A and B also relate similarly to E traveling on the train. Just when the flashes of lightning occur (as judged by O) point E, where E is sitting, coincides with point O on the ground, but it moves toward the east with speed v of the train. E (considered with reference to O) is moving toward the beam of light from A and rides ahead of the beam of light coming from B. Therefore E will see the beam of light emitted from A earlier than he will see that emitted from B. E will therefore conclude that lightning flash A took place earlier than lightning flash B.

The events which appeared to be simultaneous to O are not perceived as simultaneous when viewed by E, but rather are considered by E to be *sequential*. Thus, the time of an event depends upon the position of the observer. Time is relative to the position of the observer.

It would appear that each observer has his own time perception. Einstein (1961: 26) wrote, "...unless we are told the reference-body to which the statement of time refers, there is no meaning in a statement of the time of an event." Although Einstein believed his thought experiment of time illustrated a more accurate idea of the nature of time than the classical concept that the flow of time is experienced in the same way by all observers, he nevertheless recognized that a human observer usually perceives time from past to present to future in an irreversible direction.

Einstein's relativity logic allowed time to be considered from the perspective of two observers, instead of only one, and opened vast new horizons for the understanding of the universe. It is a logic which is difficult to apply by those who have lived a lifetime of understanding the world from one time perspective — their own. But it is a logic which continues to provide fruitful understanding in a universe which can be viewed from many perspectives. It is essential to those who measure and evaluate events occurring in the cosmos, in the atom, and in the cell.

Einstein developed the concept of the relativity of simultaneity as a *thought experiment* — much later it was used to understand the motion of sub-atomic particles in atomic experiments. As we explore the philosophical implications of this concept, we are moving from a principle which has been applied in physics and mathematics, to its broader implications. Whether we cross a theoretical line from physical *reality* to philosophical *non-reality* or to *expanded reality* — as we extrapolate to the events of speciation — is for the reader to judge. To envision the movement of sub-atomic particles is to contemplate relationships in time and space. Similar thought experiments envisioning relationships in time and space of electrical impulses, nucleotides, and speciation events are equally rewarding.

Einstein's logic borders on *paradox* — the splitting of consciousness into two or more realities which seem to be both true and false. The paradoxes of time and space boggle the mind if one thinks only in monistic logic. *Logics of paradox* are required to evaluate paradox. Einstein expanded concepts of time and space through his relativity logic. It is necessary to go further, to explore a corollary of Einstein's logic and consider other logics of paradox.

6

Logics of Paradox

Paradox has not usually been considered to be logical. On the contrary, paradox has often been considered to be *irrational*. The dictionary (Random House, 1987) defines paradox as "a statement or proposition that seems self-contradictory or absurd but in reality expresses a possible truth." This definition implies *monistic logic* as the standard of rationality but accepts a *possible truth* in paradox. *Possible truth* is in fact *real* truth if the system in which it is embedded is understood. A paradox is logical if it is evaluated by a logic of paradox.

In the past, when a paradox appears it has often been identified as such. "That's a paradox!" The mind boggles and snaps shut and the subject is changed. Paradox does not fit. Those paradoxical concepts that both *are* and *are not* cannot be considered in the Aristotelian rational that something either *is* or *is not*.

Paradox becomes evident in Einstein's concept of the relativity of simultaneity. How can time be different for different observers? How can one observer simultaneously envision one time for himself and a different time for a second observer? To do so requires a change in attitude about what is rational, a change in the basic pattern of thinking about time, and an expansion of the concept of logic.

To understand evolution, involving numerous events occurring simultaneously, in sequences, and in a time-frame lasting more than three billion years, it is necessary to gain the perspective of many observers in different times and places, not just one observer looking backward from one time and place. The timing of speciation changes within the cell requires a new perspective of time, reciprocal time, that is presented here as a logic of paradox. I have devised this logic to reason about the paradoxes of cellular events.

Logic of Reciprocal Time

If Albert Einstein had considered *two* trains, moving in opposite directions, in his thought example, he would have illustrated a logic of reciprocal time. He did not choose to do so. He believed that time was not reciprocal. Nevertheless, there are some calculations of atomic processes which can only be explained as events which occur in a reciprocal time frame. Therefore, a logic of reciprocal time, a concept of mental oscillation as one considers the timing of events, is needed. In the following hypothetical example, time may be thought to flow either forward or backward depending upon the perspective of the observers.

To illustrate the concept of the relativity of simultaneity in reciprocal time, consider two trains instead of one, on parallel tracks, moving in opposite directions, with an observer, O, standing in the area between the two sets of tracks. One train, E, is moving *east* at constant speed v, and train W, is moving *west* at constant speed v. Points E and W are in the center of the respective trains. The observers at points E and W are sitting in their bubble compartments, where they can gain a clear view of the whole sky area. Each observer, traveling in his bubble compartment moving near the speed of light, will regard all events by reference to his train. Simultaneity can be defined relative to either train, just as it is defined with respect to an observer O between trains on the right-of-way (Figure 6.1).

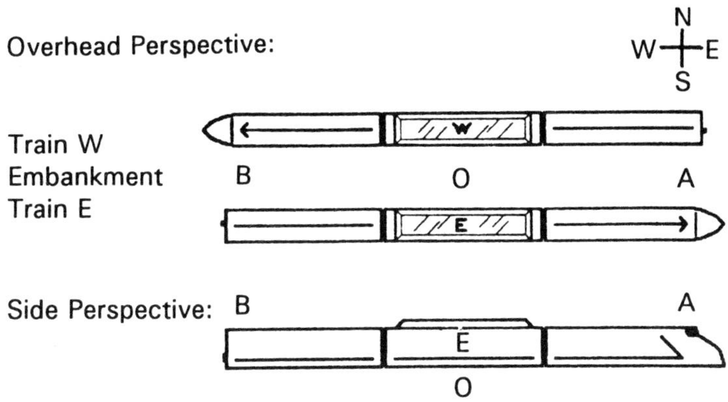

Figure 6.1 Author's concept of reciprocal time and the reciprocal moment.

The question arises: Are two events (e.g. two strokes of lightning A and B, above the trains) that are simultaneous as seen by the observer O on the rail right-of-way also simultaneous relative to the respective observers on each train? It would appear that the answer is no.

When it is said that lightning strikes A and B are simultaneous with respect to observer O on the ground, this means that rays of light from point A and B, respectively (toward the observer at point O) meet each other at O at the same time, traveling the same distance. (This assumes that light travels at a constant speed.)

The events A and B also relate similarly to positions W and E on the two trains. Let W and E be observers at the midpoints W and E of the traveling trains. Just when the flashes of lightning occur (as judged by O) point W coincides with point O, but it moves toward the *west* with speed v of the train. W (considered with reference to O) is moving towards the beam of light from B and is riding ahead of the beam of light coming from A. Therefore, W will see the beam of light emitted from B earlier than he will see the light emitted from A. W will therefore conclude that the lightning flash B took place earlier than lightning flash A.

On the other hand, just as the flashes of lightning occur (as judged by O) point E coincides with point O, but point E moves toward the *east* with speed v of the train. E (considered with reference to O) is moving towards the beam of light from A, and is riding ahead of the beam of light coming from B. Therefore E will see the beam of light emitted from A earlier than he will see that emitted from B. E will therefore conclude that lightning flash A took place earlier than lightning flash B.

Events which appear simultaneous to O are *not* perceived as simultaneous when viewed by either W or E. W has seen lightning B first, then A. E has seen lightning A first, then B. It would appear that the order in which an observer understands events to occur may be simultaneous, may be first A, then B, or first B, then A. Events, from the perspective of the observer, other than oneself, may seem to be the reverse of one's own time reality.

From this example of the concept of the relativity of simultaneity in reciprocal time, a new logic of time and space relationship emerges. The principle is that *there is no single fixed reality* such as time flowing from past, to present, to future. Events may occur in order 1,2,3,4 or in order 4,3,2,1 or in order 2,4,1,3 or in any other order, depending on the position and velosity of observers in relation to those events.

In order to understand the concept of reciprocal time, consider the moment when O, W and E each perceive the light sequence. Consider all three perspectives together. This is the *reciprocal moment*, a concept of interactional time.

For an understanding of this idea on the feeling or sensational level, think of an experience of reciprocal motion. Consider a person, E, sitting in a stationary train, looking out the window at W, who is sitting in another stationary train, on a parallel track, looking at E. One or both trains are about to move forward, in opposite directions. Assume that there is no sound, nor jerking motion, but rather a smooth and quiet beginning of motion. When one or both trains begin(s) to move, E and W will each believe with a degree of uncertainty, that it is his own train which is beginning to move, whether it is E's train, W's train, or both trains. If O were standing between the trains, he would determine the motion of the trains by means of relationship to his fixed position. This moment of perception of movement by E and W, with uncertainty about which train(s) is/are moving, would be the feeling accompanying the experience of the reciprocal moment.

Reciprocal time does not negate the reality of an individual's general experience of events occurring in the order 1,2,3,4. Rather, it shows that reality may be experienced in different order by observers in different locations. The latter perspective is not needed to understand the timing of usual events in the course of an individual human lifetime. However, it may be required to understand certain atomic and cosmic events, and *instantaneous* events occurring within a system such as the complementary movements of nucleotides within a cell, "spontaneous" mutations, or communication among cells. The logic of reciprocal time is a logic of expanded vision of close relationships in time and space.

To understand this, the observer must readjust his/her reality patterns. The experience for each observer is sequential for his/her own observation and appears to be reverse sequential for the other's observation. The reciprocal moment is the time frame of this experience.

In addition to understanding the nature of time, the seeker for the truth of evolution needs to understand the nature of the physical universe. It is no longer possible to ignore the advances of physics as they relate to evolution theory. The matter of organisms is subject to both laws of physics and logic of the quantum.

Quantum Logic of Werner Heisenberg (1901-1976)

The quantum logic of Werner Heisenberg (1983) is another logic of paradox. The paradox which Heisenberg elucidated reflects the nature of the electron, one of the basic constituents of matter and energy.

Quantum logic emerged into scientific consciousness in 1900 when the German, Max Planck discovered the "universal quantum of action," the principle: at atomic and subatomic levels of existence, the energy of atoms does not flow evenly but rather in steps or quanta, and these quanta are not continuous but are *discontinuous,* discrete packets of action.

This reasoning challenged Newtonian mechanics, the scientific system that was grounded in monistic logic. The idea of discontinuous motion could not be understood in the logic of classical science. A new logic was required. The quantum hypothesis of Planck explains the black body law of radiation, the kind of radiation occurring in the discrete discharge of energy from a heated metal object. But in doing so, it threw classical science into disarray. How could science cope with such a concept of discontinuity?

The principle of the quantum of action was virtually ignored until five years later when Albert Einstein identified *light* as quanta, later called *photons,* which move as Planck had postulated. Einstein, in 1905, devised an equation which describes the photoelectric effect. After Arthur Compton and Peter Debye independently confirmed the scattering of photons from the electron, the *photon theory of light* was recognized — and Einstein received a Nobel Prize.

In a way, this concept of light quanta seemed to be an expansion of Sir Isaac Newton's corpuscular theory of light (1672): a beam of light consists of a stream of minute *particles* which impinge on the retina of the eye to create the sensation of light. However, since 1678, there had been *two* theories of light. Christian Huygens, a Dutch mathematician, physicist and astronomer had postulated that light consists of minute *waves* traveling in a pervasive ether from a source to the eye. In the 18th Century Newton's theory had been accepted as the major theory of the physics of light. In the 19th Century Huygens' theory was persuasive in physics.

The quantum theory of Einstein replaced the corpuscular theory of Newton and took precedence in the field of physics at the beginning of the 20th Century. However, it was recognized that there was a major body of phenomena which could only be explained by the quantum theory — and a major body of phenomena which could only be explained by the wave theory (although the concept of an ether in which waves moved was later abandoned). The two theories were *contradictory* but both *necessary*, an unacceptable situation for classical science and classical logic, which have always sought to eliminate contradictions in the determination of truth.

At this point a young German physicist, Werner Heisenberg, stepped onto the stage of science history. He considered the challenge of calculating the position and momentum of a sub-atomic particle. He imagined the relationships between position and momentum of an electron moving from a source to a destination: from two slits opening into a box to a screen at the opposite end of the box.

His calculations showed that the *momentum* of the electron could be determined, but at the cost of inablity to predict exactly where the electron would be at any particular time, or where on the screen it would land. He could calculate where the electron would be or where it would land, at the cost of inability to determine its exact momentum. He found that the measuring devices needed to determine momentum actually changed the *position* of the electron, and the instruments which could determine the position changed the momentum.

The electron, like the photon of light, demonstrates a particle function in its momentum and a wave function in its position. Heisenberg learned that the greater the precision of measurement of one aspect, the hazier the precision in the measurement of the other; the more he could prove the momentum feature of the electron, the less he could identify the exactness of its wave features.

This discovery led to the amazing idea that intervention of the observer in measuring momentum actually changed the electron's wave patterning, and measurement of its wave patterning changed its momentum. The quantum function or the wave function was uncertain apart from the observation of it. Man's measuring actions affected the actual events — heady stuff for a 23 year old physicist in his first year as a Doctor of Philosophy. Nevertheless, Heisenberg's principle of indeterminacy or uncertainty was confirmed by the atomic scientists of his day and entered the mainstream of quantum mechanics, a concept precursor to atomic bombs and atomic reactors.

The logic of quantum mechanics is a logic of uncertainty, the antithesis of the supposed meaning of logic prior to the atomic age. In dealing with the world of the infinitesimal, of atoms and atomic particles, a new logic was needed. This logic requires flexibility, an open mind, not easily acquired by those who have employed inductive/deductive logic for a lifetime.

How many wish to move from a logic of relative certainty to a logic of uncertainty? Only a few! Even among physicists, there are those who disagree with Heisenberg and Niels Bohr. Uncertainty logic requires the confidence that statistical probabilities define atomic reality, that it is not possible to say that energy or matter can be measured in its exact motion or be located in its exact position at the same time.

Some physicists agreed with Einstein that "God does not play at dice" (Heisenberg, 1983: 116). They have continued the search for the certainty they hope to find beyond quantum logic. It is ironic that Einstein's concept of light quanta opened the minds of Heisenberg and other physicists to the concept of the uncertainty principle while Einstein himself could never accept it as a complete theory, as the *essence* of the way in which the atomic world really functions.

Some of those who use only logics of relative certainty, when faced with the numerous paradoxes of life, refuse to think further on the subjects. But there are those whose curiosity allows them to venture wherever their thoughts lead — even into a logic of uncertainty if that opens understanding of paradox.

Nature is full of paradoxes — of choices which may seem to be mutually exclusive, which seemingly cannot both be correct, and yet they *are* both correct in their own spheres and as two aspects of relationship in larger wholeness.

Quantum logic is available to those willing to explore the unknown as subject to statistical evaluation of its essence, as coexistent potentialities, as possibly forever indeterminate phenomena, an adventure not to be missed by seekers for truth.

The quantum logic of Heisenberg, Planck, and others is also the logic of Niels Bohr. However, Bohr used the designation "complementarity" while Heisenberg used the designation "uncertainty" to describe the basic relationship of the described events. Use of the word *complementarity* changed the perspective of the concept and made it more understandable. Complementarity logic is here considered as another logic of paradox.

Complementarity Logic of Niels Bohr (1885-1962)

The great Danish physicist Niels Bohr (1958) was 15 years Heisenberg's senior, his teacher, mentor and collaborator. Complementarity logic and quantum logic can be considered to be their viewpoints of quantum phenomena, called the "Copenhagen Interpretation." While Heisenberg described the movement of the electron in his imaginary box, as if it could be viewed from a position at ring-side or at eye level, Niels Bohr described the concept of *complementarity* as if it were possible to view phenomena from a higher perspective, above the phenomena, so to speak.

Bohr was fascinated by the idea of discontinuity of quantum phenomena. How could light move discontinuously? How could energy and matter demonstrate both quantum and wave properties? Bohr completed his doctoral thesis in Copenhagen in 1911, and began work with Ernest Rutherford in Manchester, England. He searched for answers to these questions by attempting to visualize a model of the simplest and lightest atom, the hydrogen atom, which Rutherford believed contained a nucleus orbited by a single electron.

Bohr knew that if the electron released photons of light continuously it would expend its energy and spiral, spent, into its nucleus. It had to release photons in separate packets, or quanta, in order to continue an orbit around the nucleus. Bohr conceived the idea of different orbits or shells around the nucleus. As a quantum of energy was released, the electron jumped into a new orbit closer to the nucleus. There was no continuous movement of the electron from one orbit to the next; the quantum jump was discontinuous as Planck and Einstein had theorized.

The concept of an electron, orbiting a nucleus and making quantum jumps, was most productive in identifying the process of atomic release of energy and led to the successful prediction of an atom's size. Atoms are so small that millions of them can reside on the head of a pin. If an atom of hydrogen were the size of a football stadium, its nucleus would be the size of a grain of rice, with the possibility that the electron could be located at the circumference of the stadium. Bohr received a Nobel Prize for his discovery of electron orbits.

He conceived that the discontinuous quantum jump from one orbit to another which applied to atomic phenomena, corresponds to appearance of action on the larger scale of human observations, as humans view apparently continuous actions in discontinuous pictures, like movie still-frames. This principle of correspondence between continuous and discontinuous motion seemed to Bohr to be fundamental to understand other paradoxical phenomena.

The paradox of continuity and discontinuity and the paradox of quantum and wave properties of energy and matter must be understood, according to Bohr, as *wholeness*. They need to be understood as two fundamentals of wholeness. Wholeness in this sense, is not two parts of a whole, but rather the unification of the paradox. Modern authors are seeing the correspondence of this concept to the Yin and the Yang of ancient oriental philosophy. The whole is more that the sum of its parts; it is a wholeness with the potential to join the world of previously inexplicable paradoxical relationships.

In order to visualize both aspects of paradox, Bohr developed the concept of *complementarity*. He was more aware than most scientists that it is important to translate the mathematics of new discoveries into common language. He felt that, while mathematics describes atomic phenomena in more concise and accurate terms than language, it is only through language that the ideas can become generally meaningful. He was acutely aware that classical concepts are so embedded in language that such words as matter, energy, particle, field, wave, orbit, spin, etc. have to be used to explain events which do not strictly belong in such categories.

Atomic phenomena are so different from events occurring in normal experience that the classical vocabulary is inadequate to explain them. Nevertheless, it is all that is available to make the concepts known, thus must be used, albeit with loss of mathematical precision. Bohr (1958: 29) defined his words carefully, stating, "...the lesson which we have received from the whole growth of the physical sciences is that the germ of fruitful development often lies just in the proper choice of definitions."

Let's discuss and define "complementary." The word complementary has long been used to explain relationships among colors on the color wheel. Red and green are complementary colors. Red is a primary color, and its opposite or complement is green made of two primary colors, blue and yellow. Why are red and green colors considered to be complementary? It is a relationship in which the combination of the two complements represents a wholeness, but each complement is fundamentally different from the other. Each requires a different set of criteria to evaluate its essence.

For example, to evaluate red, only one ingredient must be analyzed. The red powder is composed of all the same red element. But the green of the color wheel is composed of a designated proportion of two primary color powders, and therefore requires a different criterion for its analysis, a criterion of proportion as well as a criterion of color description. In addition, complementary color ingredients combine to create the wholeness of the color brown, whether the combination is red and green, orange and blue, yellow and violet, or combination of all the colors on the color wheel.

Another identifying characteristic of complementary colors can be demonstrated by viewing them as lights. If a person stares at projection on a white background of a red light, and the light is turned off, there is an after-image of green, indicating that there is a basic underlying relationship between the complements of red and green that transcends the physical presence of green light. Also, combination of all primary and secondary colors of light, or two complementary colors of light, create white light. This is the essence of the concept *complementary*, which Niels Bohr used to explain quantum phenomena.

Bohr expressed the relationship between quanta and waves of light as complementary. Each manifestation is seemingly mutually exclusive. Traditionally, either light was particles which could be identified as existing in certain places at certain times or light was waves which move in patterns. In classical understanding, light could not be both. Yet in a profound way, it is both. Whether light is perceived as quanta or waves depends on the way a human sees it or measures it. Light is potentially either quanta or waves; it is both quanta and waves. It is complements of quanta and waves. Expression of light's essence is complementarity.

The concept which Werner Heisenberg expressed as the *uncertainty principle* developed in a thought experiment, Niels Bohr expressed as *complementarity*, a logic comprehended in a flash of insight as a method of resolving paradox. Bohr (1958: 74) said, "The history of physical science...demonstrates how the exploration of ever wider fields of experience, in revealing unsuspected limitations of accustomed ideas, indicates new ways of restoring logical order."

In 1927, the giants of physics, Albert Einstein and Niels Bohr met at the Solvay Congress in Brussels, Belgium. Heisenberg had just completed the transition to a statistical interpretation of quantum theory, and Bohr had developed the powerful concept of complementarity. The evidence and mathematics of complementarity excluded the possibility of reverting to a causal, deterministic, pictorial description of atomic phenomena. It was henceforth to be described by statistical evaluation of complements.

Einstein remained unwilling to concede that atomic reality was fully expressed in the statistical interpretation of the uncertainty principle. He preferred to believe that it would become possible to measure *exactly* both momentum and position of atomic particles, if the right formula were discovered.

Heisenberg (1983: 117) reported that each morning for several days, after in-depth discussions of the subject, Einstein appeared at breakfast with a new and more complex thought experiment to refute the quantum logic of Bohr and Heisenberg. But each time his arguments were evaluated, it was recognized that quantum logic still held sway. In the end, Einstein and Bohr agreed to disagree, and Bohr responded to Einstein's "God does not play at dice." by saying, "But still, it cannot be for us to tell God, how he is to run the world."

Niels Bohr believed that the logic of complementarity, conceiving opposites as complements of wholes, should be applied in many fields, especially in biology and psychology. He and Heisenberg both believed that life represents wholeness and that mankind is responsible for the care of the whole of life on earth.

I combine the complementarity of Niels Bohr and my concept of reciprocal time in the following elucidation of complementarity dynamics. This is also a logic of paradox.

Complementarity Dynamics Logic

Complementarity dynamics is a logic that delineates the relationship between two entities (things), an entity and a concept, or two concepts which:

1. appear to be contradictory, opposites, antitheses, and-or paradoxical;

2. require different criteria or instruments to judge their essential characteristics;

3. are perceived to function reciprocally in dynamic equilibrium;

4. complete each other as a whole;

5. comprise one whole in a web of reciprocal unities constituting ever larger wholes.

The complements interact in an ever-moving balance between them. The interactional area, in constant change, may be understood to be, at any moment, anywhere along a continuum between the two complements, or at either end. The interactional area occurs as a function of a *reciprocal moment*, in which the complements may interact in opposing time frames. This reciprocal moment is a moment of interactional time, momentary suspension of sequential time, oscillation, a discontinuity, or as in the example of the two trains, the combined viewpoints of two observers moving near the speed of light in opposite directions. It is this recognition of the complements of time and space, in which complementarity dynamics operates.

The *dynamic equilibrium* between two complements can be illustrated by thinking of the motion of one's own train of thought in past, present, future sequence in an instant and thinking of another person's train of thought in an opposite perspective of sequence, in an instant so close as to seem that the events are simultaneous. The understanding moment may reside with either or both trains of thought. Flexible minds may experience flashes of insight at reciprocal moments.

Complementarity of Two Entities

The idea of two entities is illustrated by the relationship of two structures, such as two complementary strands of DNA, the complements of one DNA strand and its messenger RNA strand, the complements of joined egg and sperm, of nucleous and cytoplasm, of body cell and germ cell. Knowledge of complementary relationships expands as new information is gained about the natural world.

Complementarity of an Entity and a Concept

An entity such as the brain and a concept such as the mind cannot be separated, as postulated by Descartes. In complementarity dynamics, the brain and mind are visualized as interactional, integrated and whole, as well as complementary to the body and the operational ecosystem.

Complementarity of Two Concepts

To define complementary interaction of two concepts, consider the *structure* and *function* of a robin's wing. The wing has a structure of bones radiating from the skeleton of the central part of the robin. The bones of the wing are light and porous and contain blood-making bone marrow. Muscles are attached to the bones. Nerve cells are evident in the musculature. Arteries, veins and capillaries are structures of the blood system.

Skin covers the wing and holds in place the intricately designed feathers of various sizes, shapes and kinds. Each of the millions of cells in the wing has a complex structure of nucleus enclosing chromosomes surrounded by cytoplasm with various complexities in it. Altogether, the structure of the robin's wing is a marvel of nature.

The complement of the wing structure is its function. The cells operate to build and repair themselves. The bones operate to hold the wing in position. The muscles operate to move the wing in flight. The arteries, veins and capillaries carry blood containing oxygen and nutrients to each operating cell. The nerve system operates the muscles. The skin encloses the operating wing, and the feathers function to retain heat, and as air foils in flying.

The concepts of wing structure and function require different means of analysis. For structure, one must consider physical entities, their substance, their position, their pattern, their hardness or softness, their composition etc.

A different criteria is needed to consider function. The physical matter is present, but the study of function is a mental operation. One must conceive of the movement of bones, muscles, skin, feathers, blood, nerve impulses and cell operation in relation to the whole movement of the robin's wing.

The criteria for structure is static; the criteria for function is dynamic. Different criteria and-or instruments are required to measure structure and function. Each may be defined separately, but who would deny that a bird's wing is a *whole* and the complement of a larger whole, the robin?

The complements of structure and function operate together in dynamic equilibrium. Their interactions occur in a moving balance. The interactional area occurs in a reciprocal moment in which structure and function are understood as a flying robin.

Darwinians have asked the question: "Which came first, the wing or flight?" In complementarity dynamics, this is an odd question! Wings and flight are complements of a flying animal, the whole animal in its structure and function.

This logic is a reasoning pattern in which two complements are considered to interact in reciprocal moments. Each complement may be considered separately. However, their *essence* is a whole, which interacts in reciprocal moments of continuities and-or discontinuities in the network of wholes.

Complementarity dynamics consists of a choice of entity or concept and search for its possible complement. The choice is analyzed to determine relationship, looking for interactions in reciprocal moments when time no longer seems to be sequential. It is in these moments when discontinuities no longer appear to be irrational that paradox becomes clear.

The logic of complementarity dynamics is instructive to consider the implicit questions of discontinuities in evolution. Sequences of natural occurrences are obvious in nature. That discontinuities occur in reciprocal moments is not so obvious. In fact, they are so hidden that they have been, until recently, understood only intuitively, not explicitly. The term "reciprocal moment" places in our vocabulary a concept that makes discontinuities understandable.

Consideration of complementarity dynamics leads easily into a logic in which complements are conceived as systems joined together. This is an expansion from simple to complex, from individual to many, from two complements to systems of complements. It is a synthesis of Bertalanffy's ideas of systems logic and complementarity dynamics.

Complementary Systems Logic

This additional logic is essential to complete the paradox logics. Hierarchy is the essence of Bertalanaffy's systems logic, thus his reality is linear and monistic. Reality also encompasses the non-hierarchical. The complementarity, the combination of hierarchical and non-hierarchical systems is also reality. The paradoxes of complementary systems are resolved in wholeness of a multiplicity of systems.

The scientific literature gives evidence of developing understanding of the complementary interactions between separate systems, i.e. biology and chemistry in biochemistry, paleontology and ecology in paleoecology, etc. An elucidation of complementary interactions of many systems together is very complex. Nevertheless, there are many systems in nature which are combined in both hierarchical and multidimensional wholes.

The human body is hierarchical. The mind governs the whole. This is an essential and significant fact which must not be neglected. There is clear evidence that the mind is in charge. It communicates with all systems. It coordinates, makes decisions, informs and controls. Nevertheless, the body functions semi-autonomously in many of its systems. Systems function together in complementary interactions without continual intrusive, conscious hierarchical participation by the mind. A level of subconscious participation occurs. It alerts the conscious mind when problems arise, but much of the functioning of the body is not usually in conscious awareness.

The 10 quadrillion or so cells of the human body are each an elaborate functioning system which become specialized into different body systems during development from the fertilized egg in embryogenesis. Cells specialize and become the nervous system, circulatory system, structural-bone system, endocrine system, epithelial system and the other systems. Through complementary systems logic, the interactions of all these systems is envisioned.

The logic of complementary systems is a structure and process of reasoning in which systems are compared and contrasted in their interactions. It is not a logic of syllogisms and proofs, but rather a logic of interactional processing of the data of structures and functions, of relationships.

Systems can be physical systems or conceptual systems such as systems of cause and effect, competition and cooperation, continuity and discontinuity, patterning and randomness, matter and energy, innovations and constancy, space and time, in all their interactions.

The difference between this logic and complementarity dynamics is the number of systems involved. Two systems are considered in complementarity dynamics. More than two interacting systems are characteristic of complementary systems logic. For example, if all the relationships among the systems employed for space flight or all interactions of the participants in an ecosystem are studied, then complementary systems is the logic tool needed.

The stunning feat of space flight to the Moon was not only a triumph of physics and the human spirit, it was a triumph of the unidentified use of the logic of complementary systems. The rocket systems were built as complements to the space capsule. The space capsule was built as a complement of its human occupants. The internal guidance systems were built as complements to the ground computer systems. The triumph was the construction, coordination and the comprehensive functioning of a phenomenal number of complex interacting complementary systems.

A spin-off of the technology of space travel is this identification and exposition of complementary systems logic. It is interesting that this logic was used to plan for humans to leave the Earth before it was used to understand evolution of life on Earth. Its use to understand evolution is a next step.

Use of Various Logics

Of the many patterns of thought available, inductive/deductive logic is most useful when the problem at hand is considered as a structure with definable parts, and when a causal chain is an adequate concept for the actions perceived. It is deficient for consideration of complex interactions, dualistic events, spectra of events, questions of *instantaneous* time, cosmic, atomic and cellular events.

Dualistic logics are useful to explore those events which represent opposite concepts in human time frames. Each dualistic pattern has its unique approach to certain evolutionary questions. Bergson's emphasis on the importance of biological originality and innovation should not be overlooked by evolutionists. Dualistic logics help to break out of the constraints of ego to consider two or more approaches to a question. They assist in consideration of those problems which do not lend themselves to *static* analysis of parts of a whole. Dualism opens the mind to the basic *dynamics* of life processes.

The systems logic of Bertalanffy is recognized for its great potential to quantify various biological phenomena. It cannot substitute for intuitive evaluations of questions of quality but it is most helpful to grasp relationships of quantifiable phenomena. It is a logic useful to gain *macro* perspectives of reality in interactions. There is danger in considering it definitive, however. Systems reasoning should be supplemented by classical logics which consider obvious causal chains as well as *micro* events that may be lost in the massive dynamics of systems.

Einstein's logic of time and my counterpoint, reciprocal time logic, should not be overlooked when evolutionists consider the electromagnetic activities of central nervous systems, events which occur in cosmic and atomic time frames, so-called *simultaneous* events, and especially events which occur within the cell and among cells. These are difficult logics because humans are steeped in the personal experience of sequential time. It is necessary to break out of this pattern if evolutionary studies are to benefit from the insights of physics as we move into the 21st Century.

Quantum or complementarity logic is difficult because it evolved from consideration of atomic phenomena which are too small to see, difficult to visualize, and almost impossible to define accurately in classical language. Nevertheless, this logic is required to understand cosmic and atomic phenomena. Since lifeforms consist of atoms, those who discuss evolution need to be aware of the logic of atoms, quantum logic.

The logic of complementarity dynamics with its concept of the reciprocal moment facilitates comprehension of interactions between two complements. It is a system to reconcile paradox — to understand how opposites, contraries, contradictories and paradoxes are, in essence, wholes.

The logic of complementary systems is capable of bringing understanding to the complex interactions of systems such as those of the cosmos and space travel, of the body, of cellular and DNA phenonmena, and interactions of lifeforms in evolution.

To offer a variety of systems of logic to scientists already faced with vast choices of subject matter, compounds the complexity of their tasks, but it also offers kaleidoscopes of new insights that challenge the disciplines of philosophy and science.

7

Charles Darwin's Theory of Evolution

The great genius of Charles Darwin was evident in his recognition of the basic evolutionary meaning of the facts gathered by himself and others in geology, paleontology and biology. Geology was in its infancy, as Charles Lyell had only recently written *Principles of Geology* when Darwin began his Beagle voyage. The fossil record was extremely limited in 1859 when *The Origin of Species* was first published. It was mainly the study of domestic animals and the lifeforms Darwin encountered on his voyage which led him to conclude that plants and animals changed and advanced in complexity over time, that species gradually evolved.

In 1798, Georges Cuvier had published the first natural system of animal classification in France, founding comparative anatomy and paleontology. Cuvier saw a continuity of development of lifeforms; but also emphasized the discontinuities, and proposed the idea of catastrophes during which many types of life perished. Long before Darwin, Jean Baptiste Lamarck had proposed that organisms evolve, responding to environmental changes. There were others, especially Erasmus Darwin and Alfred Wallace, who recognized processes of change and advancement of life, but Charles Darwin wrote the definitive book of his time, in which he not only enumerated many facts of evolution but also posited an exhaustive theory to explain them.

When Darwin's theory became accepted by the scientific community of his day, Lamarck, Wallace and many others were eclipsed, and the catastrophism of Cuvier was rejected in the euphoria of "survival of the fittest."

Although the concept of evolution has been generally accepted, Darwin's theory of the *means* by which evolution occurs continues to be debated, challenged and blamed, as well as being supported and praised. The theory of natural selection has not been displaced but challenges and revisions are endemic and will continue. The climate is right for new challenges at this time.

Charles Darwin's theory of evolution is brilliant and fascinating but it is inadequate to explain the myriad fossils and animals he described. The deficiency is not with his observations and recording of data, but with his logic.

Darwin (1872: 367) believed that in the competition for life, the fittest survived and evolved. He was convinced that natural selection was the "main, but not the exclusive means of modification" of species, that there were other means of evolution. He maintained that he was misrepresented by those who said he attributed the modification of species "exclusively to natural selection."

If natural selection were the main means of modification, what were the other means? He said in the sixth edition of *The Origin of Species* that he formerly underrated both the direct action of physical conditions and spontaneous variations, "which lead to permanent modification of structure independently of natural selection." In addition to the competition between individuals and-or species, other effects of the environment such as climate and food were also factors in modification, and a third very important element was spontaneous variation.

He discussed these factors briefly, but their significance was generally lost in the plethora of discussion about natural selection. Discussion of spontaneous variations was very limited because Darwin had little or no knowledge of genetic principles. In 1878 Samuel Butler said: "To me it seems that the 'Origin of Variation' whatever it is, is the only true 'Origin of Species.'" W. Bateson added, "...and of that not one of us knows anything" (Haeckel et al, 1909: 107).

Even in his consideration of species relationships, Darwin acknowledged gaps in his knowledge. He said that no one ought to feel surprise that much remained unexplained about the origin of species. He thought that due allowance should be made for great ignorance about the mutual relations of lifeforms then existing, and even more so for previously existing lifeforms.

Were these mutual relations of the inhabitants of the world symbiotic (mutually beneficial) or cooperative relationships, as well as competitive relationships? Darwin hedged. He believed symbiotic or interdependent relationships were not equal in importance to competition but were *included* in competitive relationships. He said he used the term "struggle for existence" in a "metaphorical sense" which included "dependence of one being on another" and included "not only the life of the individual but success in leaving progeny" (Darwin, 1872: 52).

It is only by briefly abandoning the usually strictly constructed concept of competitive struggle that Darwin here, metaphorically, includes interdependence of lifeforms. Interdependence involves symbiotic or cooperative processes. Success in leaving progeny also involves cooperative processes, both in procreation and raising of young in many species.

Having, by means of metaphor, changed the definition of competition to include cooperation, Darwin subsequently makes assertion after assertion which again define competition in its strictest sense. He said that natural selection preserves all superior individuals and destroys all inferior individuals; that since natural selection acts by competition, it adapts and improves the species of each area *only* in relation to their co-inhabitants.

Why would Darwin muddy the definition of competition in this way? Could it be that Darwin intuitively recognized that cooperative processes are essential to evolution but they could not be considered as separate significant processes while using the deductive logic of science?

Deductive logic requires an all-inclusive major premise. In order to prove natural selection to be the main means of evolution, deductive logic requires natural selection to be the universal or all-inclusive means of evolution. This could be proven only if all other processes of evolution were considered to be a part of the struggle for existence and natural selection.

It is evident that Darwin's supporters interpreted natural selection as the universal means of evolution following the publication of the first edition of *The Origin of Species.* This explains why he found it necessary to protest most vigorously, in the sixth edition, that he was misrepresented by those who said he attributed the modification of species *exclusively* to natural selection.

The deductive logic of the science of Darwin's day apparently was not formally applied to test the all-inclusiveness of the competition hypothesis. If his main idea had been put into syllogistic form as a major premise by those who believed it to be true and all-inclusive, it might have appeared as follows:

> Major premise: All processes of evolution are competitive.
>
> Minor premise: Interdependence of living beings and success in leaving progeny are processes of evolution.
>
> Conclusion: These processes are competitive.

When Darwin's metaphor appears in syllogistic form, it becomes obvious that something is wrong. The major premise, as a universal statement, is not true. Becoming aware of this problem by the time that he worked on the sixth edition of *The*

Origin, Darwin specifically emphasized two additional factors in evolution: direct action of external conditions and spontaneous variations which lead to permanent modifications of structure independently of natural selection.

But it was too late. The word had gone out that evolution occurs by natural selection. The public had accepted this as Darwin's theory, and to this day people make little distinction between Darwin's threefold causative theory, (natural selection, direct action of external conditions, and spontaneous variations) and his main means of evolution (competitive natural selection theory). The public does not consider Darwin's concept of the action of external conditions upon speciation except to credit such a false idea to Lamarck, and does not consider the problems of spontaneous variation except to believe that genetics has proven Darwin's theory.

The deductive major premise that *all* processes of evolution are competitive was not true in 1859 and it is not true today. Even Darwin's inductive concept that natural selection is the main factor in evolution is deficient as a theory. Numerous scientific discoveries in the years since Darwin show that other factors are essential to explain evolution. Also, the other systems of logic developed in the intervening years since Darwin are more productive of understanding than are the logics of Aristotle and Descartes alone.

Before exploring the data of evolution let us consider some of Darwin's conscious and unconscious underlying assumptions about evolutionary processes and how the scientific logic of his day influenced his theory. It is necessary to understand these assumptions because they color the scientific facts that he gathered and elucidated.

Darwin's Assumptions and Their Significance

Darwin's Assumption 1: Cooperative Processes are Subordinate to Competitive Processes.

The competition of the struggle for existence and natural selection includes cooperative behavior; the concept of cooperative processes is subordinate to, and a part of the concept of competition.

One of the most telling examples of this assumption is Darwin's discussion of sexual selection. He said sexual selection depends "...on a struggle between the individuals of one sex, generally the males, for the possession of the other sex." The result is few or no offspring for the unsuccessful competitor. "Generally, the most vigorous males, those which are best fitted for their places in nature, will leave most progeny" (Darwin, 1872: 69).

Although Darwin elsewhere continuously maintained that the closest competition is between individuals of the same species, it here appears that these individuals are males (at least in two sex species). And the successful male competitors leave more progeny with little reference to the activities of the females. Females and males, although individuals of the same species, are not represented as strongly competing for the means of life.

It seems obvious today that two sex species usually cooperate in order to leave progeny. However, Darwin considered such cooperation to be insignificant in comparison to the competition between the males.

He did not entirely neglect assessment of at least one of the cooperative elements of female contributions to survival however, as he later related: "...the mammary glands are common to the whole class of mammals and are indispensable for their existence..." (Darwin, 1872: 170).

Darwin's entire presentation radiates the aura of the human male's nineteenth century perspective of the dominant male competitors and the subordinate female cooperators. Is it any wonder that the competitive processes of evolution were considered in the same relationship to the cooperative processes as the then perceived relationships between European males and females?

His concept of struggle for existence subordinated cooperative and symbiotic relationships, which today are often defined as coevolutionary. He noted that the mistletoe plant depends on birds "...and it may metaphorically be said to struggle with other fruit bearing plants, in tempting the birds to devour and thus disseminate its seeds." In these instances, he said he used for "convenience' sake the general term 'Struggle for Existence'" (Darwin, 1872: 53).

To use a metaphor that requires a plant to struggle to tempt birds and terming it "struggle for existence" for convenience does not appear to fulfill the requirements of logic.

Although Darwin's concepts of struggle for existence and natural selection depend on direct action of one individual or group competing for the means of life with another, he managed, by tortured logic, to fit ants, bees and other social insects, that exhibit instinctive cooperative behavior, into his category of natural selection. He said:

> ...I believe that natural selection, by acting on the fertile ants or parents, could form a species which should regularly produce neuters, all of large size with one form of jaw, or all of small size with widely different jaws; or lastly, and this is the greatest difficulty, one set of workers of one size and structure and simultaneously another set of workers of a different size and structure; — a graduated series having first been formed, as in the case of the driver ant and then the extreme forms having been produced in greater and greater numbers, through the survival of the parents which generated them, until none with an intermediate structure were produced (Darwin, 1872: 207).

Natural selection is simply not an adequate explanation for the speciation of the social insects. How DNA operates to develop different forms and functional patterns in insects remains open to investigation. Darwin's assumption that the evolutionary processes of these lifeforms could be covered by a competitive term is not logically sound.

Why did Darwin feel compelled to consider competitive and cooperative processes as one? Another reason, in addition to the cultural influence and the scientific logic of his day, may be the difficulty of separating the influence of each in observed natural phenomena. It seems impossible to find cases of pure competition and pure cooperation in nature because both processes are continuously interacting. The process one sees at any one moment may be one or the other because the human mind functions like the still-frames of a motion picture, in a dynamic thought process. Henri Bergson (1911) said: "The intellect forms a clear idea of the discontinuous alone." It is difficult to conceive of the interaction of competitive and cooperative processes in one still-frame of thought. If one looks for competition one sees competition. If one looks for cooperation, one sees cooperation.

Why did Darwin believe that the symbiotic or cooperative processes of nature could be (or should be) considered subordinate to, or within, a competitive process? He may not have consciously thought about this question because he was not a logician and he may have used inductive/deductive logic rather intuitively.

Various systems of logic have certain rules which, when followed, lead to persistent ways of understanding which are accepted as natural, inherent or right. The public generally may be unaware that reasoning used by different people or at different times may be one or a combination of logics. However, Darwin acknowledged his debt to Aristotle, for he said, "Linnaeus and Cuvier have been my two gods... but they were children to old Aristotle" (Rockham, 1944: 7).

Using Aristotelian logic in evaluating Darwin's understanding of the relationship of competition and cooperation, the question arises: Are they contrary concepts? Is one the negation of the other? If the definition of competition requires antagonism or neutrality toward a competitor with no help or support, and the definition of cooperation requires help or support, then in Aristotelian terms competition and cooperation are contraries.

If they are contraries, then Darwin could not believe that competition and cooperation exist in the same event at the same time. Only if they are the same thing, or one is a part of the other could they exist together. Therefore, Darwin thought they were not contrary concepts and *could not* logically think so, no matter how he might torture the concept of evolution in the process. That they could be considered to be complementary concepts was for future discovery.

Darwin's Assumption 2: Competitive Processes Are the Main Means of Evolution.

If it were possible to quantify the significant factors responsible for evolution, the competitive process of natural selection, as the main means, would account for the greatest portion of the weight of the totality of the factors. All factors, other than competition between individuals and species, would account for a minor portion of the total weight of evolutionary causes. These minor factors included influence of the external environment (in more recent years developed in the study of ecology) and spontaneous variation (in more recent years developed in the study of genetics).

For Darwin to believe that natural selection was the main means of evolution when considered together with the then unknown processes of ecology and genetics seems to have been premature, at best.

Mathematicians may continue attempts to quantify the factors of evolution to try to determine whether or not there is a main means of evolution but evolution may be too complex and dynamic to be quantified.

Darwin's Assumption 3: If There Is No Natural Selection, There Is No Evolution.

If any species does not evolve during a period of time, it is because of lack of functioning of natural selection. Darwin wrote:

> Undoubtedly if one species has any advantage
> over another, it will in a very brief time wholly

> or in part supplant it; but if both are equally well fitted for their own places, both will probably hold their separate places for almost any length of time (Darwin, 1979: 388).

He said that even though natural selection takes place over a long period of time, it is not an indefinite period; for "...if any one species does not become modified and improved in a corresponding degree with its competitors, it will be exterminated" (Darwin, 1979: 99).

Darwin assumed that a lifeform which is not evolving is either being exterminated or is surviving only because competition is temporarily stalemated. He did not recognize the possibility of a countervailing force to the competition of natural selection. Why? It is because the logic of Aristotle did not permit it. Since competition causes evolution, then if evolution is not occurring, competition is not occurring — or not effectively occurring.

Aristotelian logic, as made more explicit by Rene Descartes, requires a chain of causation, or linear inference in which effect follows cause. The logic of cause and effect is a mental habit pattern which is so deeply ingrained in the western psyche that most people consider it sacrosanct, a basic function of the normal mind.

In classical evolution theory, scientists envisioned patterns of change from one species into a different species. Darwin believed that individuals varied in their inheritable traits and, as these individuals competed for the means of life, those with stronger survival characteristics passed on their traits to larger numbers of progeny.

Darwin's study of evolution involved observation of a few fossil remains, comparisons of those with living forms, observations of differences and changes in living forms over time, and experiments with domestic plants and animals. The cause of evolution from one species into another appeared to be natural selection. The effect was new species arising from previously existing species.

This chain of causation reasoning was inherent in the study of evolution and was so pervasive that as biologists and others recognized problems that could not be addressed in this format, they developed a systems approach to study disciplines such as ecology.

Darwin's Assumption 4: The Influence of External Factors Is Unknown.

The influence of external (environmental) factors on variation and speciation is largely unknown. Darwin said that when a variation is of the slightest use to an individual, the amount to attribute "to the accumulative action of natural selection" and the amount to attribute "to the definite action of the conditions of life" cannot be determined. He noted, "In one sense, the *conditions of life* may be said not only to cause variability, either directly or indirectly, but likewise to *include natural selection;* for the conditions determine whether that variety shall survive" (Darwin, 1979: 129) [Emphasis added].

Indeed, Darwin did not know the weight to be given to environmental factors in relation to natural selection. He not only admitted the problem, but also here contradicted his emphasis throughout the book that natural selection is the main means of evolution. Aristotle would have been very unhappy with his student, for a theory that posits a main means of evolution may not suddenly include the main means in a subordinate position to a minor means and be excused due the author's lack of quantification skills. Nor can such a disaster in logic be covered up by general statements such as, "Natural selection acts exclusively by the preservation and accumulation of variations, which are beneficial under the organic and inorganic conditions to which each creature is exposed at all periods of life" (Darwin, 1979: 116).

Darwin's Assumption 5: Laws of Variation, Inheritance, and Correlated Growth Are Not Understood.

The laws of variation, inheritance, and correlated growth are anticipated but are not understood. He said that there was profound ignorance of the laws of variation: "Not in one case out of a hundred" can the reason be assigned for variation in different parts of a lifeform. However, he believed that the same laws acted to produce the differences between varieties and between species of the same genus (Darwin, 1872: 122).

He observed that there is conclusive evidence with plants that the sterility of crossed species is "...due to a principle *quite independent of natural selection*" [Emphasis added]. He believed that repeated crosses in successive generations with either parent could reduce both hybrids and mongrels to either pure parent form. (If genetics were this simple, today's biologists would not be so concerned to preserve native plants and endangered animals worldwide.)

Darwin believed that correlated growth was probably the most important of the many unknown laws of variability. It is amazing that Darwin anticipated the whole field of genetics before the pioneering work of Gregor Mendel (1822-1884) and without knowledge of chromosomes or genes, let alone DNA and RNA. The assumptions he made regarding genetic factors opened the most promising avenues toward knowledge of inheritance. They involved truly magnificent anticipations, but there were few scientifically verifiable conclusions about the means of inheritance in *The Origin of Species*.

Darwin's Assumption 6: Evolution Occurs By Chance.

The evolutionary process occurs by chance, not by plan or design. This assumption was probably one of the most revolutionary of Darwin's concepts because it changed the world view of many intellectuals of his day from a belief that species were originally created in changeless form and thus remain as they had always been, to a perception that nature provides a chance method of altering and improving species from generation to generation. The concept was devastating to some religious groups at that time and remains so today.

The basis of this conflict of perception resides in the logic used. The Aristotelian logic of both Darwin and the religious community of his day required that a choice be made:

either evolution occurs or it does not occur. Also, if it occurs, it occurs by chance or it occurs by design.

A difference between the deductive logic of a religious belief and a scientific belief is:

A religious belief is based on a major premise which is contained in a particular religious document or tradition and is not subject to change. In some religions, the belief also is not subject to interpretation, other than the interpretation of a particular leader.

A scientific belief is based on a major premise which is usually written but remains always subject to change.

People have made their Aristotelian choices over the years. Some believe, with Darwin, that evolution occurs and that it occurs by chance. Some believe that evolution occurs and that it occurs by design. Some believe that evolution does not occur. Since the latter belief is based on a religious major premise which is not subject to scientific analysis, this concept will not be discussed here.

Let's consider the question: Is it necessary to choose whether evolution occurs by chance or design? It is only necessary to choose between these two alternatives if the logic is Aristotelian. Other logics have different patterns to consider the question of how evolution occurs. If the choice is to use Aristotelian logic, which belief is the most accurate?

To determine whether evolution occurs by chance, it is necessary to study the chance mechanism(s) which is/are responsible for the process. Darwin claimed that the mechanism, the main means, is competition between individuals and species. Since it appears that there are serious questions about the weight to be given to this and his other hypotheses, and also whether there may be means, unknown to Darwin, which are just as significant, or even more significant than competition, it is premature to decide the question here.

To determine whether evolution occurs by design, it is necessary to study the question of which design or designs are under consideration. Obviously, the design would be in the lifeforms of nature, but which ones should be used as examples?

How many examples would be sufficient to establish an adequate hypothesis? The patterns in nature are so vast that many questions evoked by them must be considered open at this stage of our knowledge. The most that can be said is that there are patterns in nature that appear to be subject to laws or principles and those patterns should be considered with an open mind. Different systems of logic may be useful to explore the concepts of chance and design in nature.

Darwin's Assumption 7: Time Is Linear and Absolute.

Evolution occurs in time that is linear and absolute; evolutionary time is viewed from the lifetime experience of the viewer. Events occur in sequence from past to present to future, and time is the same for all observers.

This concept of time was accepted by scientists from Aristotle to Einstein. Einstein's perception of time as relative to position was necessary to understand cosmic and atomic phenomena. Much of the phenomena occurring in the area between the cosmic and the atomic continues to be understood very well in terms of conventional time. But living physical matter, as well as non-living physical matter is constructed from atomic particles, so it becomes necessary to understand Einstein's conception of time in order to understand the functioning of the basic structures of lifeforms.

Also, the time of evolution extends so much longer than an individual's lifetime and over so much more space than an individual can observe, that an individual's standard of time perception is inadequate to measure evolutionary events. These two factors raise the question: Is the linear, absolute, sequential time used in Darwin's theory adequate to evaluate the phenomena of evolution?

Although Darwin seemed generally satisfied about the linear time sequence of evolutionary events, he was puzzled in several important instances of sequence. The first puzzlement was in the sequence of structure and habits. He said that it was difficult to decide, and even immaterial, which changed first, habits or structure; "...or whether slight structural modifications lead to changed habits." He decided that "both probably occurred almost simultaneously" (Darwin, 1885: 141).

The second instance was in the sequence of structure and instinct. He said that he could not conjecture, in many instances, whether instinct or structure varied first (Darwin, 1885: 228).

The third puzzlement was the Lamarckian controversy over the sequence from genetic variation to somatic (body) characteristics or from acquired characteristics to genetic variation. Darwin was Lamarckian to an extent, believing that certain acquired characteristics could be inherited. He said that the technical sense of the term *variation* was used by some authors to imply a modification due to environmental causes; these variations "are supposed not to be inherited." He wrote:

> ...but who can say that the dwarfed condition of shells in the brackish waters of the Baltic or dwarfed plants on Alpine summits, or the thicker fur of an animal from far northwards, would not in some cases be inherited for at least a few generations...? (Darwin, 1979: 50)

Darwin also believed that the disuse of parts (of domestic animals) lead to their reduced size and that this was inherited. He was puzzled about the relationship between variations in lifeforms and the influence of the physical conditions of life, as indicated by his uncertain statement above. The reason for this uncertainty is that Aristotelian logic, with its cause and effect, linear time sequencing has no logical patterning for the concept of *reciprocal* relationships, those relationships which require an oscillating perception of time in order to understand their interactions. Close, reciprocal interactions require an understanding of the relativity of time and space.

Darwin's Logic

The assumptions which underlie Darwin's theory of evolution and the examples which help to identify these assumptions indicate that the theory is grounded in the logic of induction, that it does not meet the requirements of deductive logic. Inductive logic, by its nature, leads to a principle or theory which is never considered to be proven, but remains open to revision with consideration of new data.

Darwin (1964: 2) anticipated the need for expanded vision beyond monistic logic. At the very beginning of the first edition of *The Origin* he wrote:

> ...I am well aware that scarcely a single point is discussed in this volume on which facts cannot be adduced, often apparently leading to conclusions directly opposite to those at which I have arrived. A fair result can be obtained only by fully stating and balancing the facts and arguments on both sides of each question; and this cannot possibly be done here.

It is time, at last, to indulge in that exercise.

If two persons, sitting next to each other and looking in the same direction are asked to describe what they see, the descriptions will always be different. One may describe a flower in the foreground, the other, the mountain in the distance. One may see a mockingbird taking flight while the other sees only a blur of grey and white. One may describe a glorious landscape while the other sees a large area of dead weeds. Even if one person describes the scene to the other and points out the things described, the second person will not see exactly the same thing. Each person brings his own background and experience to each event he perceives.

Each person sees through his own eyes and his own experience. However, those who seek to expand their understanding, continually try to see through the eyes of others who have had expansive vision. This work attempts to add the vision of many scientists and philosophers to understanding of evolutionary processes.

8

Cooperative Processes of Evolution

The idea that evolution requires not only the competitive processes of natural selection but also cooperative or mutually beneficial processes is at least as old as Darwin. The important questions Darwin asked in 1859 and still requiring response today are:

1. What are the processes of evolution?
2. What are their relationships to each other?
3. How important are they?

Darwin's successors and challengers, using Aristotelian logic in considering these questions, are impelled to make an either/or choice: Are competitive or cooperative processes the main means of evolution?

Many authors have taken the position that *cooperative* or symbiotic processes are the main means of evolution. Until recently their work has remained relatively obscure. The world of Aristotelian linear logic has not looked for the antithesis of it's world-view, nor for a possible spectrum of evolutionary processes. Nevertheless, alternatives to Darwin's viewpoint have often been expressed.

Relationship of Competitive and Cooperative Processes

Before briefly evaluating a few of the works on this subject, it is necessary to disentangle the concepts of competitive and cooperative processes. Darwin was not the only one to fail to see the distinctions between them, nor was he the only one to rank them ordinate and subordinate.

The concepts of cooperation and symbiosis are often used interchangeably to mean mutually beneficial relationships. However, many modern biologists use the term "symbiosis" to include competitive as well as cooperative relationships. Symbiosis has been defined by modern biologists as the living together of two *dissimilar* organisms when the association is mutually beneficial. Parasitism, commensalism, and mutualism are considered to be forms of symbiosis.

A parasite is an organism that lives on or in the body of another organism, the host, from which it receives some advantage, usually nourishment, often damaging the host. Since this predatory relationship is not mutually beneficial, parasitism *should* be classified as competition (for sustenance) rather than as a form of symbiosis.

Commensalism is defined as the living with, on, or in another individual without benefit or injury to either, although some scientists consider that one individual benefits. This would seem to be a neutral area, neither competitive not cooperative.

Mutualism (or mutual aid) is a relationship between two individuals or species in which both benefit from the association, thus is cooperative.

From the above definitions and a little additional input, it is possible to visualize a spectrum of relationships among lifeforms as follows:

Beyond Competition

1. Extinction. Inability to survive competition.

Competition

1. Predator-prey relationship. Predator species or individuals are nourished, prey species or individuals die, or their numbers are reduced.

2. Other intraspecies and interspecies competition for means of life. Possibility of injury or death to one or both.

3. Parasitism. One individual or species living on, or in the other without return advantage. Possible injury or death to host and often subsequently to parasite.

Neutral Area

1. Commensalism. Living together without benefit or injury to either, or possible benefit to one.

Cooperation

1. Mutualism — Coevolution. Both of two individuals or species benefit from interaction.

2. Symbiosis, broadly defined. Mutually beneficial relationships, not only among species and among individuals, but also among organelles within individuals.

Beyond Cooperation

1. Altruism — Succession. One benefits another at its own peril.

2. Sacrifice. One gives its life for another.

This spectrum clarifies some of the major relationships among lifeforms. However, it is not entirely adequate because such a spectrum does not show the possible interactions of these processes in each individual encounter. For example, parasitism may change to mutualism, or commensalism to parasitism. Relationships among species, among individuals, and even among organelles within an individual lifeform are multifaceted and complex, requiring a logic more comprehensive than the cause and effect, either/or logic of Aristotle.

Also, overt relationships among lifeforms are only one aspect of evolution. The understanding of classical evolution needs to be synthesized with knowledge gained in symbiology, endocytology, microbiology, genetics, ecology and other disciplines to develop a more encompassing theory. The work of authors who have sought to redress the balance between competitive and cooperative processes has moved some scientists toward receptivity to new theories of evolution.

Mutual Aid — Petr Kropotkin

Petr Kropotkin (1888) directly challenged Darwin's thesis. He described the abundant animal life in Northern Asia, on the lakes and land, in colonies of animals. He described the massive-scale integrated migrations of birds and deer. In all these observations of animal life, he saw mutual aid and mutual support exhibited to an extent which convinced him that such aid is of the greatest importance to maintenance of life, preservation of species, and further evolution.

Observing the struggle of these animals against the scarcity of food, he disagreed with Darwin, writing that periods of keen competition do not lead to progressive evolution; rather such periods lead to large-scale die-offs of many individuals, yet a balance of species is maintained. He reported that Professor Karl Fedorovich Kessler, a famous zoologist and Dean of Saint Petersburg University, taught that in addition to the law of mutual struggle, there is a law of mutual aid, which is much the more important for success in life and progressive evolution of species.

The broad concept of mutual aid was identified by Kropotkin in the rearing of progeny, in animal groups providing for the safety of all and food for all, in joint efforts of certain beetles to transport and bury small animals prior to the laying of eggs in their bodies. He noted the concerted action of land crabs as they migrate to the sea, aiding those who fall on their backs or incur other mishaps.

He saw mutual aid in the association of ants. He said that the fundamental feature of ant life is the obligation of every ant to share its food with every fellow ant who needs it, food which had been swallowed and partially digested. He considered these cooperative survival techniques of much more importance than competitive processes.

Kropotkin pointed out that penguins have particular territories for resting and fishing and do not fight for them. He said that squirrels in Central Asia changed food supply from a larch forest to a fir tree forest and showed a change into a new variety, an incipient new species, without evidence of extermination among them.

Group hunting by birds of prey and other predators is given as an example of mutual aid, as is the fishing association of pelicans. He marvelled that the lapwing, given the name "good mother" by the Greeks, never failed to protect other species of aquatic birds from the attacks of their enemies.

In a state of nature, the social animals like elephants, rhinoceroses, monkeys, musk oxen and polar foxes live in associations numbering hundreds of thousands. Kropotkin noted that during three centuries of human predation, the populations of carnivorous animals were decimated, while the mutual aid/grazing animals survived in greater proportions. Kropotkin (1888: 58, 75) summarized:

> The fittest are thus the most sociable animals, and sociability appears as the chief factor of evolution, both directly, by securing the well-being of the species while diminishing the waste of energy, and indirectly, by favoring the growth of intelligence.

He concluded:

> Happily enough, competition is not the rule either in the animal world or in mankind... 'Don't compete!' Competition is always

> injurious to the species, and you have plenty of resources to avoid it! That is the tendency of nature, not always realized in full, but always present.

Kropotkin clearly chose cooperative relationships among lifeforms as the most important factor in evolution. He acknowledged the competitive processes but discounted their importance. He attempted to replace Darwin's theory with his own. Thus, he used Aristotle's either/or logic to assert that competition, essentially, is not the main means of evolution; mutual aid is!

Cooperation Among Animals — Warder C. Allee

Warder Allee (1951) asserted that instinctive cooperation is a fundamental principle of biology. His studies of animal population densities substantiate that assertion. His laboratory work with bacteria, flatworms, water fleas, sea urchins, and goldfish showed that overcrowding of animals always produced ill effects and undercrowding frequently did also.

There proved to be optimum population densities that are usually patterned by the activities of the animals themselves. In optimum densities, goldfish, both individually and collectively, were even able to survive exposure to toxic chemicals. They secreted a slime which changed the chemical into a less toxic form. Allee discovered that the slime is also a growth stimulating protein. No matter how dilute, it promotes growth when eaten by the goldfish.

Grouped water fleas survived longer in over alkaline solutions than individual fleas survived when isolated. The grouped fleas give off proportionately more carbon dioxide to neutralize the alkali. Grouping was found to decrease the respiration rates of various kinds of aggregated animals, making them more resistant to toxic materials. Allee showed that optimally massed animals survive temperature changes and ultra violet rays, which kill isolated individuals.

Marine flatworms, subjected to transfer to fresh water, lived longer if placed in water in which other such worms had previously died. The dead worms gave off calcium which had a protective effect on those that came after.

Under many conditions, a few bacteria innoculated into an animal die, while an innoculation of larger numbers cause them to grow and multiply. Bacteria battle for survival more efficiently in optimum densities. Thirty bacteria were found to neutralize at least two hundred times the amount of a toxin normally neutralized by one bacterium.

Allee found mass protection from many of life's hazards for self-perpetuating optimum populations of simple animals. He also found advantages in growth, health, energy conservation, reproduction and survival among mice, gulls and other animals when optimally grouped. He postulated a substratum of social tendencies of cooperation extending throughout the entire animal kingdom, which he believed contributes both to survival and to evolution.

Darwin, Competition and Cooperation — Ashley Montagu

Ashley Montagu (1952), reports that numerous scientists stress the significance of cooperation. He thinks that the importance of the idea "struggle for existence" has been greatly overrated, that nature is not at all like the struggle between businessmen of the nineteenth century. He saw a fundamental error by Darwin in incorporating Thomas Malthus' view that populations increase geometrically while food supply increases arithmetically. He said that Malthus applied this concept to industrial man and Darwin should not have applied it wholesale to all lifeforms.

Although Darwin considered fierce struggle to be determinant of survival between members of the same species, Montagu disagrees. He believes that harmonious or cooperative group association, not struggle, determines survival.

Similar viewpoints about the advantages of cooperation were expounded continuously in biological literature during many years. Montagu (1952: 59) quotes Chauncey D. Leak and Patrick Romanell: "The probability of survival of individual living things, or of populations increases with the degree with which they harmoniously adjust themselves to each other and their environment."

Montagu reiterates Kropotkin's view that the struggle for life requires cooperation. He believes that life is social by its nature, and all activities to maintain it are cooperative both individually and within species.

Montagu (1952: 72) especially doubts that man's combative sense ever had any adaptive value, believing that the adaptive value of human aggression is so low today that it threatens the existence of the human race. "Perhaps never before in the history of man has there been so high a premium upon the adaptive value of cooperative behavior, of peace on earth goodwill toward all men." Although he believes that natural selection may lead to the immediate survival of certain competitors, any survivors would not last long if they did not cooperate.

Montagu is correct in pointing out that cooperation is essential to evolution. However he assumes that competitive processes are usually of the death-dealing type (especially in man) when actually, competitive processes are also essential to evolution. He uses either/or logic to choose cooperation as the most significant factor in evolution, and would have preferred that Darwin make the same choice. Montagu (1952: 100) wrote:

> What a difference it would have made to the world had Darwin been able to write the *Origin of Species* and the *Descent of Man* with the emphasis placed on cooperation rather than struggle! The missing piece, in the puzzle which Darwin thought he had successfully put together, is the principle of cooperation. The next fifty years will, I believe, see that principle firmly established in scientific and in secular thought.

On the last page of Montagu's book, he anticipates a new world-view of evolution theory: competitive and cooperative processes of evolution are *complementary*. He apparently did not realize that, in this concept, he was proposing a change from Aristotelian logic to a new logic of complementarity.

Mutually Beneficial Relationships at Microscopic Levels

The word *cooperation,* in describing many activities of man is enhanced with connotations of highly developed mental operations. In describing activities of other species, we have tended to attribute cooperative action to instinct, especially in the social insects such as ants, bees and termites. However, the concept is broadened here to include mutually beneficial relationships not only among lifeforms but also within lifeforms. The object is not to obscure the meaning of cooperation but rather to expand it to avoid the necessity of using a new term for similar processes as we pass from the large to the small, to microscopic organisms, and even to the relationships of organelles within the cell.

Parts of a cell such as the nucleus, cytoplasmic organelles such as plastids and mitochrondia, and the cell wall of a one-celled organism may not be said to cooperate in a conscious decision-making or instinctive fashion, but they are parts of a cell which exhibit mutually beneficial relationships. I use *cooperation* and *symbiosis* as synonyms for these mutually beneficial relationships.

From the time of Darwin's *Origin of Species* in 1859 through the 1960s, most knowledge about small cellular lifeforms was gained by use of the light microscope. The relationships of one microbe to another or the relationships of microbes to higher animal life were studied intensively in relationship to health, mainly their destructive impact on the health of people. Dangerous and combative bacteria were fought by development of drugs to kill or weaken them. It appeared to be a struggle for existence between man and microbe — support for Darwin's competition theory.

Strange things began to happen after the invention and widespread use of the electron microscope. One-celled creatures, magnified thousands of times, exhibited astonishing behavior and astonishing structures! It was no longer so obvious that bacteria were competing with each other and with man for survival. Other things were happening. A new world of tiny creatures opened, and provides amazing new facts.

The Lives of the Cell — Lewis Thomas, and

Origins of Eukaryotic Cells — Lynn Margulis

A popular scientific author, Lewis Thomas (1974) who brought to light the wonders revealed by the electron microscope, also introduced one of the most important contributors to evolutionary thought in the 20th Century: Lynn Margulis (1970, 1981, 1993). She reviewed volumes of research by those in various scientific disciplines, and studied microorganisms. She discovered, organized and applied new data and proposed a serial endosymbiotic theory (SET) of evolution: Eukaryotic cells (with nucleus) originated by means of symbiotic processes of early bacteria (see Chapter 11).

It is now well known that prokaryotic (no nucleus) cell life preceded nucleated organisms by as many as one and a half billion years. Margulis presents extensive evidence that Precambrian prokaryotic bacteria of many varieties originally lived in an anaerobic environment.

Her most amazing theory is that these prokaryotic organisms, bacteria and cyanobacteria, with their own genophore (single ring-form chromosome) of DNA (deoxynucleic acid) some with the ability to process oxygen, joined together in symbiotic relationships. This allowed the symbionts (holobionts) to continue to exist as the environment changed from anaerobic to aerobic.

Some of these bacteria (which later evolved as plastids of plant cells) had photosynthetic ability to convert the carbon dioxide of the existing atmosphere, releasing oxygen. As oxygen built up in the environment, some of the anaerobic bacteria that could not cope with the crisis created by oxygen, died. Many bacteria burrowed into the soil and remained anaerobic but the way most bacteria became aerobic was not initially by point mutation, but by acquiring aerobic prokaryotes which provided the respiratory capacity in symbiosis with the host bacterium. Some of these prokarotes evolved into the present as mitochrondria, organelles of cells.

If Margulis is correct, a major evolutionary event, transition of early bacteria from an anaerobic to an aerobic environment, was made possible by symbiotic processes, the joining together of lifeforms which allowed the host to respire oxygen and the symbiont to be nourished by the host. Eukaryotic lifeforms evolved from separate bacteria with different abilities into lifeforms with combined abilities.

One indication that this is the case, is that the respiratory part of today's eukaryotic cells, the mitochondria, carries DNA different from that found in the cell's nucleus. The mitochondria in each of today's living body cells of plants and animals has had a different genetic history than the nucleus of that cell! The same is true of the plastids (chlorophyll containing chloroplasts) in the cells of living plants. Their DNA is different from the DNA in their nucleous and similar to the DNA of cyanobacteria. The origin of plastids is from symbiotic associations of Precambrian cyanobacteria and bacteria.

Lewis Thomas (1974: 31) realizing the importance of Margulis' endosymbiotic theory of evolution and her examples of its functioning, made "an earnest proposal:" that the nations of the world "set it as an objective of international, collaborative science to achieve a complete understanding of a single form of life." He proposed as the object of study "the protozoan *Myxotricha paradoxa* which inhabits the inner reaches of the digestive tract of Australian termites."

Myxotricha, within the hind gut of the termite, supplies the enzymes that break down the cellulose of wood into carbohydrate which can then be digested by the termite. Termites could not exist without symbiotic *Myxotricha.* Fungi which are cultivated by termites could not live without it. There would be no termite conversion of wood to soil without it.

On further examination, according to Thomas, not only is *Myxotricha* essential to the life of the termite, but is itself composed of *different* lifeforms living in symbiosis. What appears to be flagellae (hair-like projections) propelling the organism with synchronous precision are instead spirochetes (spiral bacteria) which have attached themselves over the surface of *Myxotricha.*

In addition, the interior organelles of *Myxotricha* under high magnification of the electron microscope are also found to be *bacteria*! *Myxotricha paradoxa* is not only a creature in its own right, it is also an *ecosystem*! It is an ecosystem absolutely essential to the life of the termite which is a creature absolutely essential to continuation of plant life on earth through decomposition of waste wood and creation of loam.

Such wonders of symbiotic processes are basic to evolution. Thomas (1974: 33) believes that there is an "underlying force" that causes the various creatures comprising *Myxotricha* to join together, and causes the symbiosis of *Myxotricha* and termite. He wrote:

> If this is, in fact, the drift of things, the way of the world, we may come to view immune reactions, genes for the chemical marking of self, and perhaps all reflexive responses of aggression and defense as secondary developments in evolution...

He noted that these secondary developments may be "necessary for the regulation and modulation of symbiosis, not designed to break into the process only to keep it from getting out of hand."

Following Thomas' recommendation for world-wide study of symbiosis in at least one organism, there has been rapid growth in publication confirming symbiotic processes in a large variety of lifeforms.

The work of Margulis has been prodigious not only in explaining her theory but also in bringing to American readers the broader scientific history of symbiosis research. It has only recently become known in America through the work of L. N. Khakhina (1979) that Russia has had a long history of studies of "symbiogenesis," the symbiotic origin of new evolutionary lines of many lifeforms. Lynn Margulis and Mark McMenamin (in press) will soon publish the seminal contributions of A. S. Famintsin (1935-1918) K. S. Mereschkovski (1855-1921) and B. M. Kozo-Polianski (1890-1957). Margulis (1991) considers that these early Russian biologists are the founders of evolutionary symbiology. Early symbiologists from other countries with dates of their major works include H. A. De Bary, France (1879) Paul Portier, France (1918) I. E. Wallin, America (1927) E. B. Wilson, America (1928) and P. Buchner, France (1965). Since 1980 work in the field of symbiology has expanded exponentially.

Margulis and Rene Fester organized a conference in Bellagio, Italy, 1989, and published *Symbiosis as a Source of Evolutionary Innovation: Speciation and Morphogenesis* (1991) which brought together the views of 31 symbiologists. The importance of symbiosis in early evolution can no longer be ignored. These processes are an essential complement of the competitive processes so long at center stage of evolution theory.

These authors contributed important and fascinating data to support the concept of symbiosis. Yet Aristotelian either/or logic is often apparent. To some, it appears that symbiosis is primary and natural selection is secondary.

Charles Darwin conceived a theory of evolution based largely on observation of then existing plants and animals and a very limited fossil record. Lynn Margulis conceived a theory of evolution based on microscopic examination of bacteria and protists. Darwin and Margulis came to different conclusions: Darwin believed competition between species causes them to evolve. Margulis believes that a primary basis of evolution is symbiotic processes of prokaryotes and eukaryotes.

Darwin believed that species multiplied in geometric ratios and that competition between species eliminates the excess numbers of individuals and allows only the fittest to survive.

Margulis (1970: 57) believed that all the reasons are not clear why in fresh water, marine and land environments, there exists a dynamic equilibrium of many hundreds of different species, and that an extremely stable balance is maintained. But she is convinced that "... a very specific series of symbioses led to the formation of new types of cells: our direct cellular ancestors, the eukaryotes." She believes that symbiosis is the basis of evolved cell structure, the building blocks of evolution.

Margulis has added the dimension of symbiogenesis to evolution theory. For those who pass beyond Aristotelian logic, there is understanding that competition and cooperation are dynamically interrelating processes of evolution.

9

Ecology and Evolution

Ecology may be defined as the study of reciprocal relationships among organisms, and their interactions with the physical environment. Both ecologists and evolutionists consider these relationships. Evolution theorists have taken long-term views of organism relationships with minimum emphasis on the effects of living things on environment. Ecologists, on the other hand, take a relatively short-term view, considering a specific area, or ecosystem for study of *reciprocal* life and environment interactions. Because of the relatedness of the two disciplines, an exploration of the logic of ecology brings insights for understanding evolution.

Ernst Haeckel (1866) a German zoologist, is credited with coining the word "ecology" but the subject matter had antecedents in the natural history of antiquity, in the work of Pliny and Theophrastus, who studied organisms in communities. Haeckel and other ecologists of the nineteenth century tried to mold ecology in the tradition of Darwinian mechanistic, reductionist, monistic logic. Other practitioners believed that the subject matter, which had at its base the infinite complexities of the balance of nature, did not fit into the rigid logic of cause and effect.

W. F. Ganong, Roscoe Pound and F. E. Clements, illustrious members of the biology department of the University of Nebraska in the early part of this century, called for a method of experimentation and reasoning beyond causality. These ecologists stressed attributes of quality over quantity, process over structure, concreteness over mathematical abstraction. They apparently did not realize that in making these either/or choices, they were in fact, using the very monistic logic that they were trying to transcend. They seemed unaware that they were intuitively aiming to change the logic which permeated science at that time.

With their focus on the balance of nature and the interactional processes of organisms they used an incipient systems approach. This was not the mathematically oriented logic later elucidated by Bertalanffy, but systems reasoning which challenged the orthodoxy of induction. The systems logic of these early ecologists was called unscientific by some of their colleagues, but it allowed them to suspend judgments about causality as they gathered data and interpreted interactions among plants and animals in communities.

In his masterful compendium of contributors to ecology, Robert McIntosh (1985) identified four areas of early ecological focus: Community was defined for inland water habitats by limnologists, for ocean areas by marine ecologists, for terrestrial plant ecosystems by plant ecologists or botanists, and for terrestrial animal ecosystems by animal ecologists or zoologists. McIntosh identified numerous and varied contributions by specialists in each area of interest.

The limnologist, S. A. Forbes, an early practitioner of the discipline of ecology in the 1880s, recognized that a lake is a system, with continuous origination and destruction of organisms, maintaining a dynamic equilibrium in a relatively isolated habitat.

Forbes identified the cooperative interests of predator and prey species. However, like many evolutionists and ecologists, before and since his time, he also believed that the observed equilibrium was produced by struggle for existence and natural selection. He overlooked Darwin's dictum that species competition ends in death for the unsuccessful competitors. According to Darwin's logic, the end result of competition should be displacement and succession, and yet many natural ecosystems are characterized by very long periods of relative stability of various species populations.

The belief in ecological systems with numerous species living in dynamic equilibrium, together with a belief in natural selection as the causative agency of evolution, exemplifies the dichotomy of two different logics, used but not recognized. Systems logic requires a certain timelessness, suspension of cause and effect patterning, suspension of concern for timing of events, or at least a non-linear time sequencing, because study of complex interactions of large numbers of organisms can be disorienting if logic requires consideration of chains of causation. On the other hand, the logic of natural selection *requires* consideration of causation and linear time sequencing.

The Paradox of Adaptation and Natural Selection

The dichotomy of monistic and systems logics has not been reconciled in ecological thought. The lack of precise identification of their use in discussions of ecological data, has caused great disquiet and confusion among ecologists. This is exemplified by the continuing concern over the relationship of adaptation and natural selection.

H. C. Cowles (1904) said, "If ecology has a place in modern biology, certainly one of its great tasks is to unravel the mystery of adaptation." Darwinian logic requires a linear, sequential time relationship between adaptation and natural selection. Those organisms which are best adapted to their environment are selected for survival. The selected organisms propagate those of their own kind. This logic allows change only by means of culling of maladapted individuals or species through survival of the fittest. It does not allow for changes in species based upon the influences of their life experiences. Such a concept is often considered to be Lamarckian and false. Nevertheless, ecological research documents are filled with continuing questions about the seemingly reciprocal relationships of adaptation and selection.

A. J. Nicholson (MacIntosh, 1985) reasons in monistic logic when he insists that competition regulates the density of all species populations. He, and many other ecologists over the years, did not seem to recognize that a balance of species in a community cannot be produced and maintained by competition alone. The idea of a balanced or dynamic equilibrium community requires a logical understanding of countervailing processes.

Perhaps the epitome of confusion was reached in development of the principle of competitive exclusion, or niche selection. The concept of niche is credited to Joseph Grinnel (MacIntosh, 1985) who said that competition between species leads each to adopt separate habitats, separate time patterns for eating and breeding, or different methods of food gathering within the same community. These activities exclude some species from the niche of others, thus leading to the term "competitive exclusion."

It is generally unrecognized that the end result is cooperative. The various species find a dynamic equilibrium among themselves by *not competing* for space or food supply. The idea of competitive exclusion did not go unchallenged by all ecologists, however, as C. Elton questioned the exact definition of the word, competition. F. J. Ayala wrote that the principle of competitive exclusion is invalid. However, J. M. Diamond said that the idea that processes other than competition are involved in patterns of communities "strained credulity" (McIntosh, 1985).

M. L. Rosensweig (1977: 95-96) reiterates the paradox of competitive exclusion in his study of rodents:

> ...competition remains important: natural selection to avoid competition is the mechanism that has shaped the system. Moreover, the threat of competition hovers like a ghost over the system and is necessary to its maintenance.

If natural selection is competitive, what is "natural selection to avoid competition?" It appears to mean competition to avoid competition. However, Rosensweig (1992) identifies both mutualism and competition (plus predation) within and between species, as different taxa interact to create ever changing destabilization/equilibrium of populations in their communities.

Even though the challenges of Elton, Ayala and other ecologists have had telling effects with some of their colleagues, the principle of competition continues its reign as the major cause of evolution and adaptation, but it also continues to be a matter of serious and growing dispute.

Systems Logic and Ecology

The essence of systems logic in ecology is a method to visualize and understand the infinitely numerous and diverse interactions among lifeforms in ecosystems. It is productive in understanding the relationships of producer plants, consumer animals and decomposers: bacteria, fungae and protists, soil and water-inhabiting lifeforms, and their interactions with the environment.

Since it appears impossible to distinguish causality when considering massive amounts of data of interactions, systems ecologists organize and analyze their data on the basis of related sub-systems such as:

1. Homeostasis (dynamic equilibrium) achieved through social interactions in ecosystems.

2. Influence of climate and terrain in relation to morphology of lifeforms.

3. Control of density by territorial behavior, especially among birds, fish, and mammals.

4. Organic and inorganic limits to population size, and optimum population size related to habitat.

5. Trophic levels and food webs (Who eats whom and how much?)

6. Relationship of population density
to disease, parasitism, predation, and food
quality and quantity.
7. Feeding habits related to breeding habits.

Over the years, preliminary efforts to develop systems logic have been criticized as unscientific and lacking mathematical precision. The history of ecology has therefore shown various periods of great enthusiasm for the latest mathematical models.

The wedding of systems analysis and ecology which produced such works as *Systems Analysis in Ecology*, edited by Kenneth E. F. Watt (1966) and *Systems Ecology*, by R. L. Kitching (1983) represents the wave of the future to some ecologists, who believe that computers may overcome the processing problems of vast masses of data. However, the lament of the older generation of systems ecologists is that the nuances of life are, basically, very difficult to quantify.

Need for Logic of Reciprocal Time

Systems analysis, whether the data is quantified or evaluated qualitatively, retains the element of oscillating relationships within ecosystems which requires the concept of reciprocal time. The fundamental unanswered questions of relationship between selection and adaptation, between pattern and chance, between evolution and coevolution, between succession and homeostasis, between organic and inorganic components of systems, are essentially all questions about the nature of time. They are dualities which are not addressed by monistic logic with its linear time frame nor by systems logic with its timelessness or constant time. They are questions which cry out for a logic of reciprocal time.

Forbes (MacIntosh, 1985) anticipated the need for this logic in his view that population numbers oscillate but tend toward equilibrium. Nicholson anticipated the need for a new logic to explain the seemingly irrefutable fact that populations regulate their reproduction by density induced resistance to excess multiplication. H. G. Andrewartha and L. C. Birch thought that the probability of such self-regulation is low but suggested that the controversy of population equilibrium versus density regulation awaits a *philosophical* resolution.

The perennial controversy over the relationship between selection and adaptation broke out anew in 1982 in the form of the hackneyed, paradoxical evolutionary question: which came first, the chicken or the egg? Do germ cell (gamete) mutations occur first, which are then selected by environmental conditions, or do environmental conditions, somehow, adapt organisms, followed by germ plasm changes?

August Weismann's doctrine, in support of Darwin's theory, maintains that progeny of organisms are changed by means of germ cell mutations in their parents, and it is a one-way street. The environment does not cause organisms to change and evolve, it only selects the fittest organisms as they occur, in their vast numbers and random variety (Haeckel: 1909). The theory that evolution involves only changes in nuclear DNA which is passed on to subsequent generations is now called "direct filiation."

Miles Eldredge (1982: 58) curator of the American Museum of Natural History, identified the controversy between direct filiation and symbiosis by questioning the relationship between the evolutionary concepts of Weismann and those of Lamarck. He wrote:

> *Nature,* one of the more prestigious science journals, recently published a controversial report claiming the experimental verification of the inheritance of an acquired character — in this case, one involving immunological tolerance. Though most biologists are unwilling to go so far as to resurrect Lamarck's notions of evolutionary change, it is a sign of today's unsettled times in evolutionary biology to see one of its most solid underpinnings — Weismann's doctrine — called into question.

What is the essence of this controversy? The inherent sequential framework of Weismann's doctrine, that germ cells pattern the physical body, is the antithesis of Lamarck's concept of the "inheritance of acquired characters." Weismann's position requires the assumption that since the germ cell patterns the physical body of an organism, then that germ cell precedes the altered physical body in time sequence. Lamarck's position requires the assumption that the functioning of a lifeform precedes the altered germ cell in time sequence.

Lamarck (1963: 107) a founder of evolution theory in 1809, is so often misunderstood that it is necessary to quote him directly. He wrote:

> I must now explain what I mean by this statement: the environment affects the shape and organization of animals; that is to say that when the environment becomes very different, it produces in the course of time corresponding modifications in the shape and organization of animals. It is true if this statement were to be taken literally, I should be convicted of an error; for whatever the environment may do, it does not work any direct modifications whatever in the shape and organization of animals. But great alterations in the environment of animals lead to great alterations in their needs, and these alterations in their needs necessarily lead to others in their activities. Now if the new needs become permanent, the animals then adopt new habits which last as long as the needs that evoked them.

The processes by which a body effects changes in its germ plasm (gametes) was no more known to Lamarck than the processes by which germ plasm effects development and-or change in the body was known to Weismann. However, the two scientists' theories are enhanced by new findings that nuclear DNA is a code for the structure of descendant lifeforms and that symbioses of prokaryotes and other lifeforms is a means of environmental and DNA interaction in the creation of new taxa lines.

Weismann and Lamarck lived long before the discovery of DNA and RNA and long before the logic of reciprocal time and its use to understand symbiogenesis. The viewpoints of Weismann and Lamarck are paradoxical. The perspective to consider the paradox is application of the idea of reciprocal time.

The Logic of Reciprocal Time and the Paradox of Chicken and Egg

Before Einstein, events were perceived to occur sequentially. The precise meaning of the simultaneity of events was not considered in the monistic system which was developed long before the concept of the "relativity of simultaneity."

Einstein (1961) expanded understanding of time and simultaneity with his thought experiment involving the movement of a train along an embankment. He forced consideration of the precise meaning of simultaneity relative to the location of observers. This precision has vast implications not only for atomic and cosmic events that involve relativistic speeds, but also for biology, for perception of ecological, genetic and speciation events.

With this perspective of time in mind, let us turn from Einstein's physical application to the *two* trains illustration (Chapter 6, Figure 6.1) and the biological application. Suppose train W is an expansion of Weismann's train of thought and train E is an expansion of Lamarck's train of thought. O is an impartial observer of the two trains of thought.

O observes that a prokaryote organism, a bacterium, is simultaneously both a gamete and a body cell, which divides into two gamete/body cells. W observes that the bacterium is a gamete which creates two organisms by dividing. E observes that the bacterium is an organism which creates two gametes by dividing. Is each concept equally valid for the observer involved? The answer must be yes. Just as two trains have their *own* time, so each train of thought has its own perception of time sequencing. The sequence from body cell to gamete or from gamete to body cell (or the *simultaneous* interactions of symbiotic genophore, plastids and mitochrondria) in the case of a bacterium, is *not* intrinsic to the bacterium, but represents the knowledge and focus of the observers.

To go further, assume that the bacterium undergoes a modification caused by a chemical agent. Did the body mutate? Did the gamete mutate? Did acquisition of a symbiotic organelle provide the ability to live in a new chemical environment? Did the organelle mutate at some point?

W observes: The gamete mutated in response to the chemical agent and passed on its mutated structure (having ability to produce antibodies) to two daughter bacteria.

E observes: The body, in response to the chemical agent, was damaged, developed a method of producing antibodies to the chemical agent, survived the experience and then divided into two individuals. The daughter bacteria inherited an acquired function, the ability to produce the antibodies.

O observes that the bacterium is an ecosystem which changed, developing antibodies to the chemical agent, and recovered to divide and pass on greater chemical resistance to the daughter bacteria. To the impartial observer, O, the sequence in which a mutation is passed — whether it is from gamete to body cell or from body cell to gamete — or whether symbionts were responsible for the ability to resist the chemical is not relevant. From his perspective, the occurrence is *simultaneous* because the organism is one system.

Now to return to address the original paradox: Weismann's position is that the gamete changes first and then the body changes. The controversial report in *Nature* operates from the premise that the body changes first and the gamete changes later. The perception of which came first, gamete change or body change appears to be paradoxical; however, the conflict between the positions is based on a concept of sequence that is not valid in this prokaryote case. On the fundamental level of this lifeform, gamete (egg concept) and body (chicken concept) exist together as an integrated system.

Which came first, the chicken or the egg? This is a question asked in monistic logic that has no answer. In complementarity dynamics, chicken and egg are an integrated system which evolved from earlier integrated systems.

Relationship of Reciprocal Time and Sequence

While a logic of reciprocal (interactional) time is helpful to understand the relationship of the concepts of gamete, body and symbionts of the prokaryote, a concept of sequential time is also needed to understand the functioning of the prokaryote as it gathers food and divides into two. Thus paradox continues in the need for two seemingly mutually exclusive concepts of time.

Since it is believed that life evolved from the prokaryote level to the eukaryote level, does the same logic apply in considering the structure of each? Does the concept of acausality (reciprocal time frame) between the gamete of the eukaryote and its body apply?

It would appear that the answer continues the paradox. The parts of the organism exist together as one entity, one structure, so may be considered an acausal system. However, a logic of causality appears to be required to consider the functioning of the organism. In addition, it is believed that prokaryotes preceded eukaryotes in evolution by at least a billion and a half years, thus a logic of sequence is certainly required. Paradox continues as both systems of logic are required to understand prokaryotic and eukaryotic lifeforms.

Complementarity Dynamics,

Causality, and Acausality

Synthesis of the concepts of Weismann and Lamarck can be achieved by use of the logic of complementarity dynamics. It appears that the relationship of gamete and body, in the case of the prokaryote, is acausal from the point of view of the impartial observer. There is no cause and effect relationship according to this point of view because the bacterium is one entity. However, this entity functions and as it functions, causality returns. Such a bacterium appears to be able to judge what particles are food (cause) and move to grasp, incorporate and digest those particles (effect).

The structure of the bacterium may be viewed as acausal or existing without a causal chain from gamete to body or from body to gamete. But functioning is in causal logic of sequence. Therefore a logic of acausality (logic of reciprocal time) and a logic of causality (monistic logic of sequential time) are both required to understand the prokaryote. The two complementary logics operate in dynamic equilibrium. In one moment an observer may choose to consider the bacterium in acausality logic, in another moment the observer may choose to consider it in causality logic.

Reciprocal Time and Causality in Evolution of Eukaryotes

The complements of reciprocal time and causality are helpful in analysis of the movement of evolution from the prokaryotic to the eukaryotic level of life. If the parts of the eukaryote function separately, perhaps they may be understood to function in sequence. However, if the eukaryote functions as a system, it would appear that acausality would be operative within the individual.

The eukaryote is organized as a whole, and as parts, *and* it is able to function as a whole and in its parts as it moves to capture food and as it reproduces. It is not productive to study the eukaryote exclusively in either monistic or reciprocal logic. *Both logics should be used.*

Reciprocal Time Logic and Higher Lifeforms

Causality and acausality are complementary concepts defining both prokaryotes and eukaryotes. How do these concepts apply to higher animals? Lamarck's concept of the inheritance of acquired characters was correct in recognizing that the environment effects changes in lifeforms but he did not know that the influence acted at the cellular level by means of symbiotic processes.

There is still much to learn about how various factors such as temperature, radiation, chemicals, viruses, and other environmental factors cause cells to mutate and reorganize symbiotically. However, scientists still use the terms "random events" and "spontaneous occurrences" to define probability functions.

It is still not known exactly how gametes develop from and-or interact with somatic tissues during embryonic development. It is still not known exactly what causes mitosis to operate (Chapter 11). Lamarck did not have knowledge of the genetics of DNA and RNA discovered in the 20th Century. But if one observer could observe a change in function on the cellular level, while another observed a change in cell structure, and these occurred in a reciprocal moment, what is the cause or effect of changed inheritance?

Each cell of eukaryotic organisms contains a nucleus and cytoplasm. The nucleus contains most of the DNA of the cell which is considered to be the main template for the structure and functioning of each cell. The DNA of symbiotic organelles interacts with nuclear DNA. The logic of reciprocal time can be considered to explain seemingly *simultaneous* symbiotic interactions of DNA within the cell and among cells.

Multicellular lifeforms increase substantially in organization from unicellular organisms, requiring understanding not only of the time element within the cell, between all the various cells and organs and between the brain and the body, but also between the individual and the natural environment. If prokaryotic and eukaryotic lifeforms require use of complementary concepts for understanding, how much more do multicellular lifeforms require them? The disciplines of ecology and evolution both require understanding of the time element in reciprocal interactions between complements.

Parallel Evolution, Coevolution, and Logic

Scientists have used the term "parallel evolution" to explain separate developmental pathways of such species as flowering plants and insects. They have noted conjunction in time of origination of new species which interact, but assume separate, parallel evolutionary pathways until that time. The term "converging evolution" represents the time of beginning interaction of such species.

The term "convergence" has another meaning. It is used by paleontologists to mean two similar appearing species (or genera) that have evolved from different source species (or genera). Convergence is especially evident among genera of different continents.

Scientists use the term "diverging evolution" to explain the movement of part of a population from the geographical area of its parent species into a new area, separated by a physical barrier. The species which arise in the separated areas are said to represent "allopatric speciation." In addition to divergence in geographical location of new speciation, diverging evolution is also recognized as occurring due to change in habits of species in the same geological area and this is called "sympatric speciation."

In 1965, the term "coevolution" was coined by Paul R. Ehrlich and Peter J. Raven to define the relationship between butterflies and their host plants. Those ecologists used the term to express the patterns of close interactions between two major groups of organisms: plants and herbivorous insects. "Coevolution" was later employed to define relations between carnivores and herbivores, between plants and microorganisms and between many other interacting lifeforms. Ehrlich and Raven recognized that these interactions involve complex interconnectedness of evolutionary pathways, but they did not express recognition that a systems or complementarity logic was needed to supplement the cause and effect logic of the parallel evolution concept.

By the 1990s, it has become apparent that new logics implicit in the term coevolution are required in order to understand not only coevolution and the increasingly complex interactional activities in ecosystems, but also the relationships within lifeforms, and between lifeforms and such systems as climate and geography (Schneider and Lander, 1984).

Ecologists who broke away from the mainstream of biology intuitively recognized the need to escape the straitjacket of monistic logic and strict causality. That the break was accompanied by numerous returns from a systems approach to strict induction is not surprising. Reconciliation of these two approaches has awaited identification of the logics being used. Scientists are only beginning to recognize the need for logics of paradox. Many paleontology-oriented evolutionists remain satisfied with monistic logic. A new wave of biochemists, microbiologists, geneticists, and virologists is moving the study of evolution onto a fast track to the future. The study of micromolecules, DNA, RNA and viruses has led to the speedy accumulation of so many fascinating facts that the lagging of philosophical and logical underpinnings has become obvious.

10

Beyond Darwin

An open secret of science today is that many scientists no longer believe that Darwinian or neo-Darwinian theory explains evolution. It is necessary to consider various scientists' interpretations of the fossil record and studies of living organisms in order to rewrite evolution theory.

Darwinism has been so strong in its effect on opinion, that scientific voices of dissent, as well as religious objections to the theory, have been effectively quashed or ignored by the leaders of science. There have been many dissident voices lost in the currents of history. Today, however, voices are being raised anew in a chorus which cannot be quieted.

When I wrote my first manuscript on evolution theory, *Legacy: The Land,* in 1970, there was little evidence that Darwin's theory was about to be challenged from within many different disciplines. As recently as 1942, Julian Huxley had published *The Modern Synthesis,* which had renewed Darwinism. Support for Darwinian concepts continued from impeccable sources but change was about to occur.

Niles Eldredge and Stephen Jay Gould (1972) were committed Darwinians when they sought to make a slight revision in the classical theory. Eldredge and Gould developed the concept of punctuated equilibrium: that long periods of relatively stable species are punctuated by new species arising rather rapidly. They believed that this concept did not attack the idea of natural selection, that it only added the notion that although changes in species may be very slow overall, they frequently occur in only a few thousand years.

Their theory was greeted with a rush of response. It seemed to release pro-Darwinian enthusiasm as well as pent-up dissent which neo-Darwinism had been generating. Committed Darwinians continued their defense with vigorous rhetoric. Richard Dawkins (1986: x) stated:

> ...I want to persuade the reader, not just that the Darwinian world-view *happens* to be true, but that it is the only known theory that *could*, in principle, solve the mystery of our existence. ...a good case can be made that Darwinism is true, not just on this planet but all over the universe wherever life may be found.

Many supported the opinion of Bernard Davis of Harvard, who earlier had said: "Except for those skeptics who are willing to discard rationality, Darwin's theory has now become Darwin's law" (Taylor, 1983: 37).

Darwin had proposed the concept that there was a continuous sequential relationship from the simplest one-celled lifeforms up to and including humans. He believed changes occurred by chance in small incremental steps due to the natural variation in individuals, with natural selection of the most robust individuals and species due to competition among them. He noted that the fossil record was incomplete, but thought that the patterns found in it supported his premises.

Some paleontologists, like Eldredge, held to Darwinian concepts, on the whole, accepting some seemingly minor additions and corrections. But a growing number of those studying the fossil record and microorganisms began to find major problems with Darwinism. Eldredge (1985) was disconcerted to find that some of his colleagues identified the idea of punctuated equilibrium with the theory of the geneticist, Richard Goldschmidt of the University of California, Berkeley. Goldschmidt saw major changes between taxa in the fossil record as being different in kind, not just different in degree as are the variations within species.

Goldschmidt posited that *macromutations* were the cause of large genetic differences from parental stock, and were the source of new forms in evolution. Such a view was also expressed by others, although in the 1960s it received little recognition. The problem of the jumps in the fossil record, the saltations, was solved by neither the neo-Darwinians nor the anti-Darwinians.

Another crack in the Darwinian dike was opened at a conference organized by Sir Peter Medawar in 1967. Murray Eden, professor of Electrical Engineering, Massachusetts Institute of Technology, presented the keynote speech on "The Inadequacy of Neo-Darwinian Evolution as a Scientific Theory." He said that a total of six mutations, which would be required to bring about only a small change in structure, could be expected to occur by chance only once in a billion years. Ten billion years would be required if as many as 24 genes were involved. He said that chance, or trial and error, is a totally inadequate concept to solve problems of change in the structure of animals without specific guidance (Taylor, 1983: 4).

Obviously, there was a problem with the idea of mutation, or with the idea of chance, or both. Darwinism, using monistic logic, requires the concept of small mutations or limited reorganization of germ plasm for an orderly progression from mother to son or daughter, and on down the line of generations.

John Ray said in the 17th Century, "No species ever gave birth to another species" (Taylor, 1983: 79). If species (by modern definition) cannot interbreed with another species, then John Ray was very prescient to recognize the problem of discontinuity. Nevertheless, scientists have continued to maintain that sequence from one species to a different species or genus could occur by chance variation in germ plasm if there were enough time and-or enough distance between a number of successive generations.

This could not be true in those cases in which a new *genus* was not only different in degree but also different in kind. A truly different genus, a new taxon, exemplifies a reorganization of genetic material: the chromosomes, the genes, the DNA.

Evolutionary Direction, Change, and Tempo

As late as 1977, some paleontologists were reluctant to acknowledge the paradoxes of evolution. They continued to debate these questions in an either/or fashion.

In *Patterns of Evolution: as Illustrated by the Fossil Record* (Hallam, 1977: 1-2, 1-26, 547-561) the summary by Thomas J. M. Schopf explored the questions posed by Stephen Jay Gould:

> 1. Does the history of life have definite directions... (increasing complexity of structure, or number of species, e.g.)?
>
> 2. What is the motor of organic change? More specifically, how are life and the earth related? Does the external environment and its alterations set the course of change, or does change arise from some independent and internal dynamic within organisms themselves?
>
> 3. What is the tempo of organic change? Does it proceed gradually in a continuous and stately fashion, or is it episodic?

To answer these questions, Schopf's summary analyzed the opinions of the 17 scientists who had written the book's chapters. He found that ten scientists believed there was evidence of *steady-state* taxonomic diversity through geological time and four scientists saw evidence of *increasing* taxonomic diversity in their areas of expertise. He identified nine scientists who saw directional trends (other than size) through geological time. There were 12 scientists who thought environmental factors controlled evolutionary changes and three who thought non-environmental factors controlled evolutionary changes. Eleven scientists considered that the rate of evolution was largely punctuational and four thought that it was largely gradational.

Only one author in the series, James W. Valentine, devoted more than a passing remark to the importance of DNA in structuring evolution. He devoted less than four of 28 pages to this concept.

Gould approached awareness of paradox by saying, "There are strong correlations among the issues, but all...[six] possible combinations (for two opinions on each of three questions) can and have been maintained by paleontologists." Thus, he called his opening chapter "Eternal Metaphors of Palaeontology."

The recognition that these metaphors represent complementary concepts which must be reconciled in a new logic was not expressed by these authors in 1977.

Patterns and Chance

One of the most profound beliefs of science is that nature exhibits patterns, that these patterns are governed by natural laws and such patterns and laws can become known to us through diligent experiment, study and thought. A corollary of the belief in nature's patterns and laws, which might be considered paradoxical, is the belief that nature exhibits random occurrences, that events may occur by chance, each with a certain probability of happening. However, in order for random or chance events to occur within a universe conceived as a whole, the random occurrences themselves must exist within rational parameters.

A random occurrence is defined as something which happens outside of a known pattern of purpose or reason. It is something that could not be anticipated from known patterns — being a novel, unexpected or accidental event. It is accepted by science that mathematics correlates extremely well with nature

— that random events can be identified and the probability of a random event happening can be calculated. The conventional range of probability has been considered to be 0% to 100%. Zero percent probability of an event means *certainty it will not happen.* One hundred percent probability is certainty that it *will* happen. If there is certainty that an event will not happen (0%) or certainty that an event will happen (100%) those percentages should not be considered in the continuum of chance or random events.

Chance events, therefore, range from a probability of about 99.9% to a probability of less than 1%. This is a range of uncertainty between two certainties, the continuum of uncertainty ranging between Aristotle's most certain of all principles: that something either *is* (100%) or that it *is not* (0%). This suggests a logical contrast with the uncertainty principle of Heisenberg, in which each aspect of quantum reality is certain or uncertain, depending upon perspective. In the case of the probability continuum, the aspect of certain reality (the end points of the continuum) is complementary to the range of phenomena between those two points, which is uncertain.

Some might say that a continuum of uncertainty from 99.9% to less than 1% is a numerical pattern, and certainly, mathematics is a science of the operation of numerical patterns. If mathematicians could see no coherent patterns in their calculations, they would conclude that they were not doing their work correctly. So patterns in nature are patterns of known events and chance in nature may be expressed by mathematical estimates of the degree of uncertainty, the probability that a particular event will occur. Thus pattern and chance are complements. With the wholeness of the universe thus defined as a universe of patterns which can be known with certainty or with degrees of uncertainty, it becomes possible to look at the evidence of evolution and make evaluations.

When Charles Darwin proposed the notion of spontaneous or random variations as the basis for change in species, he was, in fact, indicating a range of unknown reasons for these variations. Darwin acknowledged that he knew little about the laws of inheritance, variation and correlated growth. He was unaware of the work of Gregor Mendel who published *Experiments In Plant Hybridization* in 1866, seven years after Darwin published *The Origin of Species*. Mendel's work was unrecognized until 1901 when it was reprinted in English by the Royal Horticultural Society of London. Since then, Mendel has been recognized as the founder of the field of genetics.

Gregor Mendel first established a basis for evaluating genetic patterning by documenting the probability for variations of yellow and green color and and other variations in the garden pea. He performed and recorded a systematic study of the inherited structures of a plant species. He discovered that a species of pea demonstrates dominant and recessive color characteristics. Yellow is dominant and green is recessive.

An individual receiving two dominant characteristics of yellow or two recessive characteristics of green (the same from each parent plant) expresses that same characteristic in the first, F1 generation and breeds true in the F2 generation. (These individuals are now called homozygotes.) Pea plants (now called heterozygotes) receiving one dominant and one recessive characteristic from the parent plants appear to be like the dominant parent, yellow. This heterozygote generation passes on to the next F2 generation a pattern of individuals in the ratio of one dominant yellow characteristic, one recessive green characteristic and two combined dominant and recessive (dominant yellow characteristic) ratio: three yellow to one green.

The realization that inheritance is patterned by what came to be called the dominant and recessive alleles of genes raised a challenge to the basis of Darwin's theory. Such patterning is not caused by random variations and competitive relationships, but rather, is programed as heredity in the germ cells of organisms.

However, geneticists of the early 20th Century had been so indoctrinated by Darwin's theory that they, perhaps unconsciously, considered the relationship between two alleles to be competitive. After all, the dominant characteristic prevailed in expression in the organism over the recessive characteristic when both were present in the gene. It was thought to be survival of the dominant, even though the green gene alleles were not eliminated, remaining in ratio one green to three yellow.

Darwin's term "selection" was then appropriated in the newly developing field of genetics to mean survival only, gradually dropping the emphasis of the competition concept. It is common today to read in genetics texts that a certain gene combination is "selected for" under certain physical conditions. In the waning years of the 20th Century, geneticists still sometimes discuss selection in textbooks as though Darwin's theory were operative in their field. However, today there is also a stirring toward development of new theory based upon identification of specific genotypes of DNA and the phenotypes which they produce.

Darwin thought that evolution demonstrates a pattern of increasing complexity and fitness of organisms, but no plan to guide that pattern. Others have thought that patterns in nature demonstrate an innate purposefulness, if not a plan. To many scientists, the idea of purposefulness is anathema; it is *metaphysical* instead of *scientific*. But a number of scientists are now saying that since DNA patterns appear to be purposeful, they should be recognized as such.

Darwin was bothered by the evolution of the human eye, thinking that graduated steps from a light sensitive area on the head of an animal, to the complex structure of mammalian eyes presented some problems. However, he maintained that this gradual evolution had, in fact, occurred. Many paleontologists no longer believe that structures such as trilobite eyes, the dozen other eye forms or mammalian eyes could develop by gradual evolutionary steps.

The patterns of the eye are too complex to occur by a series of chance occurrences. And the eye is never alone. It is connected to the optic nerve and functions in perfect harmony with the structure of the brain. Many have concluded that there are patterns in lifeforms which are purposeful and there are directing principles inherent in life.

Darwin also saw patterns of change in the species he studied. He believed that evolution was constant change and improvement. But some organisms show a pattern of millions of years of constancy. Universal patterns seem to encompass systems of both change and constancy.

Darwin thought that individuals having advantageous changes of structure continue to exist and multiply but those with disadvantageous changes perish. The ammonites (Chapter 17, Figure 17.6) developed, during millions of years, successively more emphatic shell rib structures, but there is no evidence that this improved their eventual survival chances. Neither was this feature disadvantageous during their long evolution. Without evidence of *significant* change, the ammonites perished, while oysters with similar limited change lived on.

Within nature's patterns, organisms with both advantageous change and virtually no significant change live for millions of years. Organisms with significant disadvantages, such as the Irish Elk (which is actually a Siberian deer) with antlers weighing almost a quarter of a ton and measuring up to 12 feet across, survived beyond any point of expectation by Darwin's theory (Figure 10.1).

Some of the lowly bacteria have lived through environmental constancy and through massive changes since the Precambrian, showing no signs of evolving into anything else. On the other hand, bacteria are believed to have provided the source DNA which evolved into myriad advanced lifeforms.

New species constantly arise. How else could there be almost two million living species after six massive extinctions, and extermination of up to 99% of earlier species? Yet in the artificial environment of laboratories for 60 years, each new generation of fruit fly has developed in 11 days time, with numerous aberrations of development and change, but no truly different taxon has emerged.

Scientists say that no new phyla and few new families have arisen since the Permian Period; that there was a burst of placental mammals arising in the Paleocene, but little evidence of new classes or orders since then. However, many innovations which have occurred cannot be explained by existing theory.

Darwin thought that the homologous bone structure of amphibian, reptilian and mammalian appendages proved a process of evolution from one group to the next and that the analogous development of different wing structures in insects, birds and bats had no relationship to the evolutionary principle. He saw relationships of structure as evidence of evolution, but functions, such as flying, were not seen as major factors in the transformation of species. These were merely processes to be noted. The function of prey relationships, of course, was a special case and was included as part of the main process of natural selection.

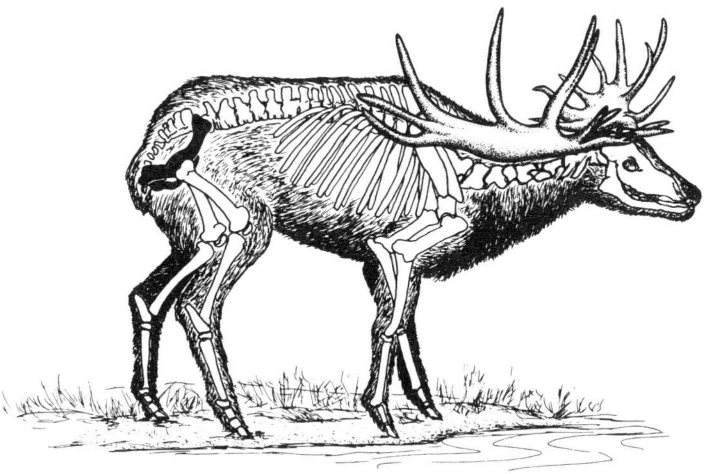

Figure 10.1 Irish Elk (*Megaloceros*, Artiodactyla). (After skeleton, University of Nebraska State Museum).

He did not consider that the symbiotic relationships of insects and flowers were a major factor in evolution. He saw symbiosis as a minor element of competition. He saw progressiveness in development of species as a *given* of natural variation, and extinction as a function of natural selection. And he considered degeneration or regression as aberrations, not as possible antitheses of noted progressions. Degeneration of organs was thought to be due to disuse.

He believed the mimicking insects such as certain moths and butterflies gradually evolved by chance, in the direction of similarity to the species they mimicked. He thought that innovative organs of species arose solely by chance as they were needed, and never before they were needed; yet he believed that their usefulness in their environment was also a chance phenomenon. Darwin's concept of a gradual sequence of individuals evolving by chance into new species is substantially challenged. It is becoming evident that major speciation events are not understandable in Darwinian terms.

Challenge to Sequential Reasoning

Until recently, paleontologists were satisfied to continue to look for the evidence which would sequentially connect the 31 or so major phyla as well as classes, orders, families, genera and species. By 1967, Alfred Romer had determined that of 43 living orders of terrestrial vertebrates, 42 orders had been found as fossils. That's 97.7%. Of 178 living families of terrestrial vertebrates, excluding birds, 156 families had been found as fossils. That is 87.8%.

Other specialists began to recognize that the fossil record is not as incomplete as usually thought. Millions of fossils have been catalogued world-wide. Some neo-Darwinians maintain that speciation events occur in such small groups that often no fossil link between different taxa should be expected; that this does not mean that connecting links never existed. However, symbiologists are now presenting verification of new taxa formed by symbiotic processes from very different species' DNA.

Some paleontologists and geneticists, as well as mathematicians, are beginning to calculate the probabilities of various natural occurrences happening together by chance, and a whole new reality is emerging. Chance is not the main impetus to evolution.

It has become apparent that major divisions of lifeforms arose in the fossil record fully formed with no evidence of transitions. There has been no adequate explanation in monistic reasoning for many of the transitions between lifeforms. Discontinuities as well as continuities exist. The discontinuity between non-life and the beginning of life was a major discontinuity, and nature exhibits numerous other major discontinuities, paradoxes which required a logic of paradox.

Repudiation of Darwinism

By the 1980s, paleontologists and others began to gather together dissenting opinions about the processes of evolution and supporting data. *The Great Evolution Mystery* by Gordon Rattray Taylor was published in 1983, followed by *Darwin Was Wrong*, by I. L. Cohen in 1985, and *Evolution: A Theory in Crisis* by Michael J. Denton in 1986. It is characteristic of these works that they point out the problems of Darwinian concepts and the evidence which refutes them without proposing an inclusive alternative theory.

This lack of a new theory to replace Darwinism is of great concern to some in the scientific community. Discussions have raised the suggestion that the collapse of Darwinism should not be expressed in public, lest religious fundamentalists preempt the field of evolution theory. The authors of these 1980s books obviously disagree. Their work indicates faith in the ability of the scientific method to bring forth theory to fit the facts, but it is evident that monistic logic of the scientific method is inadequate for this task.

Reorganization of Disciplines

By the early 20th Century, many disciplines were well defined: astronomy, mathematics, physics, chemistry, biology (zoology and botany) physiology and medicine. As the 20th Century progressed, branches of psychiatry, psychology and sociology developed — and then ecology, microbiology and virology. The subject matter of each of the disciplines was studied as a recognized system, but often induction/deduction was the underlying logic. Complementary concepts were implicit in mathematical formulas of physics and chemistry, but textbooks did not readily identify the type of logic used.

In more recent years biophysics has developed, also biochemistry, microbiology, paleoecology, geophysics and astrophysics. These movements represent reorganization of separate systems into two-system wholes. This involves recognition of complementary relationships between concepts of two fields. The amalgamation of study fields is accelerating despite the fact that this brings greater complexity as well as new insights. New efforts are taking shape to establish interdisciplinary studies to bring better understanding in many fields.

Geologists-stratigraphers have developed a system to evaluate layers of soil and rock laid down in strata over time. Paleontologists and geologists together determine the relationship of the fossils in the strata to the time of depositions. They combine with climatologists and palyologists to evaluate weather conditions and flora environment at the time of depositions. Climatologists, geologists, paleontologists and geographers are currently studying the movements of Earth's crustal plates and how these movements relate to climate — how climate and geography affect the rise and evolution of various lifeforms.

Oceanographers, climatologists, paleogeographers and geologists specializing in plate tectonics are studying how Earth's weather patterns are related not only to the processes at Earth's core, but also to the various changes in the locations of land masses and sea areas during Earth's history. Sometimes these efforts are only indirect, as when each expert writes a chapter of a book which an editor compiles as a whole. Some visionaries such as James W. Valentine (1973) J. E. Lovelock (1979) Lynn Margulis (1970, 1981, 1993) and Eric J. Lerner (1991) are beginning to write books uniting the various disciplines.

So far, however, many philosophers seem to be writing about monistic logic, the construction of systems, the reasoning of quantum mechanics, and questions of paradox, apart from the action in biology, genetics and evolutionary studies.

As numerous disciplines work together, a new world-view of evolution is taking shape. The following is an attempt to envision this new world-view. The convergence of disciplines has led to the concept of wholeness and the concept of wholeness has let to the convergence of disciplines. They are complementary.

Wholeness and Intelligence

If the universe is whole, intelligence pervades. *Wholeness* requires its fundamental structures and functions to pattern coherently.

The study of the intelligence of the universe requires examination of first principles. Monistic logicians may choose a first principle by judging whether intelligence exists only as a basic principle within the universe (science) or whether an intelligence beyond the physical universe created it (religion). Whichever choice is made, it becomes possible to consider the intelligent universe as the subject matter of science.

Logicians using complementarity dynamics may identify their first principle: "Intelligence is complementary; both within and beyond the universe." They may name intelligence which they perceive beyond the universe, or beyond the realm of human understanding, *infinite intelligence,* and intelligence within the universe, *nature* or Gaia, suggested by J. E. Lovelock (1979). In either case, the intelligent universe is the subject matter of science. By defining concepts in this way, it should be possible to discuss evolution without discord among various religious viewpoints, and without discord between religious scholars and scientists.

Intelligence has been defined as a capacity for reasoning or conscious mental activity, and some attribute this quality only to humans. However, intelligence is also defined as the ability to gather, process, distribute and exchange information and such activity is certainly inherent in life. Nevertheless, I believe we may classify intelligence as different in kind among different lifeforms.

Conscious Intelligence:

The intelligence of humans and other higher animals is a conscious intelligence, a comprehensive awareness of self in relation to other life and non-life. Conscious intelligence permits one to deliberately change perspectives. It allows free choice among alternatives. Human intelligence exemplifies an ability to understand and communicate complex theoretical concepts as well as concrete facts. A major aspect of conscious intelligence is the ability to analyze ones own thought processes, to evaluate gathered information and concepts and utilize them in the shared creation of progress.

Instinctive Intelligence:

The instinctive intelligence of lower animals appears to be limited to possibilities for action according to patterned choices. Instinctive intelligence functions with limited understanding of concrete reality and then only in terms of immediate relationship to other lifeforms and the environment. There appears to be little or no understanding of theoretical concepts. There seems little doubt, however, that animals gather, process, distribute and exchange information both within species and between species. Animals certainly make choices. Even on the level of bacteria, information passes between individuals in the form of DNA transfer and by physical, electrical and chemical signals, signifying awareness of, and interaction with, other bacteria and the environment.

Effective Intelligence:

The effective intelligence of plant life functions in coherent patterns. Plants gather and process information, but are limited, compared to animals, in distributing and exchanging information with other lifeforms and the environment. But again, DNA (information) is transferred in reproduction, and plants respond to the actions of other lifeforms, signifying a degree of distribution and exchange of information.

Is there intelligence outside of lifeforms? Is the coherent patterning of non-living systems, intelligence? Norbert Weiner has described cybernetics as self-regulating systems of communication and control. Cybernetics is evident in both living and non-living systems. There appears to be intelligence manifested in the coherent patterning of both that which is identified as living and that which is identified as non-living.

Infinite Intelligence:

On the most profound level, there seems to be intelligence which transcends individual lifeforms and human understanding. This is the intelligence which powers creation, innovation and the purposeful interrelationships among life and the environment. Whether infinite intelligence is considered an unknown force impelling the functioning and coordination of the laws of nature, whether it is considered to be an attribute of one or another of the myriad deities worshipped by humankind, or whether it is given some other definition, is an individual choice.

Gaia — James Lovelock

The Atmosphere and Life

Until the time of space flight, scientists had a tendency to view life in its relationship to other life in its rather immediate environment. A wondrous change of perspective occurred as the astronauts transmitted pictures of Earth from space. It could now be recognized in a most profound way that Earth is more than an environment. It is a source of life, a glorious whole in which birth and development are entwined with all other elements of the system.

One of the spin-offs from space travel was that new questions arose which had never been asked before. J. E. Lovelock (1979: 1) was asked to advise on an instrument design to help determine whether life had ever existed on other planets in the solar system. The first approach suggested had been that the soil could be studied to see whether any microorganisms had existed on Mars. But Lovelock proposed testing of atmospheres to see if they contained the elements necessary to support organisms.

He concluded that Mars and Venus, with an excess of carbon dioxide and little or no hydrogen, had atmospheres which could not support life. But the investigation of those planets' atmospheres suggested questions about the composition of Earth's atmosphere, questions which had not been considered adequately, and Lovelock had the opportunity to study them.

His conclusion is truly amazing. He decided that the "...only feasible explanation of the Earth's highly improbable atmosphere was that it was being manipulated on a day to day basis from the surface and...the manipulator was life itself" (Lovelock, 1979: 6). This conclusion involves a change in the concept of sequence from environment first, life second, to an interactional, systems concept of life and environment. Lovelock (1992) clarified the meaning of this concept in less poetic but more precise terms, "The system made up of all living organisms and their material environment is self regulating with respect to climate and chemical composition at a state favorable for life." (The following is my analysis of selected sections of Lovelock's work.)

The Complements of Earth and Life

Conventional science has supplied numerous theories about evolution of the early Earth. Simulations of the various influences of physical processes on temperature, climate and life appear on a regular basis. Except for the influence of plants in converting carbon dioxide to oxygen, however, little cognizance had been taken of the influences of living organisms on processes of climate and chemical balance.

The Sun provides the energy to power life on Earth. Lovelock believes its surface, with an average temperature of 5,500 degrees Celsius (C.) (9,900 degrees Fahrenheit, F.) provided 30% less luminescence about four billion years before present (4 BBP) since its diameter then was 17% smaller. He noted, that if the average temperature of Earth 4 BBP had been 30% lower than today's average, 57 degrees F., it might be expected that the Earth would have been a frozen planet for at least one and a half billion years.

However, the different forms of oxygen present in Earth's crustal rocks suggest to Lovelock that the climate has generally been close to current temperatures with intermittent ice ages apparent in only the 30% of Earth, poleward of the then temperate latitudes. This is truly remarkable since Lovelock reasoned that a 2% decrease of the Earth's average temperature is capable of bringing on an ice age (Lovelock, 1979: 19 & 23).

Table 10.1 (Frakes, 1979: 261) compares current mean global temperature to the periods of Earth's history when temperatures fluctuated warmer and colder. Ice ages occurred in the Precambrian, Ordovician, Carboniferous — Permian boundary, and in the Quaternary periods. Many of the interrelationships of events which have contributed to these changes may be unknown, but knowledge is accumulating rapidly which may allow humans in the future to exert some positive influences upon weather patterns.

Astronomers and biologists have long marveled that the Earth moves in an orbit around the Sun that is optimal for life. If Earth were about 5% closer to the Sun, it would be too hot for life as we know it. If it were about 1% further from the Sun, it would be too cold. The Earth and Sun appear to be in a dynamic relationship which makes life possible. The Sun is about 93 million miles from Earth, far enough, Lovelock noted, to provide both power and warmth, not just warmth alone. He saw the complex relationships of these factors. The same warmth could have occurred with a smaller, less luminescent Sun located closer to Earth. However, life needs power for photosynthesis, not just for light and warmth.

The Earth's axis tilts in relation to the Sun's rays, ranging from 22 degrees to 24.5 degrees, providing opposite seasons in each hemisphere. Earth's elliptical orbit has an eccentricity varying about .25% during 100,000 years. The

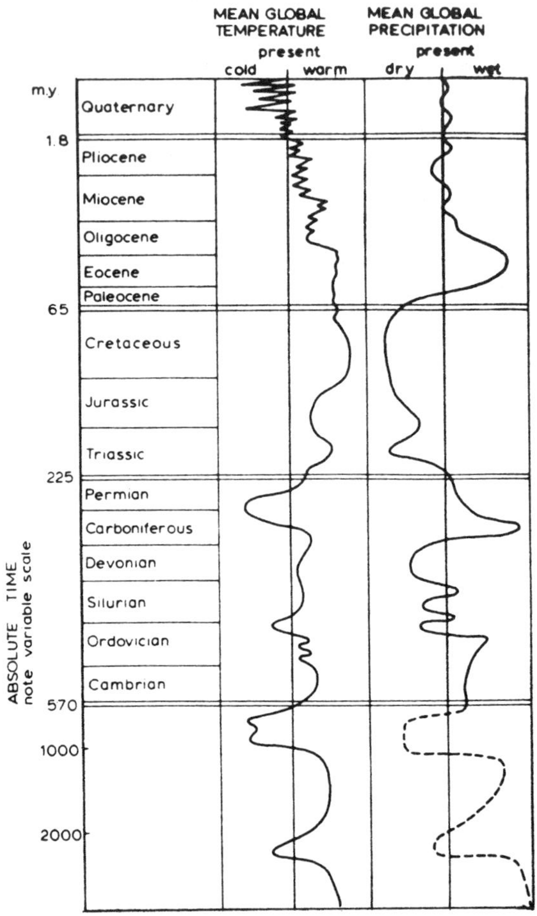

Table 10.1 Generalized temperature and precipitation history of the earth. Trends are dashed where data are very sparse. The curves are drawn to represent postulated departures from present global means, but only relative values are indicated. Note that the time scale is progressively expanded in younger time units. (From Frakes).

precession of the equinoxes (rotation of the axis of Earth's orbit) provides that Earth's approach to the Sun varies with a period averaging 22,000 years. The turning of the Earth on its own axis, providing day and night, completes the series of gross movements of Earth.

These processes cause as much as a 10% change in amount of solar radiation in different latitudes and seasons, the amount varying with time (Schneider & Londer 1984: 262). The changes in temperature from these movements would freeze any existing life if translated directly to Earth, without the influence of life processes.

Lesser known phenomena are the reversals of polarity which are believed by Allen Cox (1969) to have occurred at least 12 times during Earth's history (Table 10.2).

Normal Fields	Polar Reversals
(in millions of years before present)	
.02 - present	.02 - .03
.03 - .69	.69 - .89
.89 - .95	.95 - 1.61
1.61 - 1.63	1.63 - 1.64
1.64 - 1.79	1.79 - 1.95
1.95 - 1.98	1.98 - 2.11
2.11 - 2.13	2.13 - 2.43
2.43 - 2.80	2.80 - 2.90
2.90 - 2.94	2.94 - 3.06
3.06 - 3.32	3.32 - 3.70
3.70 - 3.92	3.92 - 4.05
4.05 - 4.25	4.25 - 4.38
4.38 - 4.50	

Table 10.2 Polar Reversals. (Data from Cox).

Earth's land mass movements, relative to polar location, known as continental drift, were first elucidated by Alfred Wegener in the 1920s. The drifting movement of the continents — or as it has developed — plate tectonics is a significant perspective. These phenomena affect climate and temperature, both broadly and in particular environments.

Whether one chooses to study the *wandering* and occasional reversal of the magnetic polar positions, or to study movement of continents in relation to each other, the equator and the magnetic poles, it is necessary to remain cognizant of the Earth's position in relation to the Sun.

That life as we know it exists on Earth at all seems to defy probability. Life on Earth has always required a range of surface air temperature no greater than a maximum of 104-114 degrees F. Average temperatures below freezing for prolonged periods, at least in the 70% of Earth's surface between 0 and 45 degrees north and south latitudes, would have prevented the continuity of life as it has existed on Earth for 3.5 billion years (Lovelock, 1979: 19).

Lovelock (1979: 36-39) estimates that an Earth atmosphere without life would be rather like Venus with 98% carbon dioxide, instead of the present .03%, with 1.9% nitrogen instead of the present 78%, with a trace of oxygen instead of the present 21% and .1% argon instead of the present 1%.

With the massive greenhouse effect of a carbon dioxide atmosphere of 98%, the average surface temperature on Earth would be 240 to 340 degrees C. instead of the current 13 degrees C. (55 degrees F.) according to Lovelock. Other chemically reactive trace gases in air today — methane, nitrous oxide, ammonia, sulphur, methyl chloride and methyl iodide — are each associated with life processes.

The composition of the oceans on Earth, with life present, is also different from the composition of a hypothetical lifeless equilibrium world which Lovelock calculated. In that world, 63% of the ocean would be water, 35% salt, mostly sodium chloride, and 1.7% sodium nitrate compared with the actual current rates of 96% water, 3.5% salt and traces of sodium nitrate and other essential chemicals. Lovelock postulated an early Earth atmosphere composed of carbon dioxide ($C0_2$) methane (CH_4) and ammonia (NH_3) which would have served to keep the planet warm, probably slightly warmer than currently.

Scientists continue to develop theories to explain the pathway from the beginning of Earth to the dawn of living organisms, 3.8 to 3.2 billion years ago. There is not yet a firm consensus about how lifeforms originated. With organisms, however, it seems evident that the chemical environment began to change. These organisms began to cycle the elements in atmospheric gases and sea water into new compounds. Early organisms probably used ammonia as food and as they did so they synthesized ammonia themselves, and produced organic compounds which affected temperature and chemical balances in the environment.

The earliest known lifeforms were photosynthesizers, blue-green algae (cyanobacteria) and bacteria, operating to change the overall reducing atmosphere of the early planet into the oxidizing environment that Lovelock reported evident in the record of the rocks of about 1.5 BBP. This set the stage for change in evolution from totally anaerobic organisms to both anaerobic and aerobic lifeforms in their separate environments.

Coccolithophores and diatoms are microscopic algae which secrete shells called "tests" and "frustules." Lovelock said that coccolithophores in the sea fixed carbon as calcium carbonate in coccolith tests, which on their demise, dropped to the sea floor as calcite and became sediments of chalk and limestone. Diatom frustules, built from the dissolved silica in the sea, became diatomaceous deposits on the ocean floor when their life spans as diatoms ended (Figure 10.2).

As these lifeforms produced the chief sedimentary alkaline and acid rocks of earth, they contributed to control of the pH balance. Life requires a pH balance near seven or eight on a scale from one, acid, to ten, alkaline. According to Lovelock, the sea has kept this chemical equilibrium by its evolving life processes during eons of time.

The change in Earth's atmosphere from a predominance of ammonia, carbon dioxide and methane, to an atmosphere containing 78% nitrogen and 21% oxygen, necessary for respiring organisms, while temperature remained relatively constant, is an amazing concept — and one that requires explanation.

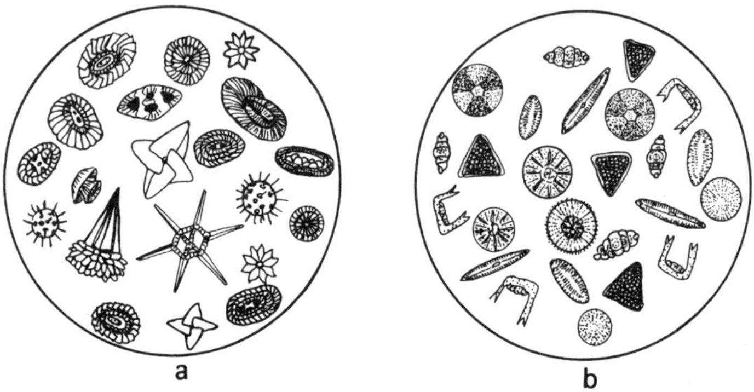

Figure 10.2 (a) Coccolithophores. (b) Diatoms. (After Bignot).

Hydrogen, Oxygen, Climate, and Life

As first life arose, some believe the seas contained about as much water (H_20) as now exists. Theories abound about the early ocean formation from outgasing of Earth core elements, but there is no adequate scientific theory to explain how it could have reached the present volume about 3 BBP and then remained about the same, while the atmosphere changed radically.

Currently oceans cover about 75% of Earth. About 30% of solar energy reaching Earth is reflected away, about half is absorbed on land and in the sea, and about 20% is absorbed in the atmosphere. The ocean, with a relatively dark surface, absorbs about 90% of the arriving radiation and most of this heat evaporates water. Over deserts, less than 1% of the absorbed heat evaporates water (Schneider and Londer, 1984: 148). The climate is driven by absorbed energy.

Stephen H. Schneider and Randi Londer (1984) in *The Coevolution of Climate and Life* described numerous interacting processes to model weather and climate, and added "The Human Connection" explaining numerous ways in which the activities of people affect Earth's climate. They made this coevolution connection at the human level, after Lovelock (1979) connected the oceans, the atmosphere, the climate and life at numerous elemental levels. It is characteristic of sequential logic that a significant new factor, such as man, may be added to a physical model of Earth and its climate. But Lovelock integrated a biological/physical interactional model in complementary systems logic, a rare achievement.

The reasoning of Lovelock suggests to me the complementary relations of living and non-living elements from the beginning of life in the sea. Organisms are composed of hydrogen, oxygen, carbon, nitrogen, phosphorus and such trace elements as iron, zinc and calcium. These elements circulate through the sea, the air, the land and living creatures.

Water consists of two atoms of hydrogen for each atom of oxygen. Hydrogen would dissipate into space if not bonded with oxygen which holds it as water, or by other chemical bonds. Hydrogen gas is very volatile and burns readily. It fires the atomic processes of the Sun.

Hydrogen is a reducing gas, that is, it functions in reverse of oxygen. It causes iron to shed its rust and supplies an atmosphere for anaerobic organisms in soil and intestinal tracts of animals. It combines with nitrogen to form ammonia to produce an alkaline environment.

Oxygen is a complement of hydrogen. It does not burn but causes other substances to burn, or oxidize, by combining with them. It rusts iron. As the complement of hydrogen, oxygen does not cause hydrogen to burn, but rather, together they become water, a fire extinguisher. Oxygen, in complementary relationship with silica (SiO_2) and other chemicals, creates an acid environment. Lovelock reported that the 98% CO_2 atmosphere of Venus is an acid atmosphere, a lifeless atmosphere.

The complements of hydrogen-alkaline and oxygen-acid function dynamically within living organisms and non-living objects to create the neutral pH balance necessary for life. Understanding of chemistry has sometimes developed using complementary concepts, but the relationships of chemicals and living organisms have often been conceived in sequential, monistic terms.

Water and Life

Oxygen and hydrogen are complements of water. Water has often been considered to be the most important component of living organisms. Monistic logic allows one to say that water is not alive and organisms are alive. By this, the monistic logician presumably means that outside of an organism, water is dead and inside an organism it is a component of that lifeform and therefore is part of life. A monist might say that water is transformed into living form as it enters the body. Or, it might be said that water remains a chemical compound until it is transformed by the body into an integral component of living tissue. In each case a sequence is postulated from non-life to life but the delineation of a transition point varies. A monist might also say that water remains a compound of H_2O wherever it is, but one finds it difficult to think of oneself as being up to 80% dead water and the rest mostly other dead chemicals.

Complementary systems do not require such either/or choices. The concepts of location of water outside of an organism or within an organism are complements of a whole, the water cycle of Earth. This water cycle can be conceived as processes of Sun and wind evaporating water into the atmosphere, precipitating it over land and sea, nourishing plant and animal life on land, picking up salt and other elements, flowing into streams and rivers, supplying water to animals, and flowing into the sea, releasing chemicals for life there.

Lifeforms participate in the cycle, incorporating water into themselves. Plants transpire water vapor, creating potential for additional rain and micro-climates like those of rain forests.

Discharges from animals provide fertilizing elements for plants on land and in the sea.

Growing plants provide atmospheric oxygen from the water (and from CO_2 of air) and provide food as an element in the web of producer plants, consumer animals, and reprocessor animals and microorganisms. In this cycle, living and dead are blurred by the realization that the cycle itself is intelligently structured and purposeful. Organisms are sustained through their own participation in nature's complementary processes.

Salt and Life

In addition to participating in Earth's generally known water cycle, organisms participate in sustaining the balance of salt in the oceans. Lovelock (1979: 92) calculated that the physical processes of the water cycle, without intervention by lifeforms, would deposit the amount of salt presently in the sea, 3.5% by volume, every 60 to 80 million years.

Thus, the sea in 3.5 billion years would be expected to be lethally salty unless some salt removal system were being employed. There is no known physical system which, of itself, could deposit salt in the massive but clearly delineated salt formations known on earth. Lovelock's hypothesis is that reef-building algae, building stromatolites; and corals, building coral atolls, established the evaporative basins in which salt precipitated from sea water and deposited in compact areas.

He postulated that the Great Barrier Reef off the northeast coast of Australia may be in development as a salt precipitation basin. If he is correct, what engineering marvels has the Earth experienced to keep the amount of salt in the sea at 3.5%? This figure is extremely important because a salt content of 6% would kill almost all lifeforms on earth, whether it is deposited internally or externally. The normal salt content of human blood is .8% to .9%.

Nitrogen and Life

The atmosphere of earth is mostly nitrogen (78%) a benign gas that moves to equilibrium as nitrate ions dissolve in the sea. If the physics of nitrogen were to take its course, apart from the activity of life, all nitrogen would eventually dissolve in the sea. This would raise both the sea's salt content and its nitrate toxicity beyond lethal levels.

Nitrogen makes up a large percentage of Earth's atmosphere because of the natural processes of bacteria and other living organisms in the soil and the sea. In the air, nitrogen regulates atmospheric density. Most nitrogen is released into the environment as ammonia (NH_3) which acts to control the acidity of the environment. Ammonia complements the oxidation of nitrogen and sulphur for the maintenance of pH near eight in the natural rains of earth (Lovelock, 1979).

Nitrogen and oxygen are complementary elements of air. Nitrogen is only mildly reactive, so instead of reacting with oxygen, it dilutes it. An atmosphere of 21% oxygen allows fires to burn with fuel up to 15% moisture content, but prevents massive conflagration which would occur with a 25% oxygen ratio (Lovelock, 1979: 37-38). The carbon dioxide atmospheres of Mars and Venus allow no fire, no lifeforms.

The microorganisms that release nitrogen, and the green plants and algae that convert carbon dioxide and release oxygen, work together to prepare and sustain an atmosphere in which lifeforms can exist on Earth. In addition to microbes that release nitrogen to air, there are bacteria growing as symbiotic organisms within the root structure of alfalfa, peas, beans and other legumes that fix nitrogen from air into a form which fertilizes these plants and establishes natural fertility in the soil.

What is the more significant intelligence of Earth? Is it represented by microorganisms and green plants which create the possibility for other lifeforms? Or is the more significant intelligence that of humankind which breathes and survives and is only beginning to learn why?

Methane and Life

There are more marvels in our air. Methane (CH_4) in the Earth's atmosphere is a whole of the complements of hydrogen and carbon. Most of it is created by fermentation of organic materials in the Earth's seabeds and muds by anaerobic bacteria. Lovelock estimates the quantity produced to be a billion tons a year, a trace compound of air. Methane preserves the anaerobic environments of Earth by clearing them of the deadly derivatives of arsenic and lead, and of oxygen that is also deadly to anaerobic organisms.

As methane bubbles up and rises into the stratosphere, it oxidizes to carbon dioxide and water vapor. The water vapor dissociates to oxygen that descends into the atmosphere and hydrogen that escapes into space. In complementary fashion, methane is oxidized in the lower atmosphere, removing enough oxygen to prevent a rapid buildup of excessive amounts. How do these seabed and mud bacteria regulate their reponse to weather and soil conditions as their release of methane keeps Earth's dynamic equilibrium of oxygen?

These anaerobic bacteria seem to have another function. The organic materials which they feed upon can remain buried rather than be eaten. It seems to be the case that the regulation of the oxygen content of air depends on the degree of activity of anaerobic bacteria in eating and rapidly multiplying, or in leaving carbonaceous material buried for their future and the future of their progeny, and ours.

Nitrous Oxide and Life

The fate of oxygen is not exclusively determined by the activity of methane producers. Lovelock indicates nitrous oxide is produced in megatons a year by microorganisms of seas and soils. The ultraviolet rays of the Sun rapidly reduce this gas, but as its production is increased or decreased, it complements the action of the methane-producing microbes in the regulation of oxygen. In the stratosphere, nitrous oxide decomposes to nitric oxide which is capable of destroying ozone, and thus may be one factor in controlling ozone concentration.

Increases in ozone are caused by changes in the atomic composition of oxygen (from O_2 to O_3) usually by means of lightning. The microorganisms which produce nitrous oxide, which depletes the ozone layer, complement the activity of algae and green plants which release oxygen. More knowledge is required to understand these complementary relationships and the influence of human activity on the dynamic equilibrium of the ozone layer.

Methyl Iodide and Life

Although produced in small quantities, methyl chloride and methyl iodide have been found to be essential to life. Lovelock reports that kelp, large sea algae, living below the low tide line on continental shelves, concentrates iodine from seawater and releases it as methyl iodide (CH_3I) which reacts in seawater and converts in part, to methyl chloride (CH_3Cl). The methyl iodide transported in air decomposes in sunlight and wafts to land where it can be absorbed by animals. Without kelp or a similarly functioning plant, there could be no healthy humans and other animals. The patterns of interrelating lifeforms and physical processes as elucidated by Lovelock are truly a spectacular insight.

Patterns — Purpose — and Process

Patterns may exemplify certainty at the ends of a continuum or a range of probability between two poles. Patterns of certainty may be considered to be purposeful. When one contemplates the purpose of the patterns within a cell, the structure and function, the purpose appears to be to maintain cell life, and also to maintain the life of the organism of which it is the whole or a part. Likewise, larger organisms generally appear to have the basic purpose of life maintenance.

Patterns are viewed by humans in relationship to themselves. Those that fulfill a human need may seem to be purposeful (a blanket, house, car, field of grain) while those that do not fulfill a need, or have not been considered or understood as to their potential to fulfill human needs, may seem to have no intrinsic purpose (a rock, microorganism, virus). It is a measure of human ignorance that purpose is often measured in this way.

Uncertainty, which can be measured in the range between 0% and 100% probability has been generally considered to be a range of purposelessness, the realm of chance, of the random, spontaneous, unknown. Even though a 50% probability retains this flavor of the unknown or unknowable, the existence of the balance of the range suggests considerable potential knowledge and purpose. Any knowledge issuing as one moves away from the 50% center (the area of the reciprocal moment) may exemplify some purpose, if only the purpose of moving toward increased knowledge existing at one of the two poles of certainty.

People often have difficulty in choosing between two certainties: Is it this or is it that? Is it true or is it false? The ability to choose on the basis of myriad probabilities is much more difficult. It is so difficult, in fact, that it is not generally taught in elementary schools and often not in secondary schools either. Monistic logic of either/or choices has not been adequately supplemented by other forms of logic.

Monistic logic is taught in the classroom using underlying concepts such as:

1. Hierarchy of knowledge:

a. The teacher is the most knowledgeable person in the classroom.

b. Pupils are ranked by the *accuracy* in verbal, written or test performance, i.e. projection of teacher's (and textbook's) viewpoint of knowledge.

2. Either/or choices:

a. Knowledge is often taught as *correct* and incorrect information.

b. Learning is commonly measured by count of the pupil's *correct* answers to true/false and multiple choice questions.

3. **Hierarchy of intelligence:**

   a. Bacteria and viruses: mindless threats to other lifeforms.

   b. Plants: limited capacity to respond to stimuli.

   c. Lower animals (such as insects): instinctive response to stimuli.

   d. Higher animals: instinct and varying degrees of reason.

4. **Evolution progression:**

   a. Movement from simple to complex, from small to large, from mindless to intelligent.

   b. Development of plants and animals through increasingly complex and more intelligent sequential stages from one-celled organisms to humans.

A teacher using complementarity dynamics might base her instruction on concepts such as:

1. **Both hierarchy and democracy of knowledge:**

   a. Each person has different life experiences and children are encouraged to evaluate their experiences.

   b. Children are exposed to many viewpoints.

   c. Experts should present a range of data for consideration.

   d. Children are encouraged to read frequently and broadly.

   e. Controversial issues are discussed.

2. **Complementary approach:**

   a. Consideration not only of those facts thought to be correct by experts, but also the complements, the opposites of those facts and the range of possibilities between extremes.

   b. Evaluation of the thinking processes of children by teachers with goal of teaching methods of reasoning.

3. **Both hierarchy and democracy of intelligence:**

   a. Intelligence is not a characteristic of humans alone; it permeates all of creation.

   b. Humans could not live if microorganisms did not regulate the elements in the seas and atmosphere.

   c. Humans have the responsibility to conserve lifeforms — not only for themselves, but also for the whole ecosystem of the planet.

   d. Humans should do no less for life than the microbes do.

4. **New theory:**

   a. Evolution is about to be envisioned from entirely new perspectives.

   b. The new perspectives of evolution via new systems of logic will carry children to expanded insights.

Perspective:

To consider Darwinian evolution as *the* theory of evolution is no longer tenable. Although it may remain as a starting point for evolution studies, Darwinian theory is crumbling under the barrage of 20th Century discoveries, theories and logics.

No doubt, present theories about the early Earth will also be displaced as more knowledge is gained. Each generation is surprised to discover the inadequacies of previous generation theories. Often, however, classical patterns of reasoning are rigidly followed long beyond their usefulness. Students have not been challenged to consider new logics. New facts are more easily accepted than are new means of interpretation. However, new means of interpretation may be at least as productive as the facts themselves.

11

Atoms, DNA, Bacteria, Viruses and Humans

Marvelous insights into evolution have been occurring in the fields of biochemistry, microbiology, symbiology and virology, virtually without public awareness. Humans have ascended to a pinnacle of evolution. So advanced are they, that they are contemplating the scientific guidance of their own future evolution by means of genetic engineering.

The human body, with quadrillions of cells functioning as a whole, is a miracle. The exquisite human mind not only controls the body, but also escapes from it in pure energy of thought.

The finite intelligence of humankind contemplates the infinite wisdom of the universe. That people identify as images, complements or essence of the infinite, is justified in complementarity dynamics. Science, philosophy and religion have all contributed to interpretations of paradox. The knowledge of science need not restrict the mind from soaring from the paradoxes of science to study of the paradoxes of philosophy and religion. It is logical to do so. How else are all the paradoxes of life to be resolved?

The sheer volume of knowledge that has been developed in the 20th Century is so phenomenal that one can only hope to grasp a fraction of it. A living generation has seen cultural evolution from horse and buggy to moon landing — evolution from neighborhood conversations across back fences to instantaneous communication across continents by means of satellites. Knowledge continues to accelerate exponentially, but unfortunately, many of the Earth's people have little access to some of the major frontiers of discovery.

As late as 1947, high school students in the midwestern USA were taught in their chemistry classes that there were 92 elements. Although radioactive elements were identified by the Curies as early as 1902, the fact that additional radioactive elements were created by atomic physicists was not taught. In 1945, before high school students were given this information in their chemistry classes, atomic tests and atomic bombs scattered thousands of curies of toxic radioactive materials into the atmosphere. Are students today similarly unprepared to deal with questions raised by genetic engineering? We must try to lay the groundwork for decision making in this field.

Molecular Basis of Heredity

In the early 1950s, university professors were teaching Darwinian evolution and Mendelian genetics, although it had been known since 1925 that the fundamental molecules of life were formed by the sharing of electrons of atoms. By 1943, Erwin Schrodinger had published *What is Life*? outlining the molecular basis of heredity and previewing the idea of a genetic "code-script" believed by some to be contained in proteins, and by others to be contained in molecules of DNA.

Recently, scientists have discovered the amazing fact that molecules of DNA, even virus particles of DNA, are active participants in evolution, while most students are still taught Darwinian evolution, and the United States Supreme Court supports the status quo.

It is difficult to grasp the significance of discoveries on the cutting edge of science, such as rearrangement of DNA to create new lifeforms. It is essential to do so if democratic peoples are to control their own evolution. The alternative is to allow scientific, military and banking elites, and ultimately, multinational corporate managers, to control the direction of planetary evolution.

The basic processes of atoms, molecules and macromolecules suggest that the logic which underlies their interactions is fundamental to an understanding of evolution of molecular life, and higher lifeforms.

When the chemist Linus Pauling, one of America's greatest geniuses, left for Europe, in 1925, to visit the physics laboratories of Arnold Sommerfield in Munich, Niels Bohr in Copenhagen, Erwin Schrodinger in Zurich and William Bragg in London, chemistry was about to be fertilized by the concepts of quantum mechanics. This was to produce, in biochemistry and biophysics, basic understanding of how the movements of electrons joined the atoms of chemical elements into molecules and into the macromolecules of life.

Basic Structure of Lifeforms

The following is an outline of the basic structures of which living things are composed: atoms, elements, molecules, macromolecules (DNA) cells, bacteria and viruses. Yes, also bacteria and viruses which have evolved into indispensable organelles and plasmids of other lifeforms.

About 98.5% of the weight of living things is composed of atoms of the elements: oxygen (O) hydrogen (H) and carbon (C). Nitrogen (N) and Phosphorus (P) are also elements essential to life. These elements combine to produce the microorganisms and fungi, more than 400,000 identified species of plants, more than 1,500,000 identified species of animals, all the living species which have not yet been identified and those millions of species which previously existed on Earth.

Atoms and Elements

The simplest atom is hydrogen, with one proton in its nucleus (atomic number one) and one electron in an orbital around it. It has a valence number of one, the number of bonds which the atom makes with other atoms. The stable state of an atom usually tends toward having two electrons in the first orbit, and if it has more than one orbit, it usually tends toward having eight electrons in the second orbit.

Although scientists first envisioned the path of the hydrogen electron as an orbit like a planetary system, it eventually became evident that electrons were in "electron cloud orbitals." These are orbits in which the electrons seem to exhibit a variety of pathways. For simplicity's sake, let us consider a hydrogen atom, as in Bohr's early insight, as an electron circling a proton nucleus (Figure 11.1).

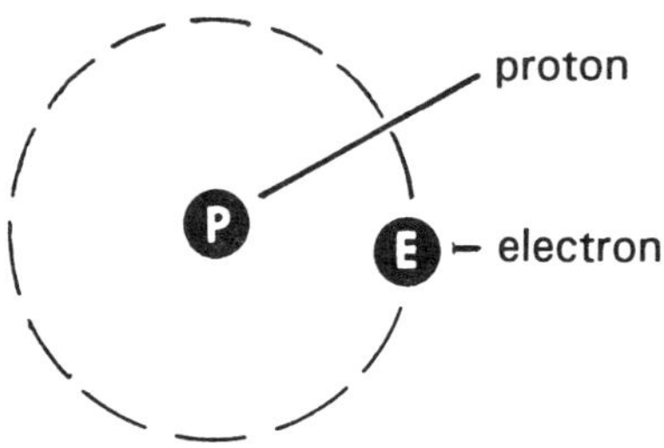

Figure 11.1 Hydrogen Atom.

Atoms tend toward having two electrons in their first orbit, one electron with a so-called "up" spin and one with a so-called "down" spin. They do not actually spin like a top but those are the names of their "flavors" their identities. An atom with only one electron in its first orbit is unstable and tends to combine with another atom.

Molecules and Compounds

Two hydrogen (H) atoms combine readily into a hydrogen molecule. While an element is made up of a number of the same kinds of atoms, a molecule or compound is usually a combination of different kinds of atoms. In the case of hydrogen, two atoms make a molecule of hydrogen (Figure 11.2). In their larger configurations, molecules are called macromolecules. Proteins are macromolecules made of many smaller units, the amino acids, which are the basic building blocks of life.

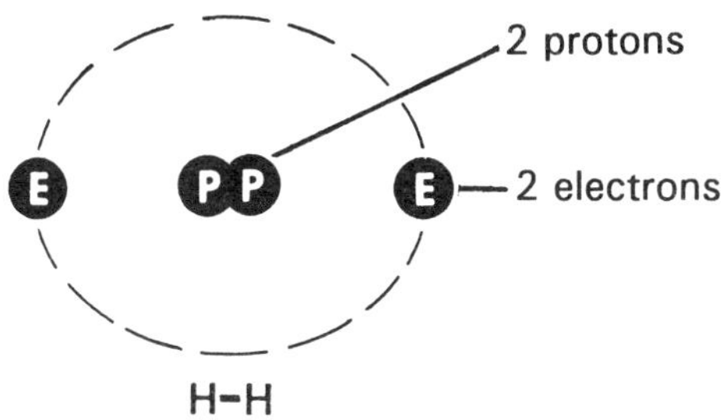

Figure 11.2 Hydrogen Molecule.

The dash between H and H in Figure 11.2 means two hydrogen atoms share two electrons. A double dash (=) between two atoms means they share four electrons.

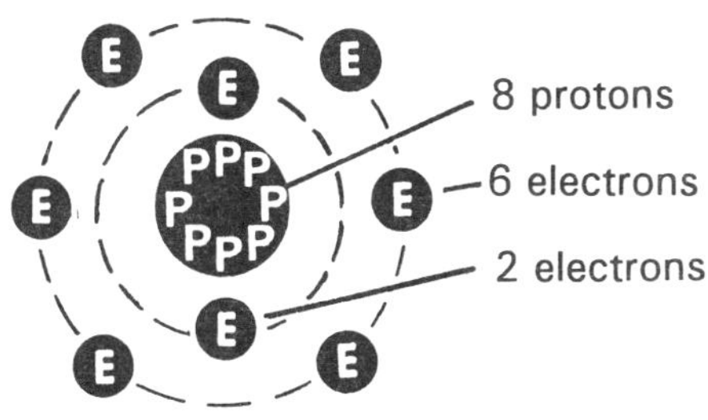

Figure 11.3 Oxygen Atom.

Hydrogen combines readily with oxygen (Figure 11.3).

Oxygen has an atomic number of eight, eight protons (and eight electrons). It has two electrons in its first orbit and six in the second orbit. This is unstable, which means that it will tend to combine with other unstable atoms. It will combine to obtain a total of eight electrons in its second orbit. Atoms move in configurations to obtain stability, therefore oxygen combines readily with hydrogen to form water (Figure 11.4).

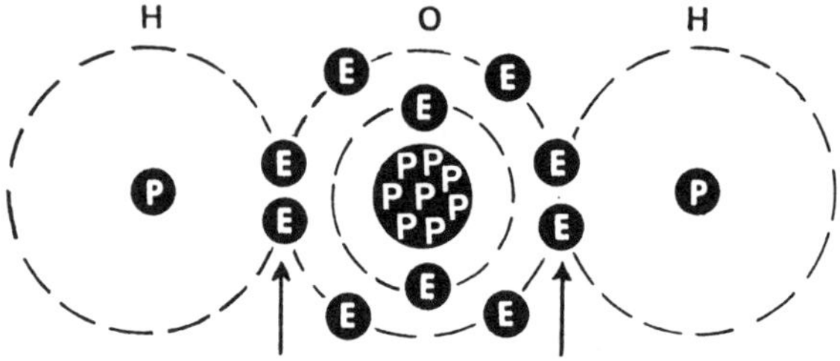

Figure 11.4 Water Molecule.

Hydrogen, with one electron, can share an electron with oxygen and effectively have two in its first orbit, and oxygen can share one electron with each of two atoms of hydrogen and effectively have eight electrons in its second orbit. Therefore a water molecule, H_2O, a main ingredient of life, is more stable than hydrogen or oxygen atoms alone.

The concept of shared electrons, in which electrons are considered to be in their orbits around two or more nuclei simultaneously (or in equal energy states) was a brilliant insight of Gilbert N. Lewis which Linus Pauling elaborated as "resonance." The philosophical concept of resonance can be perceived in reciprocal time logic. This logic is required because electrons cannot be measured in sequential patterns of motion.

The chemical element most closely identified with lifeforms is carbon. It is the basis for the field known as organic chemistry or the chemistry of life. Carbon has an atomic number of six, six protons (Figure 11.5).

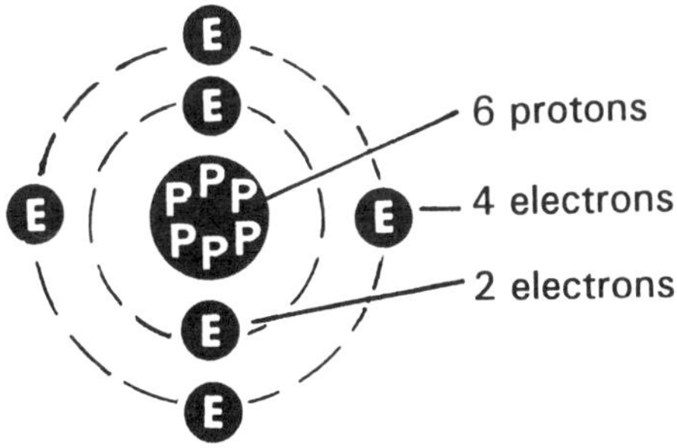

Figure 11.5 Carbon Atom.

Since carbon has only four electrons in its second orbit, it also combines readily to fill up the second orbit with shared electrons. Carbon's tendency to combine is expressed:

$$-\overset{|}{\underset{|}{C}}-$$

It easily forms a chain with other atoms. Another way to illustrate the linking of atoms is Figure 11.6.

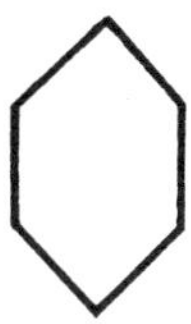

Figure 11.6 Simple Benzene Ring.

This represents the basic unit configuration called a benzene ring because six carbon atoms and six hydrogen atoms create benzene, C_6H_6 (Figure 11.7).

These figures, expressed in two dimensions, are only representations because the atoms are four dimensional. In methane, a carbon atom may be visualized as bonding to four hydrogen atoms in three dimensional space, with each hydrogen atom moving equidistant from each other in reciprocal moments of time (time being the fourth dimension) (Figure 11.8).

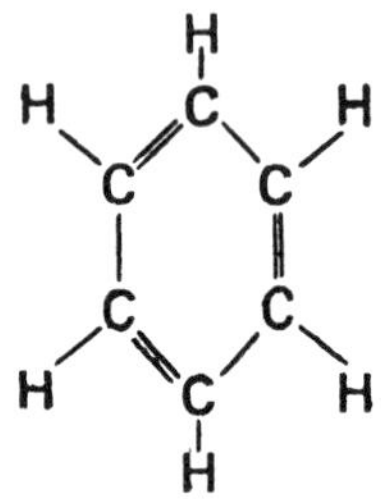

Figure 11.7 Benzene Ring.

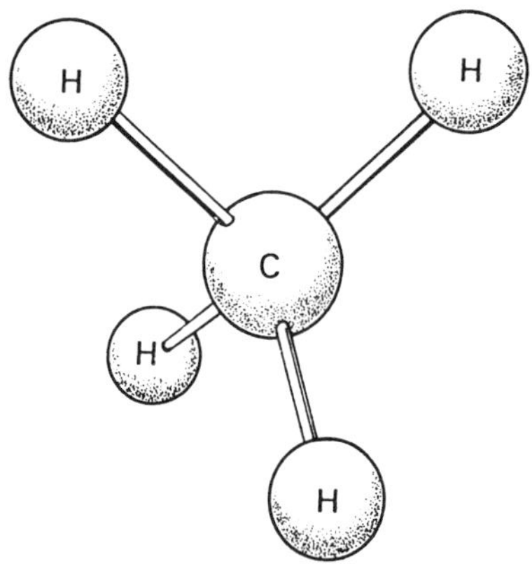

Figure 11.8 Methane, CH_4.

Electrons have a negative charge of electricity, protons have a positive charge, so they attract each other with an electrostatic force. A nucleus of protons also often has an equal number of neutrons within it, which have no electric charge. Protons and neutrons are held together by a strong nuclear force. These three atomic particles — electrons, protons and neutrons — comprise the basic atom and function together in regular configurations unless split in atomic processes of bombardment. In chemistry and biology, they can be considered to *trade* or *share* electrons to form and reform molecules and macromolecules.

The *sharing* of electrons, in reciprocal moments, is the basis for combining of atoms into different molecules. Molecules with two atoms of hydrogen for each one atom of oxygen, make the compound water. Each element may combine with other elements to form different compounds.

Macromolecules of Lifeforms

Compounds formed by the elements hydrogen, oxygen, carbon and nitrogen have been discovered to be the basic building blocks of life. Those elements and others form DNA, RNA and amino acids, the ingredients of proteins. DNA and RNA are two kinds of templates of life matter. DNA and RNA are formed in five basic molecules called bases (Figure 11.9).

As suggested by the above symbols, the cytosine molecule (nucleotide) fits with the guanine nucleotide and the adenine nucleotide fits with the thymine nucleotide to form DNA. Cytosine, guanine and adenine are present in both DNA and RNA. However, thymine is replaced by uracil to form RNA. The knowledge of these complementary patterns was used by Frances Crick and James Watson in the early 1950s to develop a model of the structure of the DNA double helix. They visualized DNA as structured in complements of AT and CG (Figure 11.10). RNA is structured from AU and CG.

The outside of the helix ladder is composed of alternate molecules of sugar (S) and phosphates (P) which hold each side of the ladder of nucleotides together as the two strands divide (Figure 11.11).

A codon of DNA is the basic hereditary code which is a fundamental structural component of the genes of all lifeforms. These codons can be visualized as triple beads on a string, triplets which reside on one strand of the double helix (reading down in this figure), and also reside within every living cell on Earth.

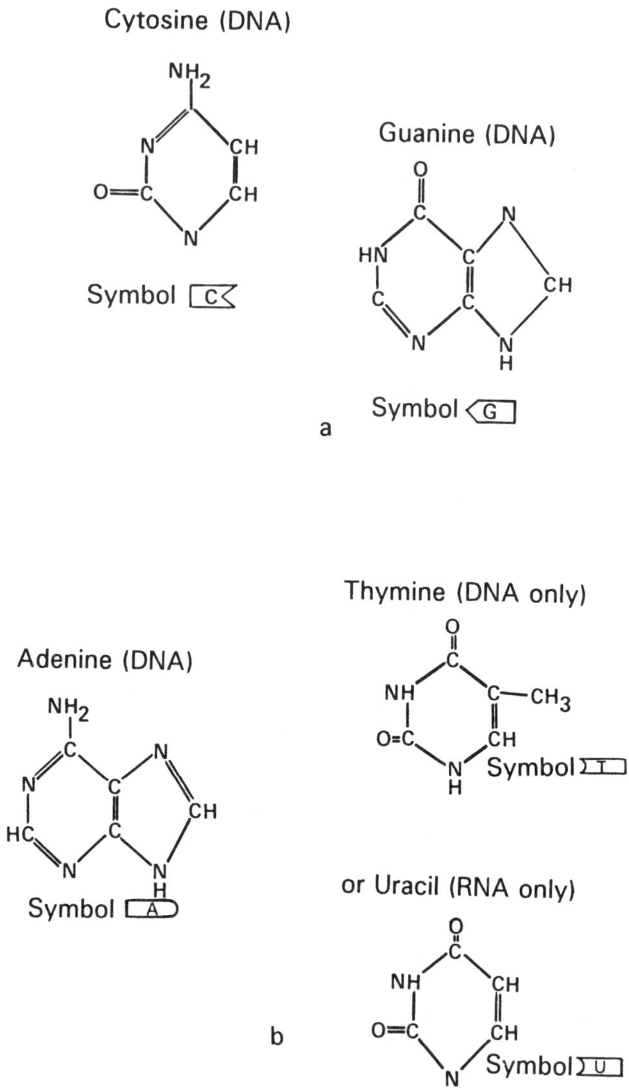

Figure 11.9 (a) Cytosine, Guanine, (b) Adenine, Thymine, Uracil.

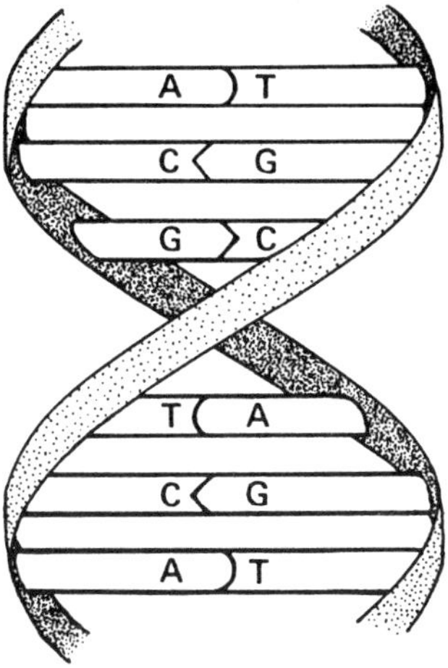

Figure 11.10 Double Helix of DNA. (After James D. Watson).

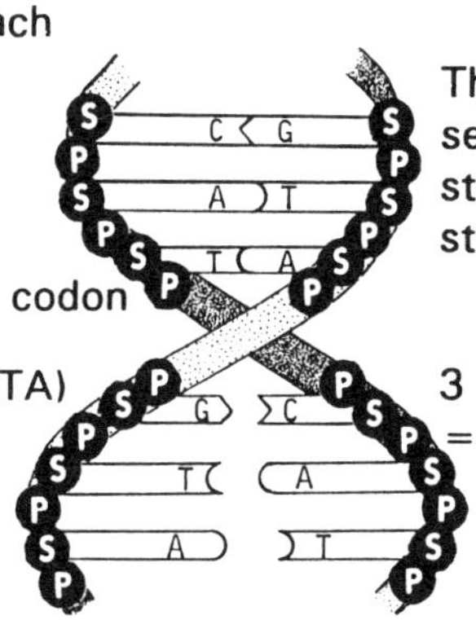

Figure 11.11 Double Helix Dividing. (After James D. Watson).

Each codon, a triplet of three nucleotides, codes for one amino acid. The order of nucleotides on the strand, including those which are start and stop signals, determines which of the 20 amino acids are created. Table 11.1 identifies the RNA code of triplets. Table 11.2 identifies the 20 amino acids of life which are determined from the RNA code.

First Letter	Second Letter: U	C	A	G	Third Letter
U	UUU -Phe UUC UUA -Leu UUG	UCU UCC -Ser UCA UCG	UAU -Tyr UAC UAA -His UAG	UGU -Cys UGC UGA-His UGG-Trp	U C A G
C	CUU CUC -Leu CUA CUG	CCU CCC -Pro CCA CCG	CAU -His CAC CAA -Gln CAG	CGU CGC -Arg CGA CGG	U C A G
A	AUU AUC -Ile AUA AUG-Met	ACU ACC -Thr ACA ACG	AAU -Asn AAC AAA -Lys AAG	AGU -Ser AGC AGA -Arg AGG	U C A G
G	GUU GUC -Val GUA GUG	GCU GCC -Ala GCA GCG	GAU -Asp GAC GAA -Glu GAG	GGU GGC -Gly GGA GGG	U C A G

Met - start
Val - start
His - terminate

Table 11.1 RNA Triplet Codon Code for Amino Acids.

The triplet codon code of DNA transcribes into RNA as U is substituted for T. This RNA code is the basis for creation of amino acids, including those which code as start and termination signals. The abbreviations, Phe, Leu, etc., are the names of the amino acids created (See Table 11.2).

Phe — Phenylalanine	Gln — Glutamine
Leu — Leucine	Asn — Asparagine
Ile — Isoleucine	Lys — Lysine
Met — Methionine (start)	Asp — Aspartic Acid
Val — Valine (start)	Glu — Glutamic Acid
Ser — Serine	Cys — Cysteine
Pro — Proline	Trp — Tryptophan
Thr — Threonine	Arg — Arginine
Ala — Alanine	Gly — Glycine
Tyr — Tyrosine	His — Histidine (terminate)

Table 11.2 Twenty Amino Acids of Life.

The Structure and Function of Cells

Every organism alive today, from the simplest to the most complex, is made up of one or more cells. Cells have characteristic structures in common and these structures are believed to be controlled by DNA/RNA as well as to contain DNA/RNA, a complementary concept.

Prokaryotic cells, with no separate nucleus, have their DNA/RNA (genophore) in the double ring formation of nucleotides in the cytoplasm. Eukaryotic cells contain DNA in the nucleus in which nucleotides are joined as codons into genes and chromosomes.

Let us consider the parts of a eukaryotic cell as an operating industrial plant (Figure 11.12). The nucleus is the head office. It contains the chromosomes, made of DNA, which are the major coded information source of the cell. The ribosome is the secretarial office where the secretary-transcriber works. The mitochondria are the power plants of the cell providing energy for operation. Chloroplasts (plastids) in plant cells only, are the manufacturing shop of chlorophyll, which convert sun energy to fuel for the cell. The centriole is the cell duplicating office which goes into operation only when the cell is about to divide into two cells. The plasmids or episomes are believed to be involved in cell control in ways complementary to the chromosomes in the nucleus. They will be discussed in more detail under the heading *Microorganisms*.

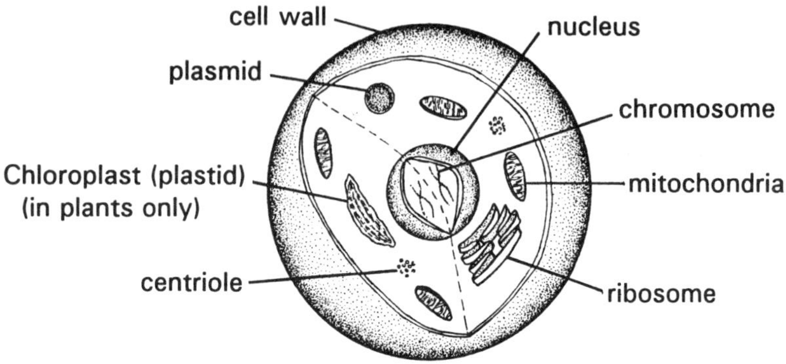

Figure 11.12 Eukaryotic Cell.

Chromosomes

Each species of animal has a characteristic number of chromosomes in the nucleus of each cell. Some plant species have a range of chromosome numbers. Humans have only 46 chromosomes. Compare the range of chromosome numbers: chimpanzees 48, rats 42, pea plants 14, fruit flies 4, butterflies 300 and single-celled foraminifera 1,500. There is no consistent pattern of chromosome numbers related to complexity of phenotype, but it is not unusual for chromosome numbers generally to be lower in more complex organisms.

Each human egg or sperm has 23 chromosomes which combine at conception to return the newly developing embryonic cell to a full complement of 46. Each human chromosome may carry approximately 30,000 genes. Although chromosomes and genes vary in length, a typical gene may be composed of approximately 1,800 nucleotide pairs (A-Ts and C-Gs). The total of 46 human chromosomes carries about three billion nucleotide pairs.

Approximately 70% of those three billion nucleotide pairs comprise single copy DNA (a string of nucleotides not repeated in the same sequence along a long chain of triplets). About 10% of this single copy DNA composes about 110,000 structural genes. Those code for proteins which build the body. Proteins may be from 40 to 500 amino acids long (120 to 1500 codons). The rest of the genes are regulator genes, operator genes (operons) or inhibitors, and genes of unknown function. Those should not be called "redundant" or "junk" genes just because they are not yet understood! It is believed that eukaryotic cells depend more on inhibitor genes for control while prokaryotic cells depend more on operons.

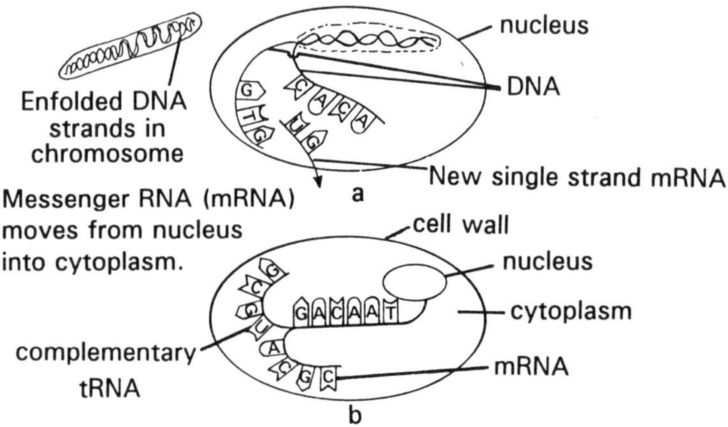

Figure 11.13 (a) Chromosome, (b) Eukaryotic Cell, mRNA to tRNA.

Chromosomes do not always appear the same under the view of an electron microscope (Figure 11.13). When they are about to divide, they are enfolded into their characteristic shapes and number. However, when they are replicating an RNA strand from a DNA strand, they appear to permeate the whole nucleus looking like a mass of threads. As the DNA unwinds from the double helix it appears to produce a complementary RNA strand which moves through a pore in the nucleus wall into the cytoplasm. This is called messenger RNA (mRNA) because it carries the DNA message into the cytoplasm.

The messenger RNA from the nucleus carries the genetic code to the ribosome where the mRNA is "read" and a complementary copy is produced, called tRNA because it is transcribed, producing itself and-or being produced from "spare parts" of atoms in the cytoplasm. The idea of producing itself and-or being produced is a complementary concept, and may explain *simultaneous* construction of the tRNA complement from the mRNA strand, in a reciprocal moment.

Each triplet of tRNA translates into an amino acid and-or a start or stop signal which form and-or are formed into protein chains (Figure 11.14).

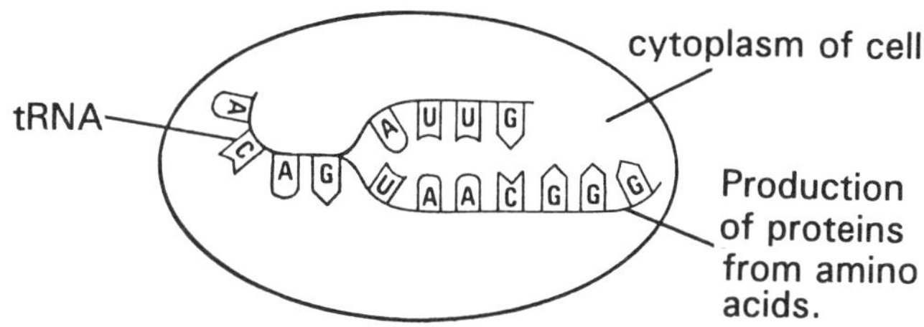

Figure 11.14 Eukaryotic Cell, tRNA to Proteins.

The triplets GGU, GGC, GGA and GGG all code for glycine, the simplest of the amino acids. The ribosome which reads this code as the mRNA passes along it composes the tRNA which composes glycine from the atoms in the cytoplasm in the following configuration (Figure 11.15):

Each amino acid contains at least one amino group (ammonia), and one carboxyl group. The other proportions of atoms of hydrogen, oxygen, carbon, and-or nitrogen are slightly different for each amino acid. It can be seen from the triplet codes that it is important that the start and stop triplets are in the correct location to produce the right triplets to create the exact amino acids needed by the cell. As the amino acids are produced, they combine to form protein chains which are the workers of the cell and the body.

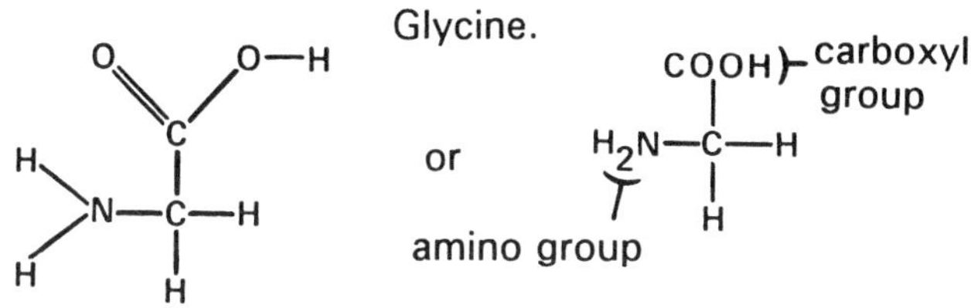

Figure 11.15 Glycine.

A common bacterium, *Escherichia coli*, contains about 5,000 kinds of organic molecules, and of these 3,000 are various kinds of proteins. The human body contains about 50,000 different kinds of proteins produced from combinations of 64 triplets which code for the basic 20 amino acids in over 200 different kinds of human cells.

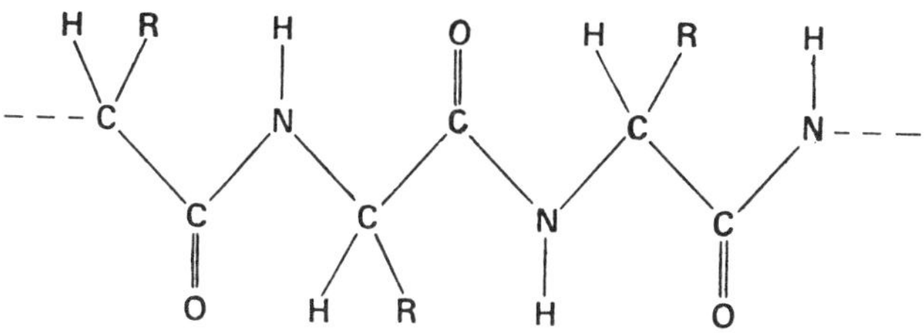

Figure 11.16 Protein ("R" represents groups of additional atoms).

A protein can be described as a chain of atoms bonded by the electrons which hold them together (Figure 11.16).

Each cell's mRNA controls the transcription into tRNA which is the structural sequences of three triplets which translate into the amino acids which build the proteins which construct cells, and as protein enzymes, control the functioning of cells in the body.

Mitosis and Meiosis

In addition to producing RNA from DNA and building proteins, the cell has an additional major essential function, reproducing itself. Prokaryotic cells replicate in a process of division of one cell into two cells which will be considered under the topic *Microorganisms*.

Eukaryotic cells divide in a process called mitosis. In this process, the two centrioles (or microtubule organizing centers) of the cell form spindles (Figure 11.17).

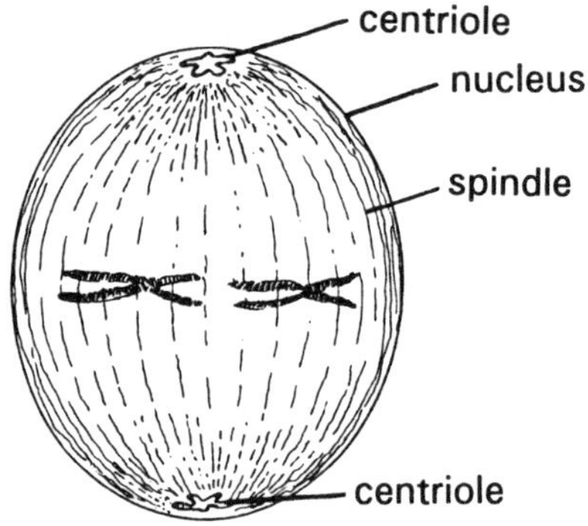

Figure 11.17 Mitosis. (After Johnson).

The chromosomes, coiled in close configuration, line up in a plane between the spindles, and one of each chromosome pair is pulled into separate halves of the cell (Figure 11.18).

After mitosis, each daughter cell contains the same number of chromosomes as the parent cell. (The chromosomes having produced their complementary strand of DNA in the process.) Mitosis is the method of replication of body cells.

In addition to mitosis, the process of meiosis occurs in two sex organisms. It is the method of replication of sex cells by division. Meiosis is a two-fold operation; the first is like mitosis, then the chromosomes divide again into one half the original number; in humans 23 chromosomes in each egg and sperm.

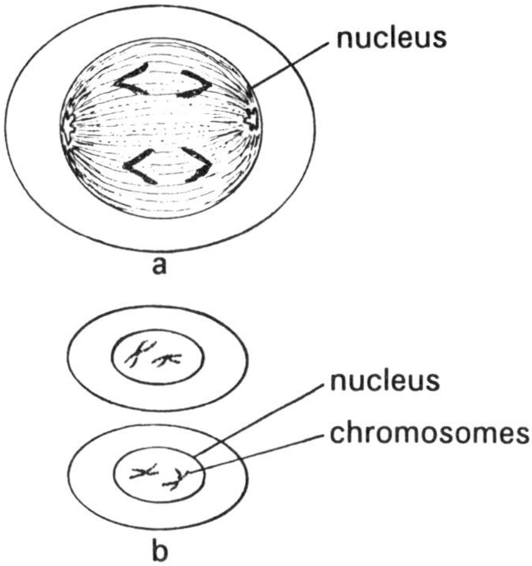

Figure 11.18 (a) Mitosis continued; (b) The cell divides into two. (After Johnson).

Body cells divide frequently in some tissue such as the lining of the stomach. There, they may divide every few days, with new cells replacing old. The neuron cells of the brain, on the other hand, are not replaced in a lifetime. Egg and sperm cells do not divide, remaining in the 23 chromosome configuration until they are joined to begin a new life.

When Crick and Watson developed the model of DNA in the early 1950s, it was thought by some that the DNA in the cell's nucleus was the only code for inheritance. As the structure and function of amino acids were determined and the genetic code interpreted as it influenced the development of various lifeforms, it began to seem that humans could learn to splice nuclear genes to assume control over evolution. But by the 1960s other genes were found in the cytoplasm. New evidence of symbioses was emerging. Inheritance was not as simple as had been thought.

Microorganisms

When Lynn Margulis (1970, 1993) studied cells and microorganisms under magnification of the electron microscope, she found not only the competition of Darwinian evolution, but also evidence of numerous symbiotic processes within cells and among microorganisms.

Her endosymbiotic theory of evolution posited that parts of the cells of advanced organisms did not all evolve together in a sequential pattern from simple to complex species. Rather, certain parts of cells had different evolutionary histories!

DNA patterns in the nucleus of cells of similar species are very similar, but the DNA of the nucleus of a cell is quite different from the DNA in the mitochondria of the same cell. The DNA of the nucleus of plants is different from the DNA of the chloroplasts (the chlorophyll container and solar energy processing organelle). The DNA of undulipodia (hair-like appendages or organelles) of organisms is different from their nucleus DNA. This led Margulis to theorize that eukaryotes originated by symbioses of different kinds of bacteria.

These amazing discoveries inspired Margulis to propose that, early in evolution, symbiotic parts of today's cells represented separate free-living organisms. These organisms joined together in symbiotic associations to solve the problems of mobility, to overcome the problems inherent in moving from an anaerobic to an aerobic environment, and to address the other problems of changing environments.

Scientists have generally accepted the symbiotic nature of the early evolutionary association of chloroplasts and other organisms and the association of mitochondria in early bacterial organisms. The jury is still out on the symbiotic association of spirochetes in the ancestry of undulipodia.

With this new view of endosymbiotic evolutionary processes, it becomes necessary to write a new theory of evolution. No longer is it possible to think of evolution as mainly a competitive process. Evolution theory must have large doses of symbiosis!

The rewriting of theory to incorporate the evolution of all the microorganisms which are known to have lived on earth is a project for decades of dedicated scientists. The vague outline of new theory is just lifting onto the horizon of thought.

This is known:

The first fossil evidence of life, 3.8 billion or more years old, is bacteria-like organisms. The blue-green algae found in the same strata is believed to be bacterial also. Margulis (1986: 119) noted, "As early as 1893 the German biologist A. Schimper proposed the idea that photosynthetic parts of plant cells came from cyanobacteria (or 'blue-green algae' as they were called at the time)."

She reported that in 1910, K. S. Mereschovski published his view that eukaryotic cells originated from various kinds of bacteria. The symbiotic concepts expressed by Margulis and her predecessors were seriously neglected by most scientists until very recently. The knowledge gained by viewing organisms through the electron microscope and the information developed about the self-organizing abilities of DNA, force reevaluation of Darwinian ideas of evolution, which suddenly seem quite archaic.

Margulis proposed "serial endosymbiosis of the eukaryotic cell" a theory dubbed SET by F. J. R. Taylor. From bacteria of various kinds (Monera) the protists (Protoctista) evolved. The protists then evolved into fungi, plants and animals (Figure 11.19).

Philosophical Speculation from Margulis Onward

If some early prokaryotes evolved a nucleus and acquired mitochondria, becoming eukaryotic about 1.5 billion years ago, as Margulis suggests, it seems probable that some mitochondria of early plants and animals was bacteria which was incorporated by amoeba-like organisms, rather than being digested by them. Perhaps the difference between symbiosis

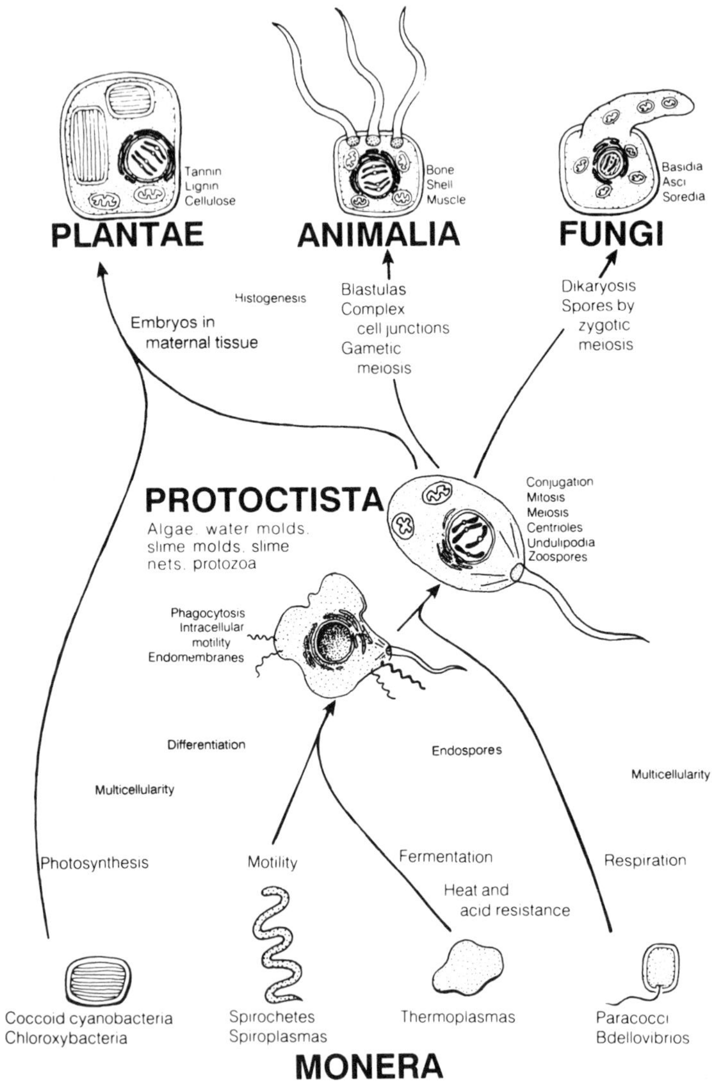

Figure 11.19 "Serial Endosymbiotic Theory of the Origin of Eukaryotic Cells" (From Margulis, 1993).

and a prey relationship was essentially a matter of the working of amoeboid-like DNA which coded for digestive juices!

Margulis estimates that humans today, in addition to having 10 quadrillion body cells, harbor 100 quadrillion bacteria, yeasts, molds, viruses, and other miscellaneous microorganisms, 10% of the body by weight. Some of the ancestors of these microorganisms may have been incorporated as mitochondria and undulipodia into the cells of larger organisms, for a billion and a half years before the beginning of humankind.

Scientists and philosophers are only beginning to speculate about the origins of such organelles and symbionts of the body. The bacterium *E. coli* is such a well-known, protective, and hard-working resident of the human intestine that it seems truly to be a part of the body though it is also clearly a bacterium, removable by scientific techniques. Has it been with us from our beginnings?

How little we know of our guests! Since the 1880s, it has been thought that all germs are villains. What a revelation to learn that as much as 10% of our weight may represent symbiotic microorganisms keeping us alive!

Since prokaryotes appear as the first fossil evidence of life, they are, or should be, a major subject of interest in evolution studies. As with early eukaryotes, limited genetic information is available directly from prokaryote fossils. It is necessary to study living bacteria in order to learn how early bacteria may have functioned.

The bacterium, *Escherichia coli* has been the subject of intensive investigation. Until the early 1940s, it was thought that bacteria multiplied only by asexual binary division and had no sexual differentiation. The circular chromosome (genophore) had not been detected. With new staining and other laboratory techniques, and use of the electron microscope, it was found that double ring-form DNA transferred in two patterns. It split into two parts, each part moving to a separate half of the cell as it divided into two new cells. DNA could also transfer from one bacterium to another in a circular motion. The transfer might be incomplete, transferring only a portion of the existing DNA.

It was also learned that bacteria are either donor cells (called males) or recipient cells (called females) (Bainbridge, 1980: 88). The sexes are not differentiated as in higher animals but contain a sex factor (F) a plasmid (a unit of replication) which transfers from one bacterium to another, sometimes separately from chromosomal transfer, sometimes attached to the chromosome strand. A female bacterium (F-) can be converted into a male (F+) by transfer of the F factor.

It was learned that the bacteria, *E. coli*, have appendages called piluses; one of which is a sex pilus, a bridge which attaches an F+ and F- bacterium together before transfer of genetic material (Figure 11.20).

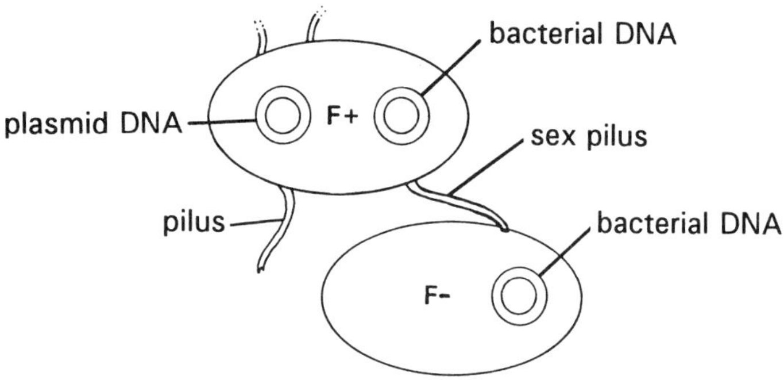

Figure 11.20 Two Bacteria.

DNA is transferred during conjugation of bacteria which may last 90 minutes or longer. It appears that a single strand of DNA is transferred and then is synthesized into a double strand in F- as F- becomes F+ with receipt of the F factor. F factor cannot be transferred if a bacterium has a resident F factor (Figure 11.21).

The circular genophore and the plasmid F factor have been analyzed. Scientists can map the DNA molecules of the genophore and also of plasmids, which vary in length from two to 60 micrometers. Similar plasmids have been found, not only in bacteria but also in fungi and other organisms.

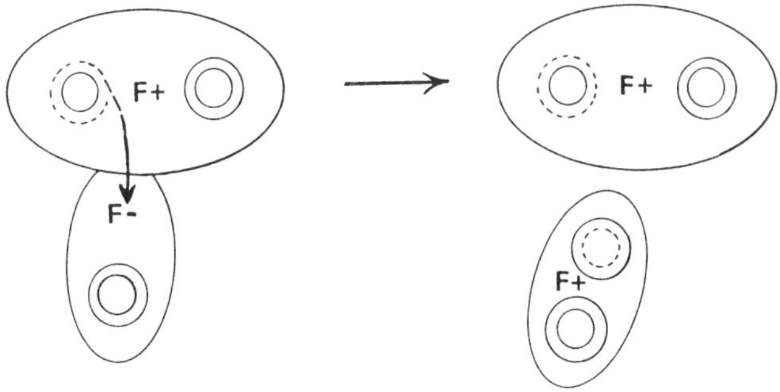

Figure 11.21 Sexduction, F Factor Transfer.

One of the most amazing discoveries in the study of bacteria has been the relationships of viruses (phages) to bacteria in the most fundamental, intimate and interrelated manner. Three male strains of *E. coli* (known as F+, Hfr, for high frequency mating strain, and F prime, [F'] a mutant male which incorporates nuclear DNA into its F plasmid) are found to interact with viruses, bacteriophages known as MS2 and fd. These phages were seen to attach to the male E. coli's sex pilus (Figure 11.22).

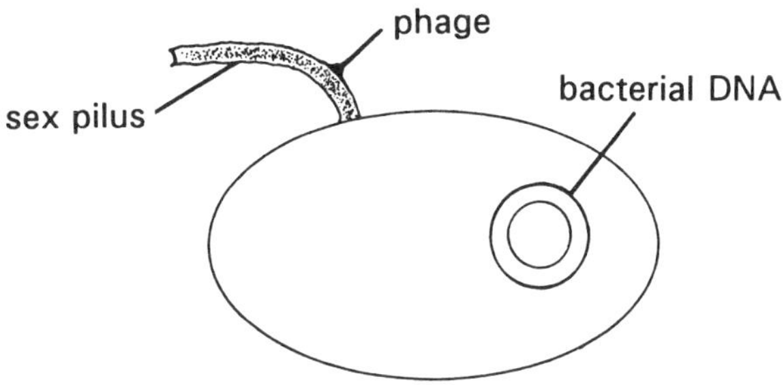

Figure 11.22 Bacterium and Phage.

Bacteria have phage receptors, specific sites where phages attach, which are chemically different for each phage type. A bacterial mutation may close a site for one variety of phage but not change the receptor sites for other types of phages. Are bacteria designed to receive and selectively incorporate viruses? A mind-boggling proposition!

It is now known that viruses of bacteria (bacteriophages) can transfer their DNA, or DNA from other organisms into bacteria or into other vectors e.g. organisms like mosquitoes, ticks, and mites. Different viruses have different vectors and different hosts.

DNA can also be transferred without phage intervention. A loose strand of DNA in a liquid medium can attach to a living cell and insert itself into that cell. It then transforms the DNA of the cell into a new DNA configuration. This is called "transformation."

When viral DNA inserts itself, and-or is inserted into a bacterium, or into a cell of a higher lifeform, it is capable of taking over the cell and putting it to its own use. A virus appears lifeless outside a cell. Within a cell, it can control the cell by converting the machinery of the cell to replication of virus DNA. A phage may use the cell machinery to create hundreds of new phages and then "lyse" dissolve the bacterial wall, to release the new viruses. A phage engaging in such activity is called a "virulent" phage. This is the process leading to disease and death of other organisms. Virulent phages always seem to kill the bacteria they infect.

A virus has other possibilities. It may enter a cell and remain dormant for varying lengths of time, perhaps for the life of the cell, apparently doing nothing. It is then called a "temperate" virus. It may appear as a plasmid under high magnification. At any time the virus may enter the lytic cycle and make more viruses. On the other hand it may persist free in the cytoplasm, or integrate into the bacterial chromosome, taking on new activities then called a "prophage." Temperate phages can incorporate their host's genes into their own and transfer them from one bacterium into another. They may transfer by generalized transduction (moving) of any gene, or by specialized transduction, in which only one or two genes are moved (Bainbridge, 1980: 107).

Among the thousands of phages there is amazing multiplicity of functions. For example, the P1 phages live individually in the cytoplasm and can move any gene. Lambda phages integrate into the genophore and move only the one or two genes next to themselves.

Viruses come in different shapes. The structure of the T4 virus gives an eye-popping clue to the nature of viruses. T4 was significant in the discovery of DNA as the genetic code of life. The icosohedran head contains the DNA. The stem attaches the head to a baseplate, to which are attached legs which latch unto a bacterium (Figure 11.23).

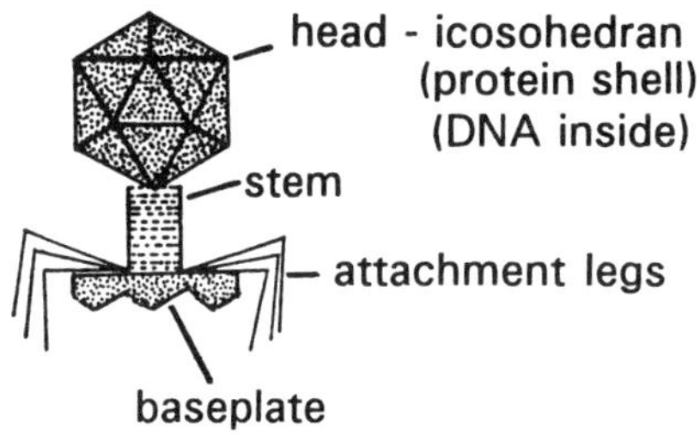

Figure 11.23 T 4 Phage.

It has been determined that T4 functions like a syringe, shooting its DNA into a bacterial cell. The outer protein coat remains outside of the bacterial cell as DNA transfers into the bacteria, becoming a plasmid (Figure 11.24).

Scientists are on the cutting edge of discovery of the essence of phages, plasmids, episomes and the F factor of bacteria. There are scientists who believe that some plasmids are degenerate phages. Reasoning in the manner of Lynn Margulis, might it be speculated that symbiotic phages became plasmid cell organelles in some early period of evolution?

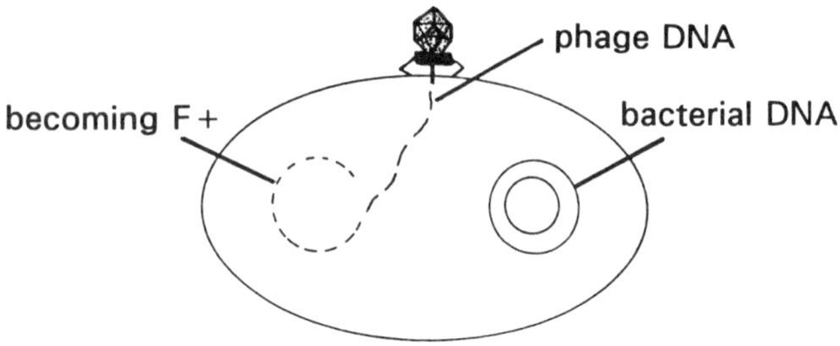

Figure 11.24 T 4 in Action. (After Nester et al).

There are nonconjugative plasmids and conjugative plasmids. They are very similar to phages, but lack a protein coat, thus are *naked* DNA. In cell division of eukaryotes, plasmids migrate in the cytoplasm and transfer with it as the cell divides. This is called *cytoplasmic inheritance.*

The term "transposons" has been coined to designate insertion sequences of genes, which are believed to interchange between phages and plasmids. The development of research on transposons may hold a key to further understanding of the interconnectedness of all lifeforms.

"Episomes" are cell organelles which seem to be equivalent to plasmids. Allan Campbell (in Smith, 1972: 534) said there are:

> ...cogent reasons for abandoning the term 'episome' altogether.... In 1958 Jacob and Wollman proposed the term 'episome' for genetic elements such as bacteriophage lambda or the fertility factor F that could replicate in two alternate modes — either as part of the bacterial chromosome or as autonomous, extrachromosomal elements.... The term 'plasmid' also includes all autonomous extrachromosomal elements, whether or not chromosomal integration is demonstrable.

Such discussions illustrate that understanding the genetics of microbes, viruses and cell organelles by microbiologists and virologists is still in an early stage of development.

Virus Function In Evolution

It is clear from the vast body of evidence developed in recent years, that viruses are present in almost all protists, plants, and animals. They are capable of transferring DNA between different organisms. They not only transfer DNA which destroys hosts, but also transfer genes which change the DNA of host plants and animals in constructive ways.

Many scientists are currently writing about the role of viruses in reorganizing DNA, and premises incorporating reorganization of DNA in evolution theory are emerging. John Gribbin (1985: 311) pointed out:

> By 1981 a meeting on evolution held at King's College in Cambridge came out loud and clear with the message that the genomes of higher organisms are in a state of dynamic change, with considerable rearrangement of the genes among the chromosomes being the norm on an evolutionary timescale, and perhaps being a driving force of evolution. It had become clear that just as human intervention using recombinant DNA techniques could take artificial genes and insert them into chromosomes, so viruses might transmit genes from one host to another during their own life cycle.

I believe this confirms that the time is ripe for a new theory of evolution.

Evolution of Prokaryotes and Viruses

Humans have wondered whether non-living atoms and molecules could evolve into complex living cells and organisms.

They have wondered whether viruses should be considered to be a stage between non-life and life. Since virus DNA can function only within the cell of a living organism, are viruses alive (perhaps in suspended animation) outside of a cell? Are they possibly a link between macromolecules and cells? Such questions of the relationship of non-life and life are posed in monistic logic demanding a sequential answer.

In complementarity dynamics, it is conceivable that atoms and molecules could organize and-or be organized into macromolecules which exhibit different traits, that these macromolecules could have joined in symbiotic relationships to overcome hazards of the environment to form both virus-like and cell-like DNA organization.

I believe virus-like and cell-like DNA have interacted in competitive and cooperative ways to initiate processes of evolution, each preserving its unique characteristics in individual segments, each also contributing to form organisms and ecosystems, large and small.

The transition from bacteria to more complex lifeforms involved numerous multicelled combinations, and varieties of reproductive strategies. Evolution appears to be not only competitive and symbiotic but also experimental. The particular ability of viruses to reorganize DNA is a major factor in the progressive salmutations of evolution, as well as a factor in cell destruction and extinctions of lifeforms.

To recap:

DNA moves within the nucleus of the cell, serving as a template for mRNA, which is transcribed into tRNA, which patterns amino acids, which form proteins. Combinations of proteins are the building blocks of life. This complex patterning of the genetic material creates the individuals which are the basis of evolution.

DNA of two sex species is generally conserved in its own characteristic configuration, but in half of original number of chromosomes as it becomes the hereditary material in the egg and sperm. DNA moves in processes which provide variations in individuals of the same species but undergoes limited numbers of mutations or salmutations from its species pattern. Deleterious mutations in humans are infrequent. Progressive mutations are even less frequent. Salmutations (reorganization of DNA) from one genus line to form a new genus line are the least frequent of all.

Evolutionary Change

Andrew Scott (1985: 13) wrote:

> ...although they can produce endless generations of near-identical organisms, the nucleic acids also have the ability to accumulate the gradual and useful changes in their structure that we believe to be the basis of evolution.

The concept of gradual change was carried over into the 1980s and 1990s as a legacy of Darwinism, although reorganization of DNA is not only a gradual process. It also may be, in a salmutation, an abrupt and discontinuous change.

Mutations and salmutations in DNA, plasmids, transposons and viruses are the basis for change in cells, organs and organisms. Scott (1985: 43, 220) calls viruses "little more than 'mobile genes' surrounded by a protective coat," and explains that "Bacteriophages...are now routinely used by genetic engineers to transfer foreign genes into bacterial cells."

Although human genetic engineers routinely use viruses to transfer genes, viruses themselves transferred genes long before genetic engineers existed. Viruses may have existed for as long as cells. Phage transfer of DNA produced repair enzymes among bacteria perhaps as long as two billion years ago. Such enzymes, produced only by bacteriophages, were required to protect bacteria from the ravages of the sun's ultraviolet rays (Margulis, 1970: 81-82). No DNA could be reorganized without repair enzymes to reintegrate the genophore in its characteristic form.

The literature on virus reorganization of DNA has expanded exponentially since the 1970s and provides amazing evidence of virus intervention in a spectrum of species. Still, the ability of viruses to reorganize DNA is severely limited by the ability of cells and organisms to block their access, and destroy virus invaders by their cell wall systems, their own DNA selectivity, their immune systems and interactional cell and neural communication systems.

J. Ponten (1971: 12) reported that normal cells communicate by facilitating the flow of ions, permitting passage between cells of molecules up to 10,000 daltons in molecular weight, and by enzymatic activity. It appears that cells communicate and operate very well on their own but are subject to virus intervention.

What evidence is there for virus participation at the beginning of evolution? Were they present in the Precambrian? Precambrian organisms are now found to be so diverse that it seems impossible to postulate any single lines of ascent. They appear to represent experimental recombinations of various molecules, some of which succeeded as organisms while others did not. The hypothesis of virus reorganization of DNA in mutations and salmutations is one explanation for observed evolutionary patterns and the interface between taxon lines. There are many other possibilities.

Mutations, Salmutations, and Logic

In reference to their studies of the fruit fly, Ruth Sager and Francis J. Ryan (1961: 39-40) said:

> Mutation does not occur unidirectionally, carrying the phenotype from one state to another in series. Rather, the phenotype oscillates back and forth among the different alternative states it can take.

Sager and Ryan then applied this concept to the bacterium, *E. coli*. This bacterium has been shown to exist in at least three states: one which is sensitive to the drug streptomycin and dies in its presence, one which is resistant to the drug and one in which *E. coli* is actually dependent on streptomycin for growth. Sager and Ryan considered the presence or absence of the drug to be the selective agent in the environment — it determined which variety of *E. coli* would survive. They believed that there was a randomness about the mutation process which made it impossible to predict which of the various mutants would arise at any given time.

A sequence, beginning with random production of different states of *E. coli* prior to a time of introduction of streptomycin, was posited by Sager and Ryan. The streptomycin killed some *E. coli* and had no visible effect on others, depending upon which variety they were. There was apparently no information available to *E. coli* about the nature, existence, location or imminence of the drug. Monistic logic thus requires that there be no causal connection from the prior existence of the drug to the subsequent existence of three varieties of *E. coli* which apparently, inexplicably, exist in forms which are destined to be affected in specific ways by the drug.

Noting oscillation in the number of *E. coli* in various states, Sager and Ryan thought that this occurs randomly. I challenge this concept. If the idea of random states were used to explain the nature of *E. coli* in relationship to every drug produced by man and nature, the random states of *E. coli* would indeed multiply prolifically and oscillate whirling dervishly! To postulate randomness as a pre-existing condition to explain a definitive life and death relationship between an organism and a drug is illogical. Sager and Ryan were on the verge of discovering a new time dimension in the relationship between *E. coli* and streptomycin, but monistic logic prevented it. If the concept of the reciprocal moment in complementarity dynamics is introduced, the interactions of *E. coli* and streptomycin may be described as follows:

In a reciprocal moment in which past, present and future are perceived as one, the *E. coli* colony becomes aware of the presence of streptomycin ($C_{12}\ H_{39}\ O_{12}$) and some individuals mutate in anticipation of, *and* in response to, its presence.

Consider a lab colony of streptomycin-sensitive bacteria (str-s) the wild strain found in nature. Each bacterium contains a plasmid which is, essentially, the DNA of a naked bacteriophage; it is called R factor (resistance factor). It has the same ability to mutate as a phage (Figure 11.25).

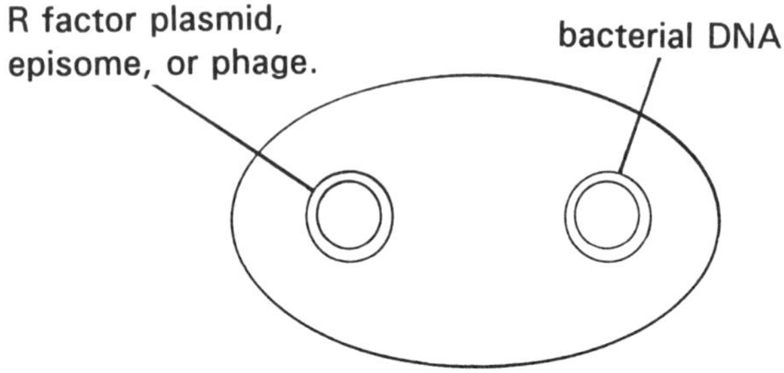

Figure 11.25 R Factor.

Consider the environment of the colony: The colony is a compact clump of bacteria with streptomycin introduced in a liquid which flows around it (Figure 11.26). Although this colony is illustrated here in a formalized configuration, the actual configurations may vary.

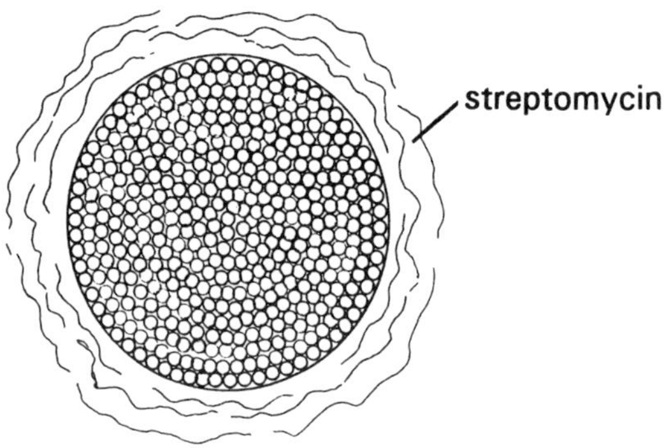

Figure 11.26 Bacterial Colony.

Three things are found to happen to different bacteria in colonies:

1. They may die. (These are called str-s, streptomycin-sensitive, the wild type found in nature.)

2. They may be found to be resistant to streptomycin. (These are str-r, a mutant.)

3. They may be found to be dependent on streptomycin for growth. (These are str-d, a mutant.) All three taxa may be found in one colony, or they may be found in separate colonies.

Application of the Principle of Responsive Salmutation

I propose here the principle of responsive salmutation, a macromutational/megamutational process mediated by a virus, in this case by the R-Factor plasmid. The bacteria in the colony salmutate in anticipation of and in response to one or more environmental factors.

I assume the colony exists in layers, for ease of explanation (Figure 11.27).

There is electro-physical and enzymatic communication between cells (Ponten, 1971: 12); thus between closely-packed bacteria. When the streptomycin is introduced, the outside layer of bacteria (str-s) communicates the nature of the chemical to the next inner layer in a reciprocal moment before they die.

The second layer bacteria, receiving the information in a reciprocal moment, communicate it within their R factor plasmids which salmutate, instantaneously reorganizing their DNA to create the gene(s) which code for resistance to streptomycin. (Others have called this a random mutation!) This salmutation is known to be inheritable by the bacterium through the R factor plasmid.

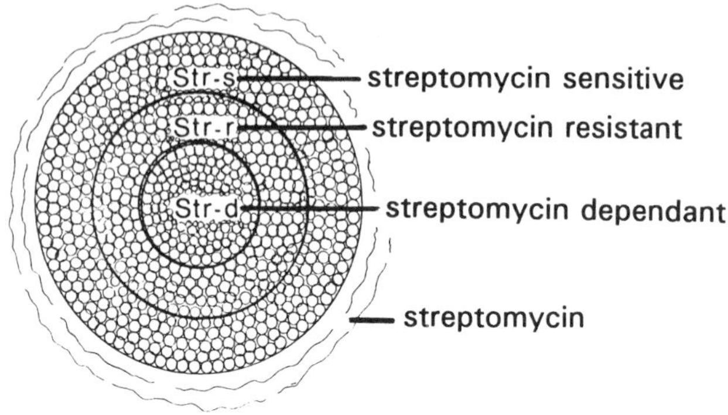

Figure 11.27 Streptomycin and the Colony.

The second layer bacteria communicate the information to the inner layer of bacteria and their R-factor plasmids, which salmutate in a reciprocal moment, into streptomycin dependent R-factor plasmids. This is known to be inheritable also. Thus, streptomycin dependent bacteria, str-d, can be isolated from the colony and will replicate streptomycin dependent bacteria progeny.

The same process may explain the salmutations of bacteria which become resistant to or dependent upon other chemical agents.

These salmutations occur in reciprocal moments of communication between these lifeforms in their changing environment. I believe this process is the basis for some of the mutations and salmutations which pattern evolution of organisms. Reciprocal time logic is required to understand the interconnectedness of the observed phenomena.

Science is at a crossroads. It must use new logics and posit new theories to explain the whole range of interactions between organisms and their environments. These can no longer be considered in classical terms.

DNA and Life

The discovery of the structure of DNA was a milestone requiring reconsideration of Darwin's theory of evolution. The discovery of the functioning of plasmids, in reciprocal time, also requires reconsideration of classical theory. As one recognizes the relative nature of time, in the beautiful complexity of creation processes, it is no longer possible to believe that random variations in organisms are the raw material upon which natural selection works.

Perhaps the blow to human ego from knowledge of viral participation in evolution will be greater than the revelation that humans ascended from ape-like forms. But in reality, it should relieve concern about heredity because there is great comfort in knowing that all nature is a unity to support and evolve life. It may be no comfort to realize that viruses are capable of evolving alternatives to humans, but that fact should unite humankind to preserve a fair place for itself among, not in place of, all the creatures of earth.

The thought that viruses were in charge of evolution might cause despair. But viruses are clearly interactional with all other lifeforms, and in harmony with intelligence beyond themselves. And humans are learning how to manipulate them.

DNA is so complexly beautiful that it requires acceptance of the concept not only of patterns, but of grandly purposeful patterns of design in nature. The basic structure of atoms and their patterning into the molecules of life create exaltation in discovery. With complementarity dynamics reasoning, one rejoices in the paradoxical oneness of the complements of structure and function. Life is process, the process of infinite becoming.

Acceptance of the idea that humans are ecosystems containing infinite numbers of microorganisms is difficult only if one does not realize that intelligence is in charge. Science is on the cutting edge of new understanding of how the human mind controls bodily functions and how it interacts with its symbionts through physical, electrical and chemical signals. Perhaps scientists and philosophers may even one day learn the secrets of the universe's immortality. There are avenues still unsuspected for humankind's future learning. Humankind is on the threshold of a new age.

12

Toward a New Theory of Evolution

Reevaluation of evolution of lifeforms through time is an exciting challenge. Facts and concepts which have appeared to be contradictory will now be considered to be dynamic complements of larger wholes. The survey of evolution to follow is not an exhaustive examination of all evolutionary steps; it is a demonstration of reasoning needed to break the shackles of monistic logic and is to serve as a starting point for theory revisions.

Human limitations must be recognized as this generation contemplates billions of years of events in both sequential and reciprocal time-frames. The timing of events depends partly on logical perspective. It also depends on the means used to determine time: whether astronomical and solar observations and calculations, evaluation of geological layering of earth in relationship to fossils, measurement of atomic decay rates, or a combination of these and other methods. It is necessary to recognize that humans, although endowed with extraordinary gifts of consciousness and intelligence, are still limited by experience in extremely short lifetimes, in limited locations on Earth, which is only one of myriad celestial bodies.

Humans think; they are curious. Questions about how the universe, the Earth and lifeforms evolved will remain fascinating challenges. This exploration blazes only one of many possible trails toward revision of evolution theory.

The Evolving Universe

George Gamow, a nuclear physicist who worked on the Manhattan project, theorized in the 1940s that the universe was created 10 to 20 billion years ago in a "Big Bang" an explosion of matter, exemplified by the expansive outward motion of celestial bodies (Eddington, 1962; Hubble, 1982). This theory is embedded in monistic logic — that the universe had a beginning and continues on a cause and effect course toward an end.

An alternate theory, reported by Eric J. Lerner (1991) in *The Big Bang Never Happened*, was proposed by the Nobel Laureate, Hannes Alfven of Sweden. His concept of a plasma universe embodies the idea that changes in particle physics, thermodynamics and cosmology are all interconnected phenomena, that not only gravity, but also electromagnetic processes are at the heart of galaxy formation.

Plasma cosmology, developed further by Lerner, envisions galaxies as disk generators shaped by plasma processes. Primordial filaments of galaxies, recently discovered by Brent Tully, appear to have been formed during a trillion or more years, thus challenging the Big Bang concept that the universe was formed about 10 to 20 billion years ago. Tully's findings and other recent evidence support the idea that the universe is without beginning.

Complementarity dynamics logic allows this complementary thought: the universe has always existed. Atomic experiments have shown that matter is neither created nor destroyed but rather, changes in form — not merely as matter but also reciprocally between energy and matter.

This does not negate the idea that major events, such as super novas, may have influenced direction and velocity of celestial bodies, but it renews the perspective of a continuously existing universe posited by Herman Bondi (1961: 12) and Tom Gold. They, however, did not recognize an evolving universe, saying, "... apart from local irregularities the universe presents the same aspect from any place at any time." It is not necessary, in complementarity dynamics, to make an either/or choice between the concepts of a beginning to the universe and a continuously existing universe. Rather, the evidence may be considered from the combined dynamic perspective.

Einstein believed that any matter which travels continuously in a straight line will eventually arrive at its starting point. Thus space is curved, an interesting paradox. This leads to the concept that the universe is finite but unbounded, another paradox.

If these principles are accepted, the movement of celestial bodies cannot be adequately explained by an initial "Big Bang" hypothesis, for this assumes not only a starting time, and that one event could explain all subsequent celestial motion, but also that matter can travel infinitely in one direction in an infinite universe. Such a concept seems more difficult to accept than the concept of a finite but unbounded universe in infinite motion. Logics beyond the monistic are needed to contemplate the meaning of the concept "universe."

It has been thought that the universe demonstrates *entropy*, with disorganization or disintegration of matter. Some scientists have proposed that a complement of entropy is negative entropy, a characteristic of life. A more positive word is needed. A complement of entropy is *organization*, characteristic of both life and non-living matter. The dynamic equilibrium of organization (the building up of pattern) and entropy (the disintegration of pattern), is apparent within all of nature. Another positive term, a complement of entropy, is *continuous creation* (patterning through change and recombination).

For example, rocks are organized systems of atomic processes, which are broken down into useful chemicals by both physical processes such as weathering and by certain lifeforms such as lichens. Rocks are re-formed by both physical processes such as vulcanism and life processes such as incorporation of minerals into the bodies of microscopic radiolarians and foraminifera which sink to the bottom of the seas and become rock strata (Figure 12.1). Minerals from rocks are incorporated into the protoplasm of lichens and plants and thereafter up the food chain.

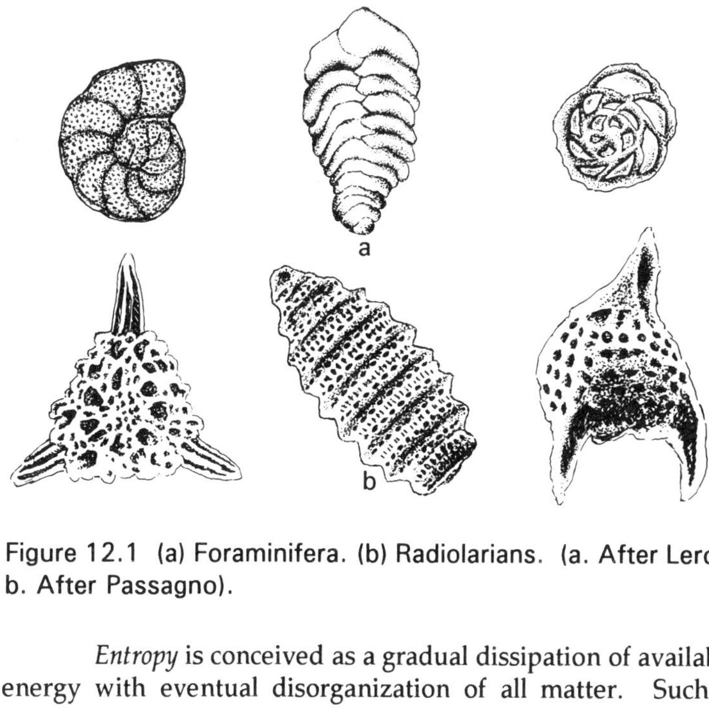

Figure 12.1 (a) Foraminifera. (b) Radiolarians. (a. After Leroy, b. After Passagno).

Entropy is conceived as a gradual dissipation of available energy with eventual disorganization of all matter. Such a dismal prospect is a misconception of the nature of the complementarity dynamics of organization and entropy. The concept of *organization* is just as evident in nature as the concept of entropy. New matter and energy evolve from dissipating matter and energy.

Since Leucippus, a Greek philosopher of the Fifth Century BC, proposed that matter consists of atoms, humans have wondered about their nature and the forces by which they move. Atoms may be considered in complementary terms: the complement of the *strong force* uniting the protons and neutrons in the nucleus of atoms, is the *weak force* holding the electrons in their orbits about the nucleus. The *electromagnetic force* exemplifies the complements of positive and negative charge of all atomic interactions. Another complementary notion is that the *gravity* of the nucleus is complemented by the orbital-spin of the electrons. The orbital-spin of electrons may be considered to be complemented by their quantum jumps. The search for complementary relationships provides changes in perspectives.

The Solar System and the Atom System

I think of the organization of the atom as complementary to the organization of the solar system. The gravity of the sun is complementary to the orbital-spin of the planets. Other energies apparent in the matter of the solar system and the atom system are: magnetic, electrical and chemical. Just as magnetism is complementary to electrical energy, electrical energy is complementary to chemical energy. This latter concept is exemplified in the alternating movement of nerve impulses from electrical form to chemical form in the neuron-synapse complement of animal nervous systems.

Einstein (1961:13) conceived that under certain circumstances different natural phenomena run their course according to the same general laws of nature. I believe that many natural phenomena run their course according to the principle of complementarity.

The concept of the four recognized forces: strong force, weak force, electromagnetic force and gravity, needs to be supplemented by the concept of a myriad of complementary forces of nature in dynamic equilibrium. Just as physicists of the eighteenth century conceived of light as corpuscles, in the nineteenth century as waves, and in the twentieth century as photons and energy waves, so too have the concepts of elementary forces and elementary particles evolved.

While some physicists study the four forces and postulate a fifth force connecting all matter, other physicists concentrate study on the elementary particles of atoms, attempting to divide all matter into its component parts.

The atom smashers of the present age have discovered 37 particles in three classifications: leptons, quarks and bosons (Asimov, 1989). Leptons come in three basic varieties: electrons, muons and tauons. Each of these has a complementary neutrino. These are considered to be six varieties of matter. Each one has a complement of antimatter, called positrons, totaling 12 leptons and antileptons.

Quarks consist of the protons and neutrons in the nucleus of atoms. Quarks are classified as up, down, top, bottom, strange and charm. Each of the six quarks has a complementary antiquark, totaling 12.

The complementary nature of forces and particles is identified in bosons: three for weak interaction, eight for strong interaction, one for electromagnetic interaction and one for gravitational interaction. This totals 13 bosons. Other bosons, such as the Z particle, are being produced currently, and still others are expected to be found when the new super collider is completed.

The concept that boson particles and fundamental forces are complementary is in early stages of development. The relationships are difficult to express in language. Even Asimov (1989) one of the best modern interpreters of physics, has not fully conveyed the wholeness of these relationships, concluding, "The bosons are particles that make it possible for leptons and quarks to interact with each other."

Perceptions of the complementarity dynamics of these relationships will expand understanding of both the atom system and the solar system. But for now, each system is studied separately, and many scholars continue to think of them in monistic terms.

The theory that the solar system came into existence about five to six billion years ago is a concept in monistic logic. It is possible to conceive that the solar system and Earth evolved from previously existing configurations of matter. It must be recognized that dating of the events in that evolution are arbitrary and subject to change with various perspectives and different interpretations of the geological and astronomical evidence.

Nevertheless, let it be assumed here that the monistic perspective of a six billion year old solar system and a five to four and a half billion year old Earth is adequate as a starting point to consider evolution. Let monistic logic carry our thoughts in a sequential pattern as long as it illuminates the facts to be examined. Let other logics be applied as needed.

Earth and Life

The Earth appears to orbit and spin in optimal patterns to support lifeforms. It is as though the objective of a habitation for lifeforms were operative in the organization and evolution of matter into the planet Earth. Humans have endlessly hypothesized about the relationship of themselves to Earth and the universe. The hypotheses continue as the early Earth is considered as a habitation for life.

Earth age dating of five to four and a half billion years is based on atomic decay data from the earliest known rock strata and from astronomical evidence. Theories of the formation of Earth involve much speculation. A current theory postulates that Earth formed from solid matter and gasses available from the explosion of a star: that as this matter concentrated, a heavy solid core formed, surrounded by a mantle. Eventually, moving tectonic plates, floating on liquid magma, supported surface land and seas.

This theory is incomplete because it does not explain the evidence that the magnetic poles of Earth have reversed at least 70 times during the Earth's history. There must be complementary electromagnetic and physical processes occurring in the core which relate to vulcanism and the movement of the crustal plates. This area of study awaits new developments.

Outgassing from Earth's core is thought to have created the ocean and atmosphere. It has been believed that the primordial reducing atmosphere consisted mostly of carbon dioxide, methane, ammonia and water vapor, but some scientists challenge this and now believe that the early Earth atmosphere consisted mostly of carbon dioxide, nitrogen and water vapor.

It has been posited that the amount of water in the ocean increased from outgassing of water vapor and torrential rains for about one to two billion years and has remained virtually the same volume since about three billion years ago. This is an incomplete hypothesis because it does not explain how such a major change from outgassing to a finite ocean volume could have occurred.

The Beginning of Lifeforms on Earth

Lifeforms are believed to have begun in an ocean environment. In the years since Darwin, many specialists have worked to fill in the details of the sequence of evolution from non-life to life. The chemist Stanley Miller proposed that available chemicals, then present in a concentration area of the ocean, could have become organized by a bolt of lightning into amino acids, the basic building blocks of life. His experiments proved that simple amino acids could be formed in this way (Cowen, 1976: 4).

Nobel Laureate Harold Urey took the next step and posited that those chemicals could have organized themselves into the various complex amino acids of life and subsequently into primitive prokaryotic one-celled organisms, the first lifeforms on Earth. He thought experiments would eventually verify this theory.

On the other hand, biologist Francois Jacob (1972: 272) wrote that such a process is inconceivable: "...there is no chance of seeing *all* these compounds being assembled correctly and a bacterium emerging fully armed from a test tube."

The Harvard biologist, George Wald (1954) proposed that the right protein molecules could have organized themselves into the first cells. He had gained information from the work of Frances Schmitt, whose experiments caused Schmitt to conclude that the main protein in the human body, collagen, could organize itself in regular patterns. Wald thought that the stability needed to organize living cells from these protein molecules would not be attained by a self-organizing process, but rather, by natural selection of precellular types of protein.

By 1975, the microbiologist Laura Hsu had produced proteinoid microspheres in the laboratory which, protected from contamination by bacteria or other microorganisms, lasted for six years. This proved that the stability necessary for evolution of protein molecules was inherent in the amino acids and was not selected from earlier precellular forms. These protein molecules demonstrated all the characteristics of organisms.

Thus the stage was set for explanation of evolution from chemicals to living cells. Wald then theorized that proteins were primary and that they self-organized as cells. This theory was a reversal of the central dogma that molecular biology had previously held: that "DNA makes RNA makes protein." Robert Shapiro (1986: 282) in *Origins; A Skeptic's Guide to the Creation of Life on Earth,* stated unequivocally that protein preceded DNA in the development of life. "Protein begat RNA and then both begat DNA."

Cells, Proteins, DNA/RNA

Scientists, in typical monistic pattern of either/or choice, have debated whether the earliest evolution from inorganic chemicals to living cells occurred in one or another of two sequences. Since one-strand RNA is simpler than two-strand DNA, Wald, Shapiro and other scientists reasoned that the amino acids which were present in Earth's early environment self-organized into RNA which then evolved into DNA.

Other scientists reasoned that chemicals present on Earth formed amino acids which self-organized into protein protocells (proteinoid microspheres, which have been produced in recent laboratory experiments, Fox, 1988). They believe that this kind of protocells evolved into the earliest living thermal bacteria, which were practically identical to fossil cells of three billion years ago. Similar but more complex forms are found today in hot springs and hot ocean floor vents as cyanobacteria or bacteria. Eventually these early bacteria evolved into modern cellular lifeforms.

This proposed sequence from inorganic chemicals to amino acids to protein protocells to thermal cells to bacteria to more advanced cellular life is currently exciting the interest of the scientific community.

It is now known that amino acids are composed of chemicals which consist of varying proportions of acids (amino combination of chemicals) and basics (carboxyl combination of chemicals). Acids have a positive electrical charge and basics have a negative electrical charge.

Therefore at the juncture of scientists' designated "non-life" and "life" positive and negative charge seem to be a key to the organization of chemicals into the basic building blocks of amino acids and proteins. In monistic logic, the question then remains: Which came first, the amino acids of positive or those of negative electric charge? And which came first, positive or negative electrical potential?

These questions can be seen as arising due to lack of understanding of complementarity dynamics. There can be no *one* sequence in evolution from certain designated chemicals of one kind of charge to certain designated chemicals of the other charge. There are, rather, dances of myriad complementary patterns which may have always existed, creating and evolving the chemicals of life as they have been, and as we know them.

This does not mean that there are no sequences in evolution; there are, indeed. However, sequences have meaning only in conjunction with the basic underlying complementary relationships. The complements of structure and information/function of amino acids cannot be divided into proteins (structure) and DNA/RNA (information/function). Proteins have both structural and functional (enzymatic) characteristics, as do DNA/RNA. The process of amino acid combination to form *both* structure and function is not sequential.

Disagreements concerning sequence: protein to RNA to DNA, *or* RNA to DNA to protein, can be seen as due to lack of understanding of complementarity dynamics. The earliest proteinoid microspheres and the nucleic acid of cells of living organisms all exhibit simultaneously (in reciprocal time) complementary structures and functions of cells: information generation (selective interaction of one system with another) membrane formation, metabolism, reproduction and communication (Fox, 1988). The evolutionary pathway from proteinoid microspheres to modern cells is a complementarity dynamics pathway.

Before the work of Sidney Fox (1988) and the other scientists who advanced the concept of evolution from "non-living" chemicals to protocells of protein, other processes were thought to be involved. Comparison between monistic concepts and complementary concepts of chemical evolution are helpful in making the transition from the opposing viewpoints of Urey and Jacob to the theories of Sidney Fox and his colleagues, and beyond.

Sidney Fox (1988) reported that verification of his research, that amino acids self-organize into protein cells (proteinoid microspheres) was expected from the Max Planck Institute of Germany. The theory that amino acids organized themselves into more complex patterns and then into living organisms should be examined. What are the assumptions underlying this theory of the origin of life? Let us start with consideration of the simplest of the twenty amino acids of life, glycine (Chapter 11, Figure 11.15).

In monistic logic, if the molecules of elements are self-organizing, there must be a cause prior in sequence to the existence of organized glycine. Urey believed the causative force leading to the organization of macromolecules such as glycine, from existing molecules was an electrical charge, lightning. Fox's work shows that heat also (by means of Brownian motion of particles) energizes the formation of amino acids.

In monistic logic, this assumes that there is a principle of organization which is either activated by the elements themselves, *or* by lightning or heat. An observer might assume that the organizing principle is within the elements, and the lightning bolt or heat is the energy to start the activation.

If the elements operate within a principle of organization which patterns their activity, it must be a flexible principle that would be inherent in the organization of all twenty amino acids. How is the decision made to organize glycine rather than another amino acid from the same ingredients? Can elements make decisions? Can lightning make decisions? Is the production of glycine a random event?

What organizing principle could be invoked to explain the organization of the specific genetic code system, the DNA, without knowledge of that code? How could the organizing principle for amino acids be invoked to prepare them for the life of a cell without a concept of cell or of life? Monistic logic isn't a means to explain the organization of elements into the amino acids of life, much less explain the salmutation from amino acids to living one-celled organisms.

Complementarity dynamics might allow that the organizing force is both within and beyond the elements. It might be posited that the intelligence which organizes glycine cannot be localized in the elements, in the lightning, or even in a combination of the two. It is necessary to posit a process, an interaction of intelligence, both finite and infinite, in order to conceive of movement from "non-life" to lifeforms.

Glycine is a product composed of the elements hydrogen, oxygen, nitrogen and carbon in an organized configuration. It is the specific *organization* of these elements which is the definitive criterion of glycine. Glycine is the same whether it is composed from chemicals organized by an electric spark, by the processing by DNA and RNA within a cell of a living organism, or by extraction from cellular material in a human industrial operation.

It is now generally accepted theory that specific amino acids originate from chemicals assembled by a physical process, energized by an electric spark or heat. Specific amino acids can also originate by the self-organizing function of DNA and RNA in a living cell. Both processes imply intelligence; they both organize specific products essential to lifeforms.

The difference in the two processes is that one seems to begin from a disorganized group of chemicals, plus a bolt of physical energy, while the other seems to begin with a highly organized, possibly self-activating, and living substance, the DNA within a cell. Because both processes produce the same product, glycine, both processes may be assumed to be orchestrated by the same intelligence. There exists intelligence for the organization of glycine and presumably also, intelligence that orders glycine, in conjunction with the other amino acids of life.

A so-called "non-life" process and a "life" process produce the same product. This is not unique with the formation of chemicals into an amino acid. Both "non-life processes" and "life processes" form oxygen, carbon dioxide, hydrogen, carbon and nitrogen, in numerous patterns. The complementary interactions of chemicals and lifeforms are legion. To understand these complementary interactions is to define the artificial borderline between "life" and "non-life" in the reciprocal moments of evolution. The beginnings of life on Earth remain a miracle and a mystery that may never be fully understood, but use of complementarity dynamics clarifies conceptional options.

The gap between "non-living" matter (the complex amino acids) and living bacteria not only involves a few sequential steps, but also involves complementary interactions. The DNA of all living things is too complexly patterned to be considered to be organized by bolts of lightning.

For Fox and his colleagues, the question remained: Was the process of self-organization of protein random *or* was it determined? If the process were random, it would agree with Darwin's concept of random variation, extrapolated to the molecular level. If the process were deterministic, it would agree with Einstein's belief: "Everything is determined, the beginning as well as the end, by forces over which we have no control. It is determined for the insect as well as the star" (Fox, 1988: 164).

This monistic form of the question, random self-organization *or* determined self-organization, threw the problem into the limbo of paradox, where it has remained until today.

Sidney Fox (1988: 167) enumerated the theoretical options: "The essential positions to be compared do not ease the problem: total determinism, 'almost' determinism, 'almost' indeterminism, and total indeterminism."

These positions are posited in unexamined Aristotelian logic. Let us examine them:

1. All the self-organization of proteins (S) is determined (D).
2. No self-organization of proteins (S) is determined (D), (thus, self-organization is random).
3. Some of the self-organization of proteins (S) is determined (D).
4. Some of the self-organization of proteins (S) is not determined (D) (Plate 12.1).

The problem is in the logic. In monistic, deductive logic, a choice must be made between (1) and (2). If (1) is chosen, a syllogism may be devised with a minor premise and a conclusion may be drawn. If (2) is chosen, a syllogism may be devised with a minor premise, and a conclusion may be drawn. Neither (3) nor (4) can be used as a major premise because they are not all-inclusive. Cause and effect sequences, beginning from (1) or (2) may be constructed to reach additional conclusions.

Either/or choices are required and cause and effect sequences are allowed, but such choices and sequences are made in a logic that is incapable of patterning the dynamics involved. Deductive reasoning prohibits consideration of reciprocal moments of dynamic interactions, and therefore does not structure an adequate paradigm for modelling nature.

In complementarity dynamics, truth is kaleidoscopic. It may be found at any locus in reciprocal moments (Plate 12.1 b). Truth may appear wherever one looks. To find the whole truth, one must consider the whole of all possibilities and all the interactions among them. Therefore perception of truth is relative.

(a) Deductive Logic:

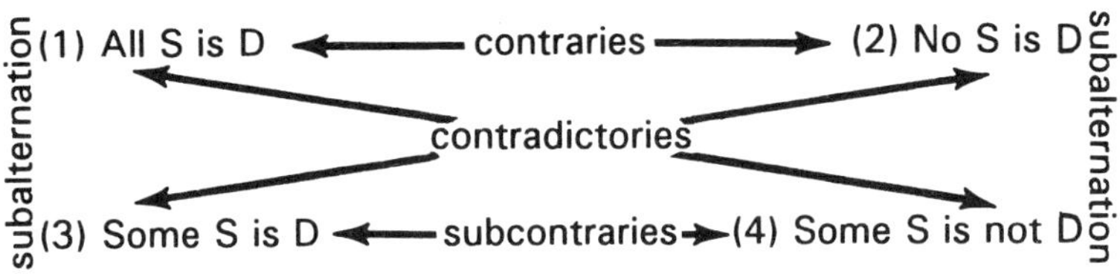

(b) Complementarity Dynamics:

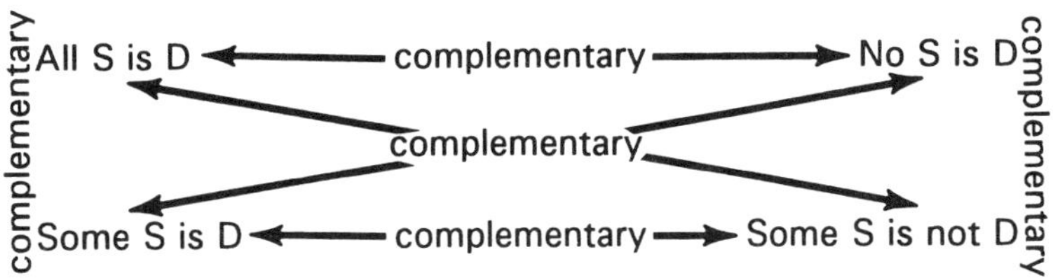

Plate 12.1 (a) Deductive Logic. (b) Complementarity Dynamics.
Deductive logic requires either/or choices and consideration of whether a thing is part of the whole. Complementarity dynamics requires consideration of relationships and probabilities.

Determined self-organization and *random* self-organization are two complements of a larger whole. The positive and negative electric charges that perform the dance of life in movements of particles and-or forces of atoms of the elements and their products, including protein, RNA and DNA, are complementary. In complementarity dynamics, electricity functions as a whole in two charges, both sequentially and discontinuously.

Assessment of the processes which occur in the interface from chemical evolution to biological evolution is only part of the larger question: Is the intelligence of life and the universe, which is exemplified here, inherent, pervasive and infinite? Complementarity dynamics leads me to think that it is.

Assuming that protein protocells self-organized in complementary dynamics with their environment, and evolved into thermal cells, thence into cellular organisms such as archaebacteria and bacteria, and into higher lifeforms, we need also to explore the role of DNA in creating the higher lifeforms which have appeared in evolution.

New Forms of Life and Salmutations

Scientists are beginning to believe that new lifeforms are created by reorganization of "superfluous" DNA into "new regulatory gene patterns" (Augros and Stanciu, 1987: 181). I see superfluous DNA as DNA which scientists don't yet fully understand. Minor reorganization of DNA which does *not* produce an identifiable new taxon may be of deleterious or neutral significance, or it may be a mutation recognized in the phenotype. The process whereby DNA is reorganized, producing a new taxon line, is a *salmutation*.

There are several processes capable of effecting reorganization of DNA including: radiation, action of certain chemicals, heat and activity of viruses. It is also believed that DNA *within cells* is not only self-organizing but also self-repairing. Radiation, chemicals and heat seem often to produce deleterious changes. The self-organization of DNA within cells, appears to be purposeful, as does the recombination of DNA by viruses, within and between cells.

Beginning about 1864, viruses (then known as small particles of living animal matter) were thought to cause only disease and death. But recent amazing discoveries lead to the conclusion that certain viruses not only reorganize DNA to produce more viruses (and-or disease) in plants and animals, but also reorganize DNA in progressive ways, and are capable of reorganizing DNA upon manipulation by humans (Eron, 1981: 5-6).

Gene splicing is not just man's cutting and reassembling of DNA. It is human intervention in the acquisition, transfer and recombination of DNA, often accomplished by means of viruses. Before man intervened, viruses were reorganizing DNA based upon other than human intelligence. Perhaps they have always been in the ultimate feed-back loop in which organisms change to cope with their physical environment.

Viruses have been considered to be non-living because they do not carry on metabolic processes. They do not eat. They do not grow. They do not age. They do not produce energy. They cannot move on their own volition, nor do they seem to respond to external stimuli outside of a cell. They are identical to all other viruses of their type (Pauling, 1975: 443). Viruses crystalize like ordinary chemicals. They *do* have DNA or RNA, within their protein structure. One of the functions they perform, which is similar to living beings, is reproduction. To reproduce, they require use of the genetic machinery of a living cell (Hughes, 1977: i).

However, viruses do something much more important in evolution than mere reproduction. They reorganize not only their own DNA but also the DNA of their hosts. They can incorporate foreign DNA into their own. They carry it as they are passed, by various vectors, from one host to another, from bacteria to insect, from insect to plant or animal, from plant to plant, from plant to animal, from animal to plant, from animal to animal. Different viruses have different modes and mediums of transport. They can also lie dormant in crystallized form for unknown periods of time, to come to "life" upon incorporation into a living cell, perhaps having passed by means of air, water, or soil.

Then they can incorporate the foreign DNA they are carrying into the DNA of a new host, changing the expression of body cells, perhaps initiating cancer, or reorganizing the germ cells, changing the expression of the genotype, the actual genetic structure of the host. In such a reorganization of the germ cells, a salmutation, a new taxon may be created.

Viruses can also reorganize existing DNA in a cell without transporting it from one organism to another. This, in fact, seems to be their usual method of operation. However, the complementarity dynamics between viruses and cells is not entirely understood. The capability of cells to reorganize themselves and the capability of viruses to reorganize cells are at least as difficult to understand as the complementary relationship between photons and waves of light.

Viruses have been classified in various systems. One major classification system is based upon their close relationship to the organisms in which they are usually found (Fraenkel-Conrat, 1976). Thus, scientists have learned that viruses, generally, do not move readily from one lifeform to another.

It must be recognized, however, that this classification system has been found to be inadequate, and other systems are being advanced which are based on the structure and function of viruses, rather than on the usual host species. I believe no classification system will be able to keep viruses in their place, however, because it is in the nature of ever-changing viruses frequently to defy the status quo, in reciprocal moments of dynamic interactions.

For example, in the case of crown gall in plants, DNA is transferred between prokaryotes and eukaryotes by a prokaryotic plasmid (a form of virus?) which consists of DNA sequences that become functional genes in eukaryotic cells. Patricia Zambryski et al (1983: 505) wrote:

> It is interesting to speculate that other such natural instances of DNA translocations,

involving widely different organisms, can be expected to occur in situations involving a close, long-term symbiotic or parasitic association between these different organisms.

Perhaps, this speculation will lead to new insights into the process by which early bacteria evolved into new organisms of various kinds, also suggested by the symbiotic relationships discovered by Lynn Margulis.

Scientists are making new discoveries every day involving the DNA relationships of viruses and cells. Many physicists since Delbruck have transferred their interest from physics to microbiology and virology, where the frontiers of knowledge are currently extremely volatile and exciting. Complementarity is evident in the processes they are observing.

Transitions Between Genera

In the years since Darwin, many biologists have worked to fill in the details of evolution. Scientists who specialize in the study of different animal taxa have had serious disagreements over numerous questions of sequence. Most scientists have not taken a definitive stand on the question of saltations between major sequences of genera. The more that is learned, the more questions arise. It is impossible to find elucidation of a complete sequence of evolution from early organisms to present lifeforms. One might expect that in more than 130 years since Darwin's book, paleontologists would have been able to delineate this sequence, if the concept of sequence were adequate to express the process.

Since scientists disagree about which lifeforms are in the line of sequence at numerous points along the way, it seems that theory requiring sequence has failed. No matter which species are projected to have lived at the interface between other species, there is disagreement about their relationship as connectors between the prior animal and the subsequent animal, or between prior plant and subsequent plant. My discussion will emphasize animals, as did Darwin, but I recognize that evaluation of microorganisms and their interface with fungi, plants and animals needs to be elucidated more fully.

In spite of the problems encountered, scientists continue to try to develop the evolutionary sequence of the animals they are studying. Scientific literature abounds with evidence and opinions from many experts. Let us now consider new concepts of evolution, in conjunction with classical evolutionary sequences.

Within the field of paleontology a new discipline called cladistics emerged as a significant force for change in the 1980s. It is a method which classifies organisms by what purports to be an estimate of the recency of common descent. Decisions about relationships are made by considering the structure of living and fossil forms, and by comparing the DNA of living forms. Cladistics has produced opinions about classification of organisms which are substantially different from the opinions of Darwinians. It has led some scientists to the conclusion that no fossil species can be known to be ancestral to, or descendant of, any other!

These cladists see an inherent contradiction between the ordered, distinct divisions of lifeforms and the concept of evolutionary biology that organisms form a sequence of types. Michael Denton (1986: 140) concluded:

> Cladism is merely making explicit a fact enshrined in classification schemes ever since Aristotle, but which has lain dormant for most of the past century — that in the final analysis nature's order is not sequential.

This conclusion again illustrates the monistic, either/or logic of science: Either nature's lifeforms are sequential or they are not sequential. Paradox logics allow a conclusion that nature's productions are sequential in some respects and discontinuous in others, a concept blocked by the strictures of monistic logic.

The logic problem, involving the links between major classifications of animals, requires careful consideration. If classical logic requires sequence, and Darwin's theory, based on this logic, requires the concept of sequence, then sequences will be postulated from the facts which are available. This is not to deny that there *are* sequences in the fossil record. Sequence is evident in the movement from simple to more complex forms. It is evident in changes within species and between species, but discontinuities are also evident, and it is now possible to explore the relationship of continuities and discontinuities.

The "inherent contradiction" identified by Denton, between the distinct divisions of lifeforms and the concept of sequence from one type to another can be addressed by asking the question: How is the DNA of one generation reorganized to produce a new taxon *in the next generation*? This question has meaning only when one is free from the shackles of the concept of "random" variation of germ plasm, and the obsolete dictum that "no species ever gave birth to a new species."

This is, essentially, the same kind of question that is asked in order to understand the process of metamorphosis from caterpillar to butterfly. In this case, the process takes place within the chrysalis, so the evidence is dramatic and demands immediate attention. It has been found that the caterpillar disintegrates into a protoplasmic "soup" with only brain, heart and hind gut retained. The caterpillar protoplasm then reintegrates into a butterfly, which emerges from the chrysalis fully formed, as though a miracle has occurred, and indeed it has! (Figure 12.2.)

The scientific community does not yet understand how DNA relates to metamorphosis. Scientists are only in the initial stages of understanding the reorganization of DNA which occurs when the body of a bacterium (or the fertilized egg of an animal) is penetrated by a virus. I have used the word "penetrated" instead of the word "infected" because viral penetration or insertion is then considered in its positive aspect of salmutation, not in its deleterious effect connoted by the word "infection" which is the subject matter of medicine.

Viral insertion occurs naturally in many different ways. This process has only recently been initiated by human manipulation in gene splicing. Humans have learned to alter lifeforms by adding and subtracting DNA strands, both by using viruses and by other methods.

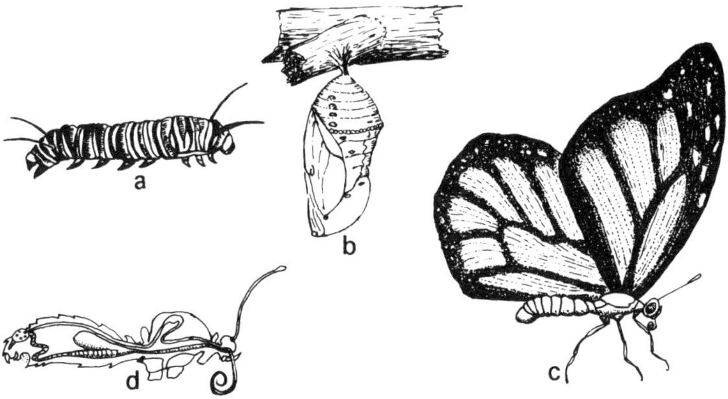

Figure 12.2 (a) Caterpillar. (b) Chrysalis. (c) Monarch Butterfly. (d) Cutaway of butterfly.

The power of gene alteration was brought forcefully to the attention of the public on April 24, 1987, when genetically engineered ice-minus bacteria (*Pseudomonas syringae*) were released into the environment for the first time. The new bacteria, minus the gene which makes them sensitive to frost, were sprayed on strawberry plants in Brentwood, California, USA, so that they could invade the plants and create frost resistant strawberries (Cowen, 1988).

In this case, a gene was deleted, not inserted. However, it is important to note, that in the case of human manipulation of bacterial DNA using viruses, humans do not always make the final decision about where, on the existing bacterial genophore, a viral insertion of DNA will be located. Viral DNA seems to "choose" that location for itself! Thus, the final genetic production may be, to some extent, a viral "choice" not entirely a human determination. Much more needs to be learned about the interaction of bacteria and viruses before more experimental organisms are widely released into the environment.

Scientists are now making comparisons of DNA from various species, in order to learn possible junctures in the evolutionary pathways they have taken. In my opinion, the cladistic comparisons between DNA sequences of species with closely similar skeletal structures may provide little information about the *time* of DNA differentiation. Rather, some of the DNA sequences of the designated species may describe viral DNA involved in salmutations, which may have occurred at different times and in different places. For example, F4 viruses have DNA sequences very similar to the E. coli bacteria of humans. Tobacco Mosaic viruses have DNA sequences very similar to tobacco plants. Simian viruses have DNA sequences very similar to chimpanzees.

The first scientific classification of viruses was based upon their identification as residents in the first host in which they were found. I believe that the reason identifiable viruses were found in particular host species, is that the viruses evolved in conjunction with particular hosts and reorganized their host's DNA in salmutations, along various evolutionary lines.

The evolutionary process has previously been interpreted as *Mendelian sequences,* with mutations leading to new taxa. Mendelian inheritance is not overturned here, but rather supplemented. The Mendelian principle involving dominant and recessive genes and homozygous and heterozygous individuals, is especially needed to explain the several phenotypes in one species. An example is: the various kinds of ants such as workers, soldiers and drones. (However, a prescribed diet transforms a worker pupa into a "queen.")

Another example is: the three varied patterns of females, and a fourth pattern of the male in a Malaysian butterfly, identified by Alfred Wallace (Figure 12.3). These different wing patterns can be explained by Mendelian sexual recombination processes.

Another example of different Mendelian expressions of DNA within one phenotype is the various patterns of human hemoglobin. However, I believe that hemoglobin from a DNA segment of 486 base pairs may have evolved (by means of viruses) from the hemolymph of invertebrates and the legoglobin of certain plants. Thus, both the Mendelian principle and the viral insertion principle are needed to explain inheritance of the identifiable characteristics of blood and its evolution.

Figure 12.3 Malaysian Butterflies (*Papilio*) three females, one male.

The implications and ramifications of these concepts are so vast that many scientists and philosophers will study and write on the subject before society will be able to make all the choices necessary to control evolving bioengineering technology. Here, we proceed to implications of the viral insertion principle for evolution theory, as basic groundwork.

Scientists such as Niles Eldredge have acknowledged that genetics must guide the conclusions of paleontology. However, scientists have not found all the keys needed to release the secrets of DNA from the fossil record. Thus fossils have been gathered and evaluated apart from the evidence becoming available from the sister discipline of genetics. This has led to interesting criticisms of the classical theory of evolution from different perspectives.

In the following discussion, data from various disciplines and disparate sources are considered, in order to posit new evolutionary pathways, incorporating salmutations as well as sequences.

Radiations and Extinctions

Some of the most puzzling and fascinating of the phenomena found in the fossil record are recurrent evidence of rapid radiations and massive extinctions of both marine and terrestrial lifeforms. These phenomena cannot be explained as occurring due to random mutations and competition between species. A number of complex interrelating factors are evident.

The speed of change in the record of nature ranges from hundreds of millions of years of relative constancy in the body form of many species to rapid change in others species, such as modern canines, which radiated into various domestic varieties during human history.

Would modern dogs be considered to be in different genera or species instead of different varieties or breeds if only fossil evidence of their skeletons were available? I think perhaps so. There are more anatomical differences between living varieties of dogs such as Lhasa Apso and Russian Wolfhound than between many different genera or orders in the fossil record. For example, the differences among living dogs are much greater than the differences between different orders of trilobites (Figure 12.4).

The changing evolutionary pattern of dogs may be contrasted with the constancy of ancient trilobites, which appeared in the fossil strata of the Cambrian, 570 million years ago, radiating into hundreds of taxa called species, genera, classes, and orders, from 1/8 inch to 24 inches in size but remaining identifiably and consistently trilobites. During 300 million years they changed in the most minute manner, becoming extinct 230 million years ago in the Permian period. (Nevertheless, there is evidence that the trilobite DNA configurations did not become extinct with trilobites, as we shall see in Chapter 14). Rapid change, long-term constancy and discontinuities are all characteristics of nature.

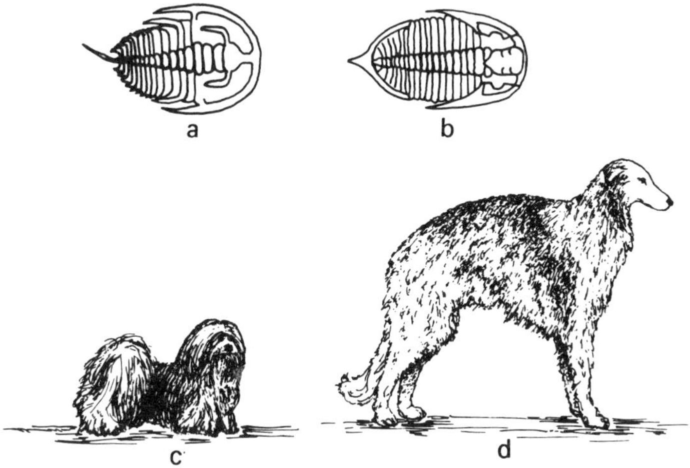

Figure 12.4 (a) Cambrian trilobite. (b) Permian trilobite. (c) Lhasa Apso. (d) Russian Wolfhound. (a, b, After Clarkson; c, d, After Hutt).

Darwin (1872: 49-50) despite his many postulations about species, believed that species were not definitive entities. He wrote:

> Finally, varieties cannot be distinguished from species...for two forms, if differing very little, are generally ranked as varieties...but the amount of difference considered necessary to give to any two forms the rank of species cannot be defined.

He finally decided it was up to experts (like himself) to judge what was, and what was not, a species. Modern scientists continue to reclassify species of plants and animals when they think these plants and animals no longer clearly fit into previously identified categories.

Nature shows continuous existence of organisms such as oysters and mussels with little change — for more than 400 million years. It also gives evidence of massive extinctions during relatively short periods of a few million years. Of the estimated 31 major phyla which have existed on the planet, nine are believed to be extinct (Taylor, 1983: 19).

It has been estimated that 60%-99% of species that have lived on Earth, are extinct. Almost two million existing species have been identified. Estimates of living species which have not been named, range from 3-30 million. The fossil record suggests that as many as 50 billion species may have lived on Earth. It also shows that massive extinctions occurred in the Cambrian, Permian, Triassic, and Cretaceous periods, with less extensive extinctions in the Silurian and Devonian.

Cuvier's concept of catastrophism, long neglected by evolution theorists, is evidenced in the massive extinctions in the fossil record. Darwin's emphasis on survival through chance variation and natural selection is substantially challenged by new support for Cuvier's position. I believe that perhaps being in the right place at the right time is as meaningful a key to survival as "fitness."

Many theories have been advanced to account for radiations and extinctions. Perhaps one of the most significant recent theories is presented in the work of James W. Valentine (1973) Valentine and Eldrege M. Moores (1982) and Bjorn Kurten (1982). They correlated major changes in lifeforms, both massive extinctions and radiation of new species, with major environmental changes due to movement of the earth's tectonic (crustal) plates. These correlations may become more understandable when considered in the logic of both reciprocal time and sequential time. Some changes appear to be instantaneous, while others appear to occur during a span of millions of years.

Plate Tectonics and the Fossil Record

The changes during geological time, in the configuration of the Earth's continents is illustrated on Plate 12.2. These tectonic changes were accompanied by the rise and fall, origination and demise of seas with complementary changes in the height of land above sea-level. Because the theory of plate tectonics and the correlations made here are so recent, revisions will, no doubt, be necessary as new evidence accumulates.

The fossil record contains much more data than can be indicated here. Plate 12.3 shows *Complementary Systems Evolution of Six Kingdoms of Life.* I feel impelled to add one taxonomic kingdom to the five kingdoms delineated by Whittington (1969) and supported by Margulis (1970, 1981). This kingdom is Virusa, initiated and named here as a new category. This is a category for all viruses and viral configurations. Because viruses consist of DNA or RNA (and interact with all lifeforms which have been studied as their hosts) they should be considered in a kingdom connecting life with non-living matter. Therefore Virusa and Monera begin the six kingdom system which I propose here (see Taxonomy).

Eras, periods and epochs of geologic time were devised as time standards to organize data from geologic and fossil evidence lying in various rock strata around the world (Plate 12.4). The eras are:

1. **Archeozoic,** the oldest time frame of known life, beginning perhaps, 3.8 billion years before present (3.8 BBP);

2. **Paleozoic,** beginning perhaps 570 million years before present (570 MBP) — the era of invertebrates and primitive vertebrates;

3. **Mesozoic,** the middle time — beginning perhaps, 230 million years before present (230 MBP) — the era of reptiles;

4. **Cenozoic,** beginning perhaps 65 million years before present (65 MBP) — the era of mammals.

The word "perhaps" is used because there are differences of opinion among experts about the evidence in such

1. Precambrian

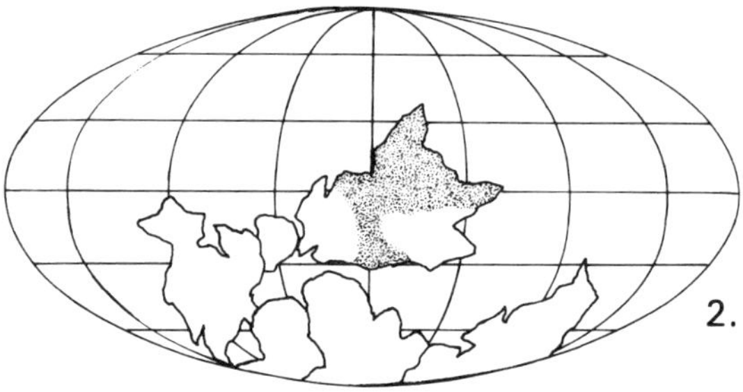

2. Cambrian

3. Devonian

Plate 12.2 a

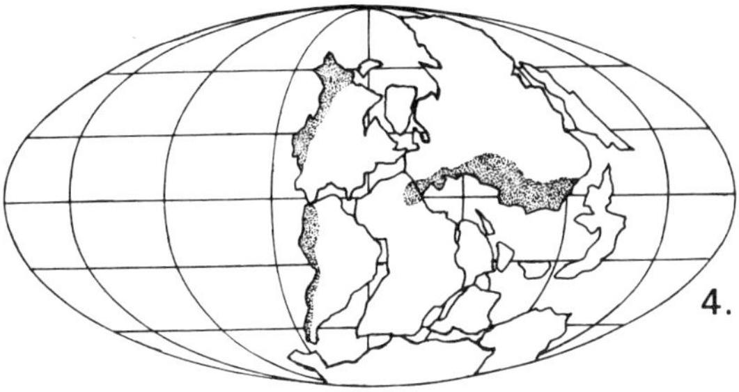

4. Permian

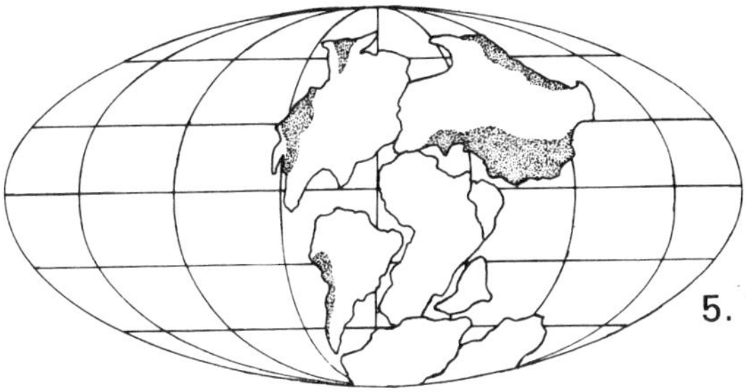

5. Triassic

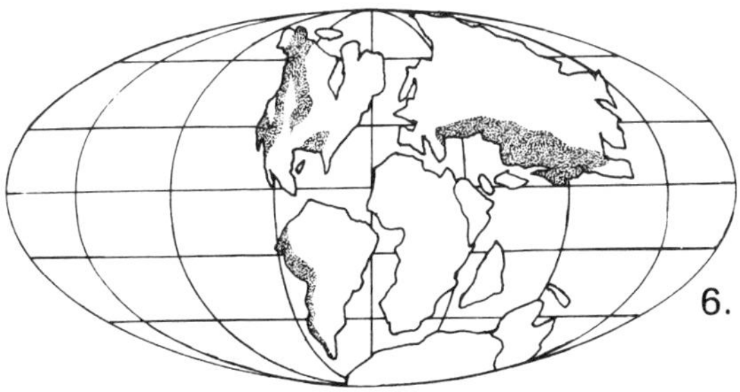

6. Jurassic

Plate 12.2 b

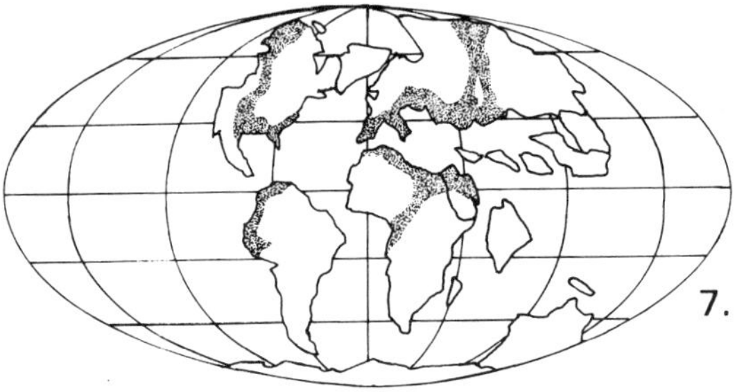

7. Cretaceous

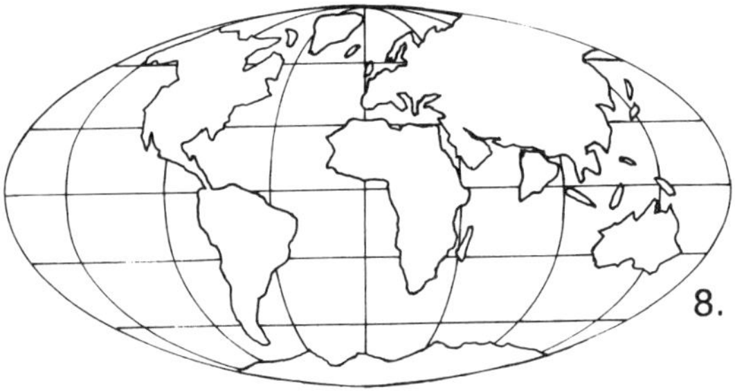

8. Quaternary

Plate 12.2 c

Plate 12.2 Configuration of Continents in Selected Geological Periods. Epicontinental seas are indicated by stippling.

1. Precambrian Period, 3.2 BBP - 570 MBP. Land was configured as one continent: Pangea. North America was adjacent to South America in the Southern Hemisphere.

2. Cambrian Period, 570 MBP - 500 MBP. Pangea fractured into the continents of North America, Eurasia, and the southern continent, Gondwana.

3. Devonian Period, 400 MBP - 340 MBP. The world was divided into three continents: Euroamerica, Asia, and Gondwana.

4. Permian Period, 280 MBP - 230 MBP. Pangea was again configured as one continent.

5. Triassic Period, 230 MBP - 180 MBP. The Tethys Sea began to separate Pangea into Laurasia in the north and Gondwana in the south.

6. Jurassic Period, 180 MBP - 130 MBP. Massive splitting of continents continued, forming the North Atlantic Ocean. Gondwana split into Antarctica, Africa, and South America.

7. Cretaceous Period, 130 MBP - 65 MBP. The earth was configured as six continents, with Australia joined to Antarctica, and India, an island.

8. Quaternary Period - Holocene Epoch, 10 TBP Present. Present configuration of earth's continents.

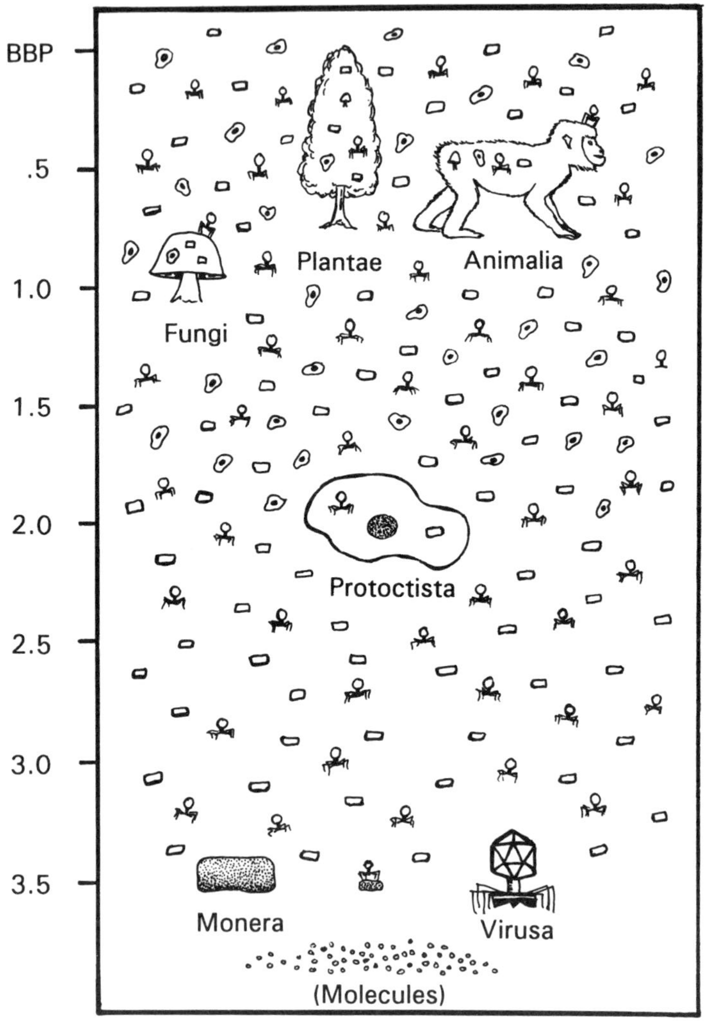

Plate 12.3 Complementary Systems Evolution
— Six Kingdoms of Life

ERA	PERIOD	EPOCH	BEGIN DATE
CENOZOIC	QUATERNARY	HOLOCENE (RECENT)	10,000 YEARS BP
			MILLIONS, YEARS BP
	QUATERNARY	PLEISTOCENE	2
	TERTIARY	PLIOCENE	5
		MIOCENE	25
		OLIGOCENE	40
		EOCENE	60
CENOZOIC	TERTIARY	PALEOCENE	65
MESOZOIC	CRETACEOUS		130
	JURASSIC		180
MESOZOIC	TRIASSIC		230
PALEOZOIC	PERMIAN		280
	CARBONIFEROUS	PENNSYLVANIAN	310
		MISSISSIPPIAN	340
	DEVONIAN		400
	SILURIAN		450
	ORDOVICIAN		500
PALEOZOIC	CAMBRIAN		570
ARCHEOZOIC	PRECAMBRIAN		3,200
AGE OF EARTH			4,500

Plate 12.4 Geological Time.

huge time frames. There are variations in the estimates of beginning and ending dates for the various rock strata. As more evidence has accumulated and new dating methods have been devised, the time frames have been — and continue to be — adjusted.

Each era is divided into periods; the two most recent periods are divided into epochs. Although scientists have divided these epochs into ages, these will not be used in this work. Eras, periods and epochs will be the divisions used, and it is recommended that the reader refer to Plate 12.4 as needed to keep the time of evolutionary events in perspective.

As continents moved in relation to seas, far-reaching changes occurred on the planet. In the seas, major changes include: shifting depth, sea floor spreading and mountain range building — changing relationships of sea to land forms — changes in temperature and salinity, in mineral, ion and oxygen content, in currents and turbidity and in amounts of light — solar radiation and penetration. Evaluation of all these factors over eras of time is an immense scientific task.

The changes of land configurations were as complex as changes in the seas — even mind boggling — when coordinated with lifeform changes, but scientists have made magnificent progress in describing them. Changes on land include: configuration of continents in relation to other continents and seas, changes in weather and climatic zones, in elevation, in rock and mineral formations, in soil types, in seismic activity and volcanism, in solar radiation, in varying seasons and lengths of day and night. There were changes of plant and animal beginnings, varying evolutionary patterns, interactions among lifeforms, and extinctions.

A survey in these pages of some of the tectonic movements of continents and their relationship to the fossil record, together with evolutionary analysis of selected species occurring during a period of 3.8 billion years, is only suggestive of the mass of data available in libraries around the world.

Norman D. Newell (1982) contributed to knowledge of the rise and extinction of lifeforms, although his work was done before the general acceptance of plate tectonics in the late 1960s. Tectonic correlations with the history and structure of vertebrates would be almost impossible without reference to the superior work of Alfred Romer (1966).

Richard Cowen (1976) was a pioneer in coordination of tectonic and fossil data, and Stehli and Webb (ed., 1985) have made coherent the interaction of tectonics and paleontology in the interchange of lifeforms in the Western Hemisphere during the Cenozoic. Some of the information in the following chapters is drawn from these sources.

This sketch of fossil, geologic and plate tectonic evidence in relation to evolution, is written with some trepidation. To select key information from those fields and combine it with knowledge gained from the study of genetics and logic is a pioneering undertaking.

I have given common names to certain fossil animals to facilitate reading flow. The genus is usually the basis for the common name, with some exceptions to avoid confusion. Scientific names and certain particulars of classification follow the common name in parentheses.

These are the key points of contrast between classical and complementary principles which will be demonstrated in considering the fossil record:

Key Points of Contrast Between Classical

and Complementarity Dynamics Principles of Evolution

Classical Theory

Darwinians believed the fossil record has shown:

1. Simple species preceded more complex species.

2. Simpler members of one taxon preceded more complex members of the same taxon.

3. Smaller organisms preceded larger organisms of the same kind.

4. Regressions in development have been permanent.

5. Soft-bodied organisms preceded hard-bodied organisms.

6. The age of fossils is determined in order to assess the common ancestors and lines of descent of plants and animals.

7. Linnaenian classification is a basis for constructing Darwinian evolutionary patterns.

8. Key homologous structures employed in classification can also be used to distinguish evolutionary progress; i.e. type and placement of fins, type and location of cartilage or bone, head bone structure and openings, type of hip bones, development of ear ossicles, configuration of bones of hands and feet; epidermal bones, horns, antlers, scales, feathers, skin, hair. The line of development of homologous teeth is a key criterion for determining classification and descent, especially the difference between carnivores and herbivores.

9. Series of small mutations fill saltations between lines of descent.

10. Mammals evolved on land and some have become sea-living animals; e.g. porpoises, seals, sea cows, whales, etc.

11. Key fossils (called by their Latin genus and species names to preserve scientific accuracy worldwide) show evolutionary patterns.

12. Local environments can be defined for identified species. Predator and prey

relationships are judged to be of primary importance.

13. Preadaptations are evident, but are ignored in theory, or are considered to be random.

14. Parallel evolutionary pathways are acknowledged but are not explicable within the theory.

15. Coevolution or convergent evolutionary pathways are considered to be random.

16. Divergent evolutionary pathways are generally considered to be movements of evolving species into a new location, separated from the niche of the parental species.

17. Key functions can be used secondarily to distinguish evolutionary progress; e.g. reproductive processes, eating habits, aquatic or land living.

18. Evolution proceeded from simple and primitive species to invertebrates, to jawless fish, to jawed fish, to amphibians, to reptiles, to birds and mammals, to primates, to humans.

Complementarity Dynamics Principle Applied

The fossil record supports many of the above principles. It also supports their complements. Using complementarity dynamic logic, I believe that the fossil record shows that:

1. Although there is a general trend toward more complexity, complex creatures with definitive features occurred very early, and many simple lifeforms have unusually long histories.

2. Some taxa have shown simplifications or regressions over time. Complex organs of simple taxa have often preceded simple organs of taxa destined to evolve into more complex lines of organisms; e.g. the eye is vastly more cómplex in Cambrian trilobite species, than the sense organs of chordates such as Pika (Chapter 14, Figure 14.15 b).

3. Some taxa have remained virtually unchanged in configuration as well as size over millions of years; e.g. certain mussels and clams. Large animals have often arisen without linear evidence of a previous smaller kind; different sizes of the same taxon have arisen together in the record; e.g. large and small saurischian dinosaurs in the Jurassic, such as *Brachiosaurus*, weighing over 80 tons and chicken-size *Ornitholestes*. Some taxa have large-size members in the fossil record long before smaller members have arisen. Examples include ground birds, sloths and a gigantic ape line.

4. There are some examples of regressions, considered permanent by Darwinians, which may be temporary DNA modification in adjustment to environments, such as blind cave fish.

5. Soft amoebae and organisms with hard tests may have arisen about the same time. Bony fish appear to predate cartilaginous fish.

6. The age of fossils is determined to assess the possible ancestry and ascent of plants and animals. Complementary relationships

of numerous characteristics suggest possible DNA connections.

7. Orthodox classification does not always represent successional relationships. Notoungulates and ungulates may be related by similar DNA although they apparently arose on different continents. The relationship of marsupials and placentals may or may not be by successional descent. Not all early amphibians preceded early reptiles. Various means of classification suggest different routes of evolution. All should be considered.

8. Key structures of one kind may suggest a different lineage than key structures of another kind, e.g. dinosaur hip bones and foot bones. The complementary relationships of homologies (similar structure) and analogies (similar concepts and-or functions) are both valid in considering classification and ascent of plants and animals. Unusual characteristics and combinations of features should be explored as possible keys to discovery of recombinations of DNA in lifeforms which may appear far apart in time. The most important criteria for assessing classification and ascent are age, the wholistic configuration and concept of the lifeform, its DNA and the DNA of its symbionts and viruses, its relationships to other lifeforms, and its interactions with the evolving environment.

9. Saltations in the fossil record are evidence that reorganization of DNA has created phenotype expressed mutations and salmutations.

10. Sea mammals have generally evolved in the sea. A number of land animals of various kinds have evolved in separate lines, some originating from different sea-living animals, coming to land at different times.

11. Key fossils show many evolutionary patterns not previously examined. (Some of these will be considered in the next chapters.)

12. The concept of environment is expanded to include the changing configuration of the continents, assessment of relationships of water, land and weather, and interactions of various lifeforms in the assessment of key fossils.

13. Preadaptations are considered to be evidence of intelligence and purposefulness and are significant in the assessment of evolutionary patterns.

14. Parallel evolutionary pathways are considered to be evidence of similar DNA configurations.

15. Coevolution, and convergent evolutionary pathways are considered to be evidence of intelligent coordination of lifeforms and processes, perhaps involving experimentation.

16. Divergent evolutionary pathways are considered to be evidence of reorganization of DNA to produce different lifeforms, often to fill a new niche opening in response to a changing environment.

17. Key functions, such as living in an aquatic or land environment, are not clear cut. There are many movements between the two

environments in patterns which are not yet clearly understood, e.g. Desmo (Chapter 19, Figure 19.10).

18. Evolution generally proceeded from simple to complex but retained the simple. The lines of ascent involved reorganizations of DNA which did not merely follow even-flowing, sequential pathways, but also exhibited salmutations. Evolution shows discontinuous jumps in characteristics of lifeforms by means of DNA reorganization, while still maintaining a continuous flow of reproduction from one defined group to the next.

Classification Systems

Perception of the differences between the classical Darwinian theory of evolution and the evolutionary implications of complementarity have not only occurred in flashes of insight, but have developed over time in many different disciplines. Emerging systems of classification are evidence of scientists' divergence from conventional theory.

One of the first evolutionary "trees of life" was drawn by Ernest Haeckel in 1866 (Plate 12.5). Paleontologists are now drawing diagrams which show only parallel progressions of lifeforms or mostly parallel progressions from a common, usually unknown (assumed) ancestor (Plate 12.6). Cladists use either/or choices to represent junctions between species lines in cladograms, which show divergent patterns, but not "trees" (Plate 12.7). These two latter approaches are evidence of change to new evolutionary perspectives.

Evolution theory has been refined since Darwin in the classification of newly-found fossils and living organisms. Comparison of characteristics, which may be similar, may be present or absent, or may vary widely, are used by scientists in classification.

Generally, smaller and simpler fossils have been found in earlier strata. The exceptions to this pattern, although

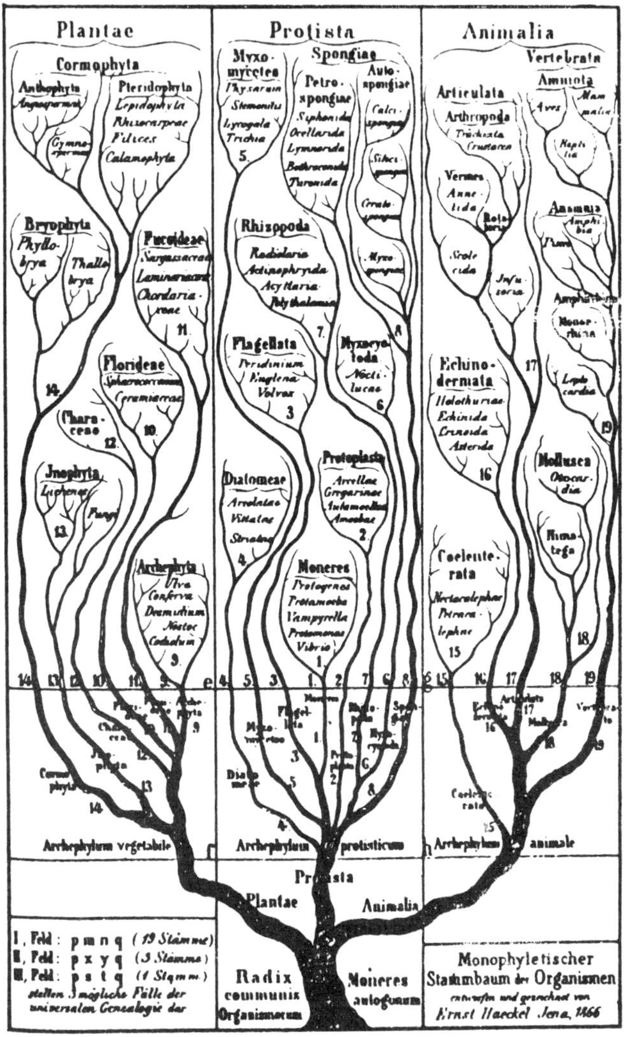

Plate 12.5 Ernest Haeckel Tree of Life. (From Haeckel 1866).

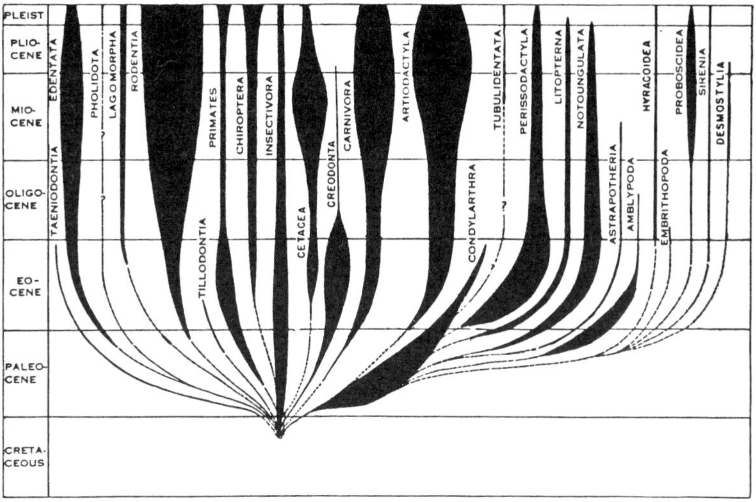

Plate 12.6 Parallel Evolutionary Pathways. (From Romer 1966: 208 "The chronological distribution of the placental mammals.")

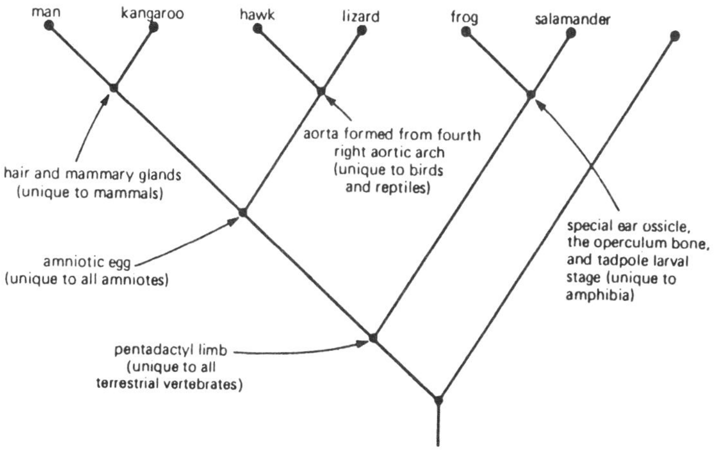

Plate 12.7 Cladogram. (From Denton 1986: 130).

numerous, have often been ignored. Rather, they should be highlighted to frame the questions about how and why these exceptions occurred.

The complexity of the Precambrian and Cambrian fauna suggests very early diversification into at least seven phyla which evolved in both separate pattern lines and in interrelated patterns over time. The frustration of scientists who have tried to categorize organisms along sequential lines, oozes from the pages of many taxonomies. Reclassification of organisms from one category to another is becoming endemic as scientists now proclaim many source (polyphyletic) origins for numerous organisms, but have no polyphyletic taxonomic system.

The discovery of DNA and its function in the patterning and repatterning of genetic expression of lifeforms requires a new taxonomic system. However, it is premature to devise such a system before the DNA patterns of numerous organisms become well known.

In the meantime, it may be wise to avoid extensive restructuring of classifications based upon external and internal bodily characteristics of the organisms apart from DNA configuration. I believe it will be more productive to begin to identify those characteristics which may represent the same or similar DNA configurations in organisms in which they appear. This should make it possible to study the relationship between phenotypic characteristics and DNA configurations as a beginning point for new taxonomy and new evolution theory.

Therefore, in the following discussion of selected lifeforms which lived in different geologic eras, I will suggest possible characteristics which may represent similar or the same DNA patterns. These similarities may be significant whether or not the pathway by which the DNA traversed between lifeforms can at present be identified.

An ongoing, one-hundred-thirty-year-long, perspective of evolution as chance occurrences of variation in individual organisms is monolithic in its concept. To change to a perspective in which each individual organism is coded by DNA and each species is identified by its organization of DNA, as well as its chemical and physical attributes, is not easy. It becomes necessary to consider the possible meaning of this changed perspective in the case of a number of specific organisms through time.

A new perspective cannot be adopted as a general approach without a significant exercise, citing examples, using the new perspective. Therefore I use complementarity dynamics logic to consider the possible operation of DNA reorganization in examples of lifeforms through geologic time.

It should be recognized that this is not a complete scientific overview, but rather a philosophical exercise meant to stimulate the thinking of scientific experts as well as general readers. The fossil record will now be examined using a complementarity dynamics perspective.

13

Archeozoic Era — Precambrian Period First Lifeforms on Earth

Life first arose in the Archeozoic, the earliest era of designated time following formation of Earth. The land was configured as one continent, Pangea, during at least part of the Precambrian Period, the time span from 3.8 billion years before present (3.8 BBP) to 570 million years before present (570 MBP) (Chapter 12, Plate 12.2 # 1).

It is believed that the ocean was near its current volume, but much less land was exposed as it was relatively flat. There were no mountains, and neither high altitude nor polar ice. The climate was warm, not only from the rays of the sun but also from the greenhouse effect of the gasses in the stratosphere.

Before the invention of the microscope, it was thought that the seas were practically devoid of life during this time span. Scientists had been searching for large fossils which seemed non-existent. In 1954, Elso Barghoorn used the electron microscope to identify numerous microfossils associated with silicified algae deposits in Africa, giving a new Precambrian starting date to the miniscule organisms which were studied by micropaleontologists.

The largest of the microorganisms, foraminiferans, have been known since the time of the Greek geographer Strabo (63 BC-24 AD) who observed them in Egyptian pyramid limestones and wrote their description. In 1758, Linnaeus classified a few species of foraminifera in the genera *Nautilus* and *Serpula.* (They were subsequently classified in the Kingdom Protista, Phylum Sarcodina, Class Rhizopoda, Order Foraminiferida).

Gerard Bignot (1985) called C. G. Ehrenberg (1795-1873) the true founder of micropaleontology, as he was the first to understand the rock-forming role of microfossils. When it was recognized that these organisms were of stratigraphic importance to oil drilling (1874) the field became dominated by oil interests. The emphasis on identification of foraminifera and other microorganisms in different strata took precedence over the study of their evolutionary relationships, although these were secondarily considered.

Micropaleontologists have based their investigations and classifications mainly on *hard* parts of organisms retained or imprinted in the fossil record. While this is understandable, it is inconclusive for comprehension of evolution, which now requires comparison of the fossil record with knowledge of the DNA of *soft* organisms, and the soft parts of organisms which secrete tests and shells.

Some scientists believe there is not enough fossil evidence available to fit the pieces of the puzzle of Precambrian life into a coherent picture of early evolution. The extensive movements of the tectonic plates may have obliterated much of the hard evidence which once existed. The great elapsed time since this period may also prevent the gathering of data sufficient to construct an adequate Precambrian theory. But why not speculate about the puzzle pieces which are currently available?

First Known Bacteria and Blue Green Algae

Evidence was discovered by Elso Barghoorn (1982) and J. William Schopf that Precambrian strata of approximately 3.2 BBP, known as the Fig Tree cherts of Swaziland, Africa, contain fossils of what appear under magnification of the electron microscope to be bacteria and columns (stromatolites) of blue green algae (Cyanobacteria or Cyanophyta). Those were considered by Barghoorn and Schopf to be the first evidence of life on earth, single-celled organisms (prokaryotes) creating oxygen by photosynthesis even in that early time (Figure 13.1).

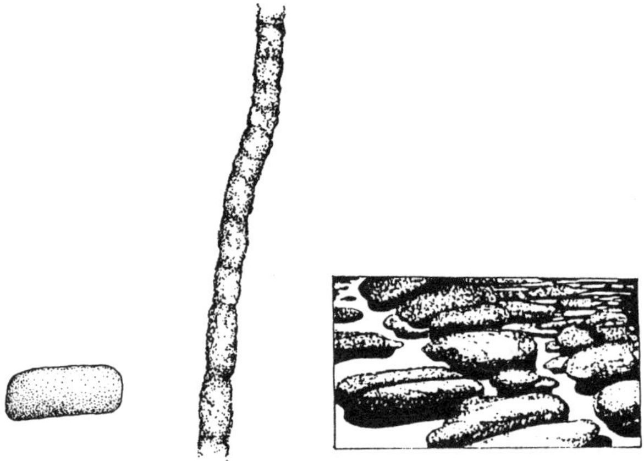

a b c

Figure 13.1 (a) Bacteria. (b) Blue-green Algae. (c) Stromatolites. (c. After Cowen).

The nature of those fossil organisms was judged by comparing them with similar lifeforms existing today. It has been thought by a few scientists that some of those early fossils were protein molecules, a transitional form between non-life and life. Lynn Margulis has "fossilized" protein molecules in the laboratory, which seem identical to some of those real fossils.

However, weight of opinion now seems to be that those cellular forms were bacteria and algae.

It is impossible to go back in time to evaluate the DNA patterns of Precambrian organisms which are known to us through fossilized mineral replacements of their bodies, fossil imprints and pathway traces in stone. The only secreted parts of organisms which have been found in Precambrian deposits have been microscopic shells (tests), worm tubes and reefs of algae and coral deposits.

However, much may be learned about early evolution by comparing the DNA of various living organisms, which are similar to lifeforms whose fossils appeared in early geologic strata. Although it is difficult to judge the nature of the internal parts of those early fossil organisms, let's assume with their discoverers, that those bacteria and blue green algae were constructed similarly to modern bacteria and algae, with ring-form DNA.

The columnar structure of the fossil algae suggests a beginning of cell association in a fixed configuration, an initial structure of organisms ancestral to plants. The freely moving bacteria, even though some were also photosynthetic, suggest the concept of animals. Were these fossil creatures prototypes for plants and animals to come?

By 1985 some scientists thought that bacteria and algae have a history even earlier than the Fig Tree cherts. H. D. Pflug found stromatolites in Greenland dated 3.8 BBP. Were viruses in the bacteria of the African Fig Tree cherts and the algae of the Greenland stratas? Could they be detected, if there?

Principle of Complementary Origination of Bacteria and Viruses

I hypothesize complementary origination of bacteria and viruses. Apparently when the first viruses on earth arose, they required living bacteria, or other living organisms, in order to reproduce and thus survive. Those viruses must have arisen very early in evolution. It is possible that the first bacteria had no means to evolve apart from the functioning of viruses.

In complementary systems logic, it is possible to theorize that the early bacteria and early viruses arose as a system, a complementary system of life and non-life, a system capable of supporting life, of damaging or destroying life, and also capable of producing evolving new patterns of lifeforms. The verification of this principle depends on future development of techniques to detect whether viruses were embedded in the bacteria of the African or Greenland formations, or in the bacteria of other early strata.

The assumption that both bacteria and viruses arose before those fossil formations is warranted. Modern evidence suggests that bacteria and viruses may always have been indispensible complements. The relationship between bacteria with F+ factor (containing ring-form plasmid DNA) and bacteria with F- factor, (containing circular DNA but no F plasmid) shows that life exists with more than one system of DNA. Two-system DNA: genophore DNA and cytoplasmic DNA, such as the F+ plasmid may have evolved from an original complementary system of bacterial and viral DNA in the Precambrian Period. The F plasmid may be found in bacteria today, either as a part of the genophore or as a separate structure. Perhaps it was always so.

The F plasmid of a living bacterium may be the same DNA as that of a virus. Modern evidence that the body wall and sex pilus of bacteria have specific receptor sites for specific viruses suggests a long evolution of intimate relationships. Could this relationship have started in the Precambrian?

A virus, attuned to a receptor site on the sex pilus and capable of injecting its own DNA into the bacterial DNA passing through the sex pilus may be an agent capable of changing the genetic form of an F- bacterium. Many same viruses in an area could change a number of F- bacteria into the same new taxon of F+ bacteria. As noted earlier, only F- bacteria or their sex piluses, can be inserted by viruses. F+ bacteria are immune from viral insertion.

These hypotheses have mind-boggling implications. It may be that before male and female organisms arose in evolution as separate entities, each with slightly different DNA, viruses (as F plasmids) were already present on earth, reorganizing DNA of bacteria. Viral recombination of DNA may be considered a precursor of sexual reproduction, which became evident in early organisms of the Precambrian.

Processes of DNA Reorganization

Salmutation as the process of recombination of DNA may have occurred in the Precambrian, creating new and different lifeforms from existing bacteria and algae. The principle of salmutation could apply in the evolutionary interface between the earliest known bacteria, and later, between genera of other organisms. Some of the possible interface points will be suggested in this and later chapters.

The scientific literature identifies seven processes of *salmutation* (DNA reorganization and transfer to produce new species and new genera). It remains to be learned when these processes first arose, but some of them may have begun with the earliest bacteria. The study of Precambrian organisms is still in the stage of identification and classification. Consideration of their possible DNA and evolutionary affinities remains for future scholarship. DNA of living organisms is reorganized by processes which probably evolved during eons of time, beginning in the Precambrian:

1. Point mutations — exchange of two nucleotides between complementary strands of DNA.

2. Deletion — removal of nucleotide(s).

3. Insertion — addition of nucleotide(s).

4. Duplication — replication of a segment of DNA.

5. Inversion — reversal of a segment of DNA on a strand.

6. Translocation — movement of a segment of DNA between non-homologous chromosomes.

7. Chromosome reorganization — DNA, in the form of genes recombining to become new chromosomes, or chromosomes multiphying, joining together or splitting, leading to an increase or decrease in their number. Some of these processes may also involve "repair" of reorganized DNA strands. This is facilitated by repair enzymes produced by amino acids.

Processes of DNA Transfer

DNA is transferred among living cells, single-celled organisms and viruses, in seven processes, some probably occurring as early as the Precambrian:

1. Fusion of two or more cells or organisms. Fusion may involve recombination of DNA in the fused cell. This may be the method by which early symbiotic prokaryotes became eukaryotic. This process, in its early stages, may have been similar to that represented today by the ingestion process of amoebae.

2. Division of a cell into two or more cells. The DNA is divided between the cells as they split: in asexual species by such processes as binary fission, budding and spore formation; in later sexual species, usually by mitosis and in the case of sex cells, by meiosis.

3. Sexduction. A plasmid transfers from one to another bacterium during conjugation.

4. Viral infection and transfection. Virulent phages insert DNA into bacteria where it is

transformed by the cell's processes into new phages.

5. Viral transduction. The bacteriophage introduces viral, bacterial or other DNA into a host where it may remain inactive for a time. It may be incorporated into the host cell chromosome or remain as a plasmid.

6. Transformation. Adsorption of naked DNA by bacteria or other lifeforms leads to incorporation of foreign DNA in a body cell or germ cell.

7. Human intervention. Some of the above processes have recently been mediated by humans.

How do the processes of viral transfer of DNA lead to changes in taxa lines? Scientists are just beginning to explore this question. It can be theorized that viral DNA insertion into bacteria began processes of evolution. Later, viral DNA could have inserted into fertilized eggs and transformed various taxa DNA in small population explosions. It would be necessary for a number of individuals to salmutate together if a new sexually reproducing genus line were to result.

This evolutionary proliferation could occur from direct transfer of DNA by viruses into the eggs of a genus line about to be transformed, from virus crystals dormant in the environment for a period of time and then coming into contact with eggs, or by human manipulation. But where is the intelligence to produce a viable new genus? Is it in the virus, the virus particle, the provirus, in the plasmid? Is it in the virus crystal? Is it in the whole DNA, in the base pair, in the codon? Where is it?

It is revealed in the process. Intelligence does not reside in a structure in the sense of having physical parameters. No intelligence can be found in the brain nor in the computer — intelligence is not a component of the hardware. Nor is it in the software apart from the hardware. We cannot see intelligence; we comprehend it. Intelligence becomes evident in the processing itself, in complements of structure and function. Salmutation into a new genus manifests intelligence, directed not by the virus alone nor the germ cell alone. It is an event in harmony with the symphony of the universe. Salmutation is a finite form of infinite becoming.

Two Billion Years Before Present

First Fossil Eukaryotes

More than a billion years after first fossilized life, in the two billion year old strata of the Gunflint cherts (very hard black quartz) of Ontario, Canada, eight genera of aquatic plants are evident (Barghoorn, 1982: 47). Some of the fossils exhibit structures which indicate that they may have been eukaryotic, the first evidence of life with a cell nucleus, the cell structure evident in subsequent multicellular organisms.

The salmutation from the double ring DNA of a bacterium to a fully formed nucleus within a eukaryotic cell was a major salmutation. It was an essential preadaptation to allow lifeforms to evolve beyond single-celled organisms. Did this innovation occur many times? We don't know. Cytology and virology are very young sciences that are just beginning to demonstrate evolutionary perspectives.

But it would appear that no viruses and no cells could be considered to possess the degree of intelligence required to preadapt for the whole future of myriad multicellular organisms and myriad reproductive strategies known on Earth. The salmutation from DNA of a prokaryote to a fully formed nucleated cell seems to call for the concept of an organizing intelligence.

Sexduction with circular transfer of DNA from an F+ bacterium to an F- bacterium might have been a satisfactory processes of gene transfer for single-celled organisms, but how could multicelled creatures have evolved in this fashion? Impossible. A new method for transferring DNA was required.

It would appear that the division of DNA into two parts must occur in both asexual reproduction between cells (cell division) and sexual reproduction between organisms. If F- bacteria could have continuously divided asexually without a method for acquiring plasmid DNA, life might have escaped transformation by viruses, and F- bacteria could have smothered the earth in a mass of uncontrolled protoplasmic growth.

Therefore it seems likely that the processes of mitosis and meiosis came into existence as complements. The complex and beautiful movement of newly arising parts of a cell (centrioles and spindles) which serve to organize and distribute DNA from the nucleus of one cell into two new cells, each with a nucleus, is a truly creative innovation.

Meiosis arose in complementarity dynamics with mitotic division, providing a second division of the two complements of DNA into two chromosome forms: male and female, each with slightly different configurations of DNA composing one of the sex chromosomes. This was a preadaptation which appeared in myriad forms as evolution unfolded. It seems likely that very early in the fossil record, as early as two billion years ago, as eukaryotic lifeforms appeared, both asexual reproduction and sexual reproduction arose in the alternating patterns exemplified in early microorganisms.

Although Lynn Margulis and Dorion Sagan (1986: 116) postulated that symbiosis of two different organisms began to occur 1.6 to 1.4 BBP, I believe that those fusions must have occurred before two billion years ago, prior to the salmutations to the nucleus form of eukaryotic cells and the processes of mitosis and meiosis. Scientists believe that those early symbiotic organisms lived by photosynthesis and-or by "eating" inorganic matter, and that the atmosphere changed gradually as they produced oxygen.

Some early organisms functioned well in the oxygen build-up while others died from its effects. Margulis believes that evolution from anaerobic to aerobic organisms was exemplified by symbiotic fusion of two or more lifeforms. On the other hand, perhaps there were also salmutations at that time.

Food Intake and Photosynthesis

There are those scientists who believe organisms ate organic matter *before* they ate the inorganic. Scientists educated in monistic logic continue to debate the issue, which may be sequential and-or may be in reciprocal time. In any case, producer, consumer and recycler organisms are found very early in the fossil evidence of the Precambrian, and in every subsequent level of evolutionary time. It will be interesting to compare the DNA of predator and prey organisms which often seem to arise together in the fossil record. Is there a predator-prey DNA system evident in formation of genera? Are there plant-animal DNA complements which arose together?

An example of combination of phenotype characteristics is the modern protist *Euglena*, which alternates between the processes of animals and plants: it eats when food is available and photosynthesizes when it is not. Such a lifeform does not fit readily into Darwinian theory and Linnaean classification. It has DNA configurations in its organelles of both plants and animals. Could it also have originated in this very early time period, although it left no known fossils? (Figure 13.2.)

Organisms may have been photosynthesizing and eating both inorganic and organic materials more than two billion years ago. As the ability to incorporate a variety of organisms into the diet of eukaryotes grew, so also, in complementary fashion, organisms may then have been incorporated in symbiotic relationships. It is not difficult to imagine that some early organisms in amoebic form, moving as blobs, engulfing bits of organic matter, underwent a salmutation, producing change in digestive enzymes. These enzymes could have allowed an aerobic organism, later called mitochondria, to exist in symbiosis within the amoebae and perform the function of breathing and energy production for the combined holobionts.

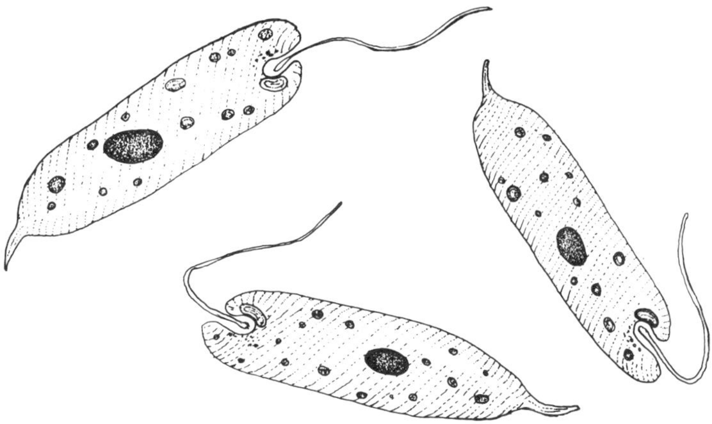

Figure 13.2 *Euglena* (Protista). (After Johnson).

It became necessary for organisms to become aerobic, become symbiotic with aerobic organisms, or define an exclusive anaerobic niche in order to survive the major change to an oxygen environment. Those organisms which clung to anaerobic niches, burrowing into silt and mud, or into the tissues of other organisms, were preadapting for a time to come when their descendants would live in the intestinal tracts of animals, to create the possibility for animal and human life. Such preadaptations indicate a majestic unity in nature.

One Billion Years Before Present

Impressive Evidence of Evolving Eukaryotes

In the Bittersweet cherts of Australia, dated almost one billion years before present, 20 kinds of filamentous blue green algae and three species of bacteria have been identified, also four groups of lower aquatic plants, two species of fungi, and two genera of green algae, definitely eukaryotic, with impressive evidence of the process of mitosis. Two unclassified lifeforms remain to be identified (Barghoorn, 1982: 54).

If lichens were not present as early as one billion years ago, algae and fungi, lichens' source symbiotic species were evident then. I believe algae and fungi were preadaptive for the

13,500 species of lichenized fungi known today (Hawksworth and Hill, 1984: 8). Lichens, which live in myriad symbiotic forms, reproducing asexually and sexually in a multiplicity of patterns, are impressive examples of the power of symbiosis.

As the oxygen content of the atmosphere increased, the stratospheric ozone shield began to form. Near the end of the Precambrian, the oxygen production of proliferating algae rose precipitously. The concentration has been estimated to be 1% to 10% or more of the present level of oxygen in the earth's atmosphere. The ozone shield, as well as the increased level of atmospheric oxygen were preadaptations of the environment for life on land. No land-living lifeforms could have evolved without oxygen and the ozone shield.

Seven Hundred Million Years Before Present

Early Eukaryotic Life and Rise of Multicelled Organisms

It has been postulated that in the late Precambrian, about 700 million years ago, the continent of Pangea was still devoid of life. The sea crashed on the shores of the dead land. But there is evidence that a myriad of different types of protists inhabited the ocean of late Precambrian times. It is amazing to discover evidence also of diverse multicelled life much earlier than previous theory had indicated. The Protoctista Kingdom classification is in a state of flux as a number of scientists strive to work out the sequential evolutionary pathways from bacteria of various kinds to nucleus-celled life and multicelled lifeforms.

Organisms which have been classified as Protozoa lived in the Precambrian Period. Moore et al (1952: 39) classified one-celled lifeforms in four main classes: Flagellata and Sarcodina, some of which have left hard parts in the fossil record, and Sporozoa and Ciliata which have only soft parts and, along with Flagellata were not then identifed as Precambrian fossils.

After taxonomist R. H. Whittaker (1969) classified one-celled lifeforms in two kingdoms, Monera and Protista, the four classes of Protozoa discussed by Moore were no longer considered relevant. The invertebrate paleontologist M. A. Sleigh (1979: 23-54) suggested that the Protista Kingdom be considered an aggregation of lower eukaryotes, and it is regarded by him as one of the five kingdoms formulated by Whittaker.

Sleigh pointed out, however, that botanists and zoologists have included in the Protista Kingdom, organisms such as euglenoid flagellates, ciliate protozoa, algae and lower fungi. He noted that these include groups of both unicellular and multicellular forms. It no longer seems possible to draw exact division lines between single-celled organisms and multicellular ones. Colonial aggregations appear among both groups, suggesting many source (polyphyletic) evolutionary foundations for the higher plants and animals.

Sleigh listed 23 groups of lifeforms as Protista. Although he separated dinoflagellates, collar flagellates and ciliates into separate groups, it is now known that both cilia and flagella have a common (9+2) microtubular structure (Figure 13.3 d). This structure, which may have salmutated from a simpler tubular structure of bacterial spirochetes, is the same in cilia of single-celled organisms such as paramecia, and cilia of cells of higher lifeforms. The (9+2) structure is found in the spindles of eukaryotic cells in mitosis, and in the structure of the axions and dendrites of nerve cells (Figure 13.3). The evolution of the (9+2) structure calls for new concepts.

A new classification seems to be needed for organisms characterized by this structure. Indeed, Margulis (1978) identified the (9+2) structure as "undulipodia" but the evolutionary ramifications of this structure are complex and remain in dispute. Because undulipodia are not just organisms, but also cell parts, the concept of *individual* becomes clouded. Perhaps this requires a moratorium on reclassification of organisms pending more exact and full DNA data, in order to avoid the confusion caused by ever-changing classifications based on structural data alone.

Wonderful Ediacara Organisms

The Precambrian Ediacara strata of Australia contained more than 1,400 specimens of fossil organisms classified in 30 genera. The majority of them were segmented worms, jellyfish and corals. The formation also contained fossils of unique organisms such as Trib (*Tribrachedium,* Figure 13.4 a) organisms which do not appear to have direct antecedents or to be direct ancestors of any later developed lifeforms. Other ocean organisms which appear to have a plant-like structure, *Charnia, Rangea* and *Arborea* (Figure 13.4 b, c, d) suggest possible progenitor DNA, or preadaptations for later appearing (Silurian Period) land-based plants. Such discoveries open a new beginning chapter in the story of life and one that demands an open mind in the development of evolution theory.

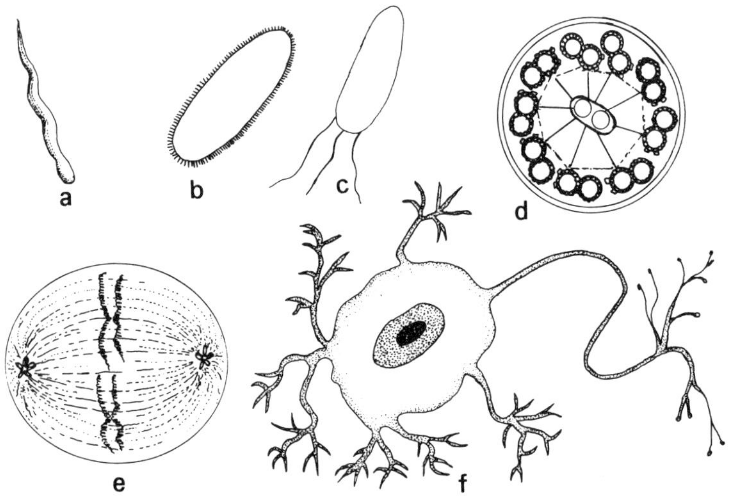

Figure 13.3 (a) Spirochete bacteria. Undulipodia: (b) Ciliated organism. (c) Flagellated organism. (d) Microtubular 9 + 2 cross section. (e) Spindles of cell in mitosis. (f) Nerve cell with axion and dendrites.

Early Amoebae and Other Microorganisms

Two groups of protists generally with soft parts — amoebae and paramecia — may have arisen in the Precambrian. The evidence is quite speculative, but their very early existence now seems feasible. Two groups of fossil organisms with hard parts were first identified in Ordovician deposits (foraminifera and radiolaria, Class Sarcodina, Chapter 12, Figure 12.1) and they may have evolved from Precambrian microorganisms.

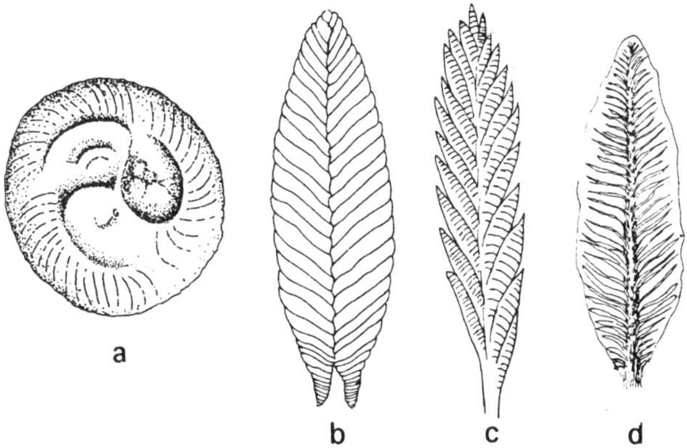

Figure 13.4 (a) Trib (*Tribrachedium*). (b) *Charnia*. (c) *Rangea*. (d) *Arborea*. (After Graessner and Wade).

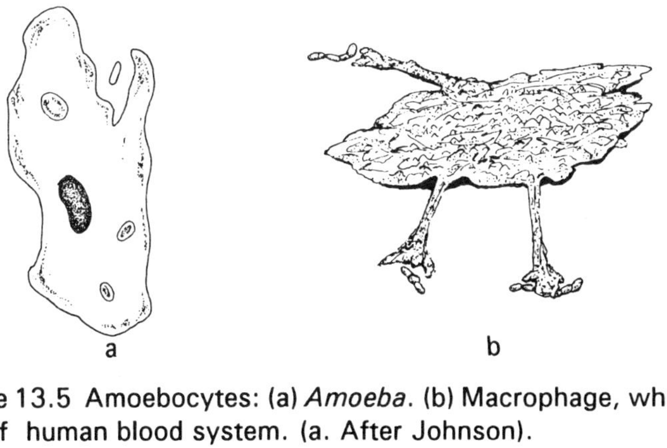

Figure 13.5 Amoebocytes: (a) *Amoeba*. (b) Macrophage, white cell of human blood system. (a. After Johnson).

It is now believed that the sea floor was crawling with life in the late Precambrian. Sea water probably teamed with soft-bodied microscopic life such as amoebae which apparently left no fossils (Figure 13.5 a). Still, there is evidence that amoebae existed at this time. Interface points between the microorganisms of the Precambrian and the larger lifeforms evident in the fossil record are now being studied by scientists.

In the 1940s, biology students learned that amoebae are microscopic, single-celled organisms without a rigid cell wall. They live in an aquatic medium, moving by flowing motion of their cytoplasm, engulfing food particles, digesting them within the cell, and reproducing by asexual division.

One of the methods that can be used today to gain insight into the evolution of amoebae, is the comparison of DNA of modern amoebae and the DNA of organisms such as sponges. Numerous hard skeletal parts of sponges (spicules, Chapter 14, Figure 14.1 a) are found in early Cambrian deposits, so it is known that sponges lived at that time. Modern sponges contain wandering amoeboid cells, so it may be assumed that Cambrian sponges were similarly constructed. Therefore, I believe amoeboid cells probably existed within sponges in the Cambrian.

If we accept that simple organisms joined together in symbiotic relationships to form more complex organisms, then it would appear that amoebae preceded the sponges and existed as individual organisms, possibly in the Precambrian Period, prior to their incorporation as body parts of sponges.

Today it is known that amoebae are not only organisms in their own right, but are an integral component of symbiotic life processes in many lifeforms, and participants in the evolutionary pathway to many higher organisms. Amoebocytes moved freely within the basic structure of early eukaryotes and, in my opinion, they eventually evolved into macrophages, white blood cells of animals and humans, a basic component of the human immune system (Figure 13.5 b).

The concept of this evolutionary pathway of amoebocytes is indeed mind boggling, for it implies that an essential preadaptation for the human immune system originated and developed in the long history of microscopic single-celled organisms. This is another essential preadaptation which cannot reasonably be considered to be a matter of chance. It requires the perception of infinite intelligence as integral to evolution.

"Simple" Paramecia and Radiolarians

From primitive eukaryotes, early multicellular organisms arose. Perhaps some of these evolved from the DNA of paramecia-like lifeforms (Figure 13.6 a). Paramecia are ciliated soft-body protists which live in aquatic environments. They are considered to be one of the simplest organisms. They have mouths for intake of food and expelling of waste, but no specialized circulatory, digestive or nervous systems. They function as one cell. In contrast to amoebae, they have defined cell walls and move by rapid movements of cilia (hair-like, 9+2 structured projections). These cilia may have originated in a salmutation from symbiotic spirochetes, primitive bacteria, which joined with the basic paramecia body.

The cilia of paramecia act as oars propelling the body in a spiraling forward motion (Johnson, 1983: 830). When the organism swims into an object, the point of contact on the paramecium is depolarized and impulses are passed along the plasma membrane, causing a reversal in the cilia motion. The organism backs up for a few seconds, and the membrane becomes repolarized. The organism angles slightly and resumes a forward spiraling motion. It is known that one gene mutation can cause a delay in repolarization, thus a longer than normal backward oaring motion in a "paranoic" paramecium. This factor breeds true as the paramecium reproduces in asexual division.

A single DNA gene mutation is thus shown to involve the most minute function of a simple one-celled creature. As coded life patterning, the full evolutionary significance of DNA can no longer be ignored. It is the initiator of life structures and functions; it even provides the directional patterning of one of the simplest lifeforms seen moving in nature! What is the relationship between the paramecium and its viruses? Is this a creature which can provide evidence of salmutations?

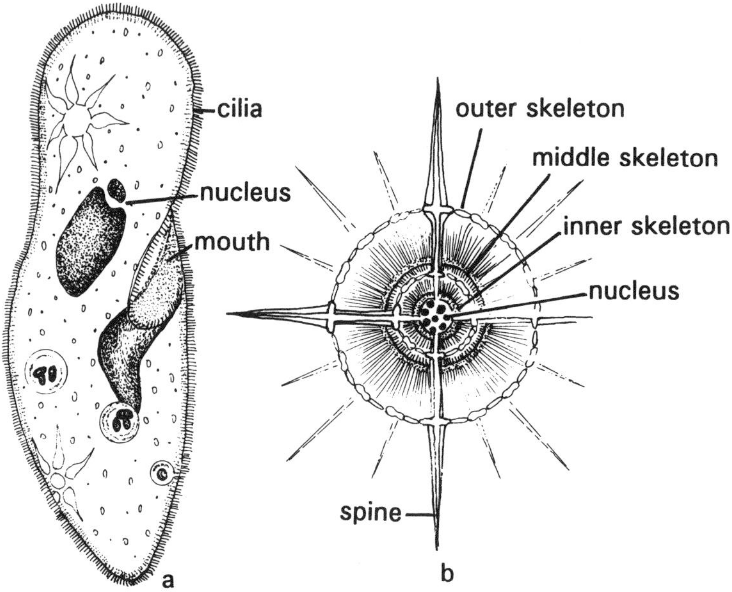

Figure 13.6 (a) *Paramecium*. (b) *Radiolarian* structure. (a. After Johnson; b. After Moore et al).

Early eukaryotes may have been ecosystems of symbiotic relationships of the paramecium, its cilia, and its endosymbiont photosynthesizing bacteria, as suggested by Lynn Margulis (1970: 174) and Lewis Thomas (1974). (Figure 13.6 a.)

Organisms like microscopic paramecia may have evolved through salmutations into macroscopic primitive organisms such as worms, perhaps flat worms, a larger form of the basic body structure represented by paramecia. The Precambrian fossil record shows several different kinds of worms. It will be interesting to compare the DNA of paramecia with the DNA of various species of worms, to try to detect possible evolutionary salmutations.

Protists such as radiolarians (Figure 13.6 b, also Chapter 12, Figure 12.1 b) left numerous fossils, tiny shells known as tests, embedded in and creating the shales and limestones of Earth. Radiolarians are minute silicious lifeforms which may be spherical or may be laterally or radially symmetrical. A typical body consists of a nucleus or many nuclei within a central membrane, enclosed by an inner spherical skeleton. A second spherical skeleton encloses a protoplasmic layer where reproduction occurs. A third skeletal layer encloses protoplasm where food is digested in vacuoles. This food is microscopic protozoans and diatoms (algae) which are collected by thread-like pseudopodia within a third outer skeletal sphere. Spines of various kinds stabilize the animals in their planktonic environment. Reproduction is thought to be by asexual fission, with one animal retaining the skeletal sphere, the other building a new one.

Some radiolarians do *not* collect their own food but live in symbiotic relationships with algae, supplying carbon dioxide and nitrogenous wastes in exchange for oxygen, fats and carbohydrates from the alga symbiont. The 800 or so genera of radiolarians which had been described by 1952, are a small fraction of the microscopic sea life which needs to be understood more fully to assess patterns of evolutionary development. Radiolarians have been so numerous that their tests form the principal constituent of rock layers in some Devonian, Jurassic and Tertiary formations. Radiolarians give evidence of the complexity of the microscopic fossil record and the importance of symbiotic processes in evolution.

Comparison of the DNA of modern microscopic, test-producing organisms with the DNA of animals like mollusks may give evidence of possible evolutionary salmutations. This might explain why there is structural similarity between certain Precambrian tests and later mollusk shells.

Ancestral Worms?

Numerous fossils of multicelled organisms such as Ana (*Anabarite,* Figure 13.7 a) a hollow shelled burrowing worm, Dick (*Dickinsonia,* Figure 13.7 b) an annelid worm and Sprig (*Spriggina,* Figure 13.7 c) a simple symmetrical worm are found in Precambrian strata. These may be ancestral to subsequent organisms.

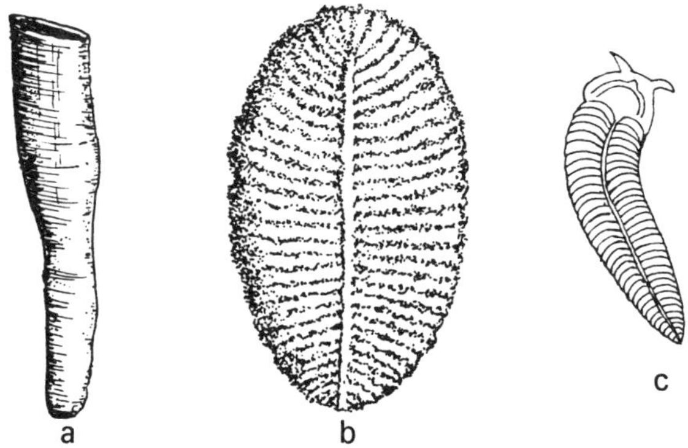

Figure 13.7 Marine worms: (a) Ana (*Anabarite*). (b) Dick (*Dickinsonia*). (c) Sprig (*Spriggina*). (After Matthews and Massarzhevsky).

Sea anemones also lived in the shallows of the Pangean continent. Scientists are amazed by the richness and advanced stage of development of some of the lifeforms of the Precambrian that have been discovered only in the last few years. A new theory of evolution must take into account the new evidence of this period. The theory posited here is that the evolutionary steps between those early animals and their successors were often salmutations.

Segmented worms may have salmutated from the DNA of flat worms. Early round worms had a coelom, an open space within the cell structure, functioning for storage, digestion and elimination. Segmented annelid worms reproduce asexually. If one is divided into two equal parts, each part can regenerate into a whole worm. They can also reproduce sexually. They are hermaphroditic, having the organs of both sexes. They cross fertilize with either, or both sexual structures of another worm. Perhaps worms were the first terrestrial animals, boring from the mud under the ocean into the mud underlying the land, preparing the soil for aquatic plants to salmutate and grow up the mud banks onto the land.

Advancement of Jellyfish and Corals

Organisms, considered by some scientists to be on the next step up the evolutionary ladder from worms, are jellyfish and corals (coelenterates). However most scientists now say that cnidarians (coelenterates) evolved first — then flat worms — and later, round and segmented worms. Jellyfish and corals are metazoans with a three layer body wall composed of epidermis, gastrodermis and mesenchyme (except for Hydrozoa) with radial or bilateral symmetry. The epidermis layer secretes skeletal elements by crystallization of calcite.

Coelenterates have a single body cavity open to the aquatic medium in which they float. Some species are attached to the sea bottom. They have a mouth but no digestive tract, no separate excretory system nor anus, and no circulatory system. They feed by filtering organic particles from seawater, waving tentacles to direct seawater into their mouths. Most coelenterates reproduce in polyp and medusa forms in alternating asexual and sexual reproductive stages. They are represented in solitary forms, and in colonial forms as hydroids and corals.

Jellyfish have been among the most numerous of early coelenterate fossils. Kim (*Kimberella quadrata,* Figure 13.8 a) appeared in the Precambrian Ediacaran deposits of central Australia. Its gastric filaments are believed to be similar to modern jellyfish. Its unpaired gonads projected into the lumen of the bell (Jefferies, 1975). Jellyfish were among the first organisms known to swim efficiently in the water medium. They had soft body parts, and were formed of definite tissue layers.

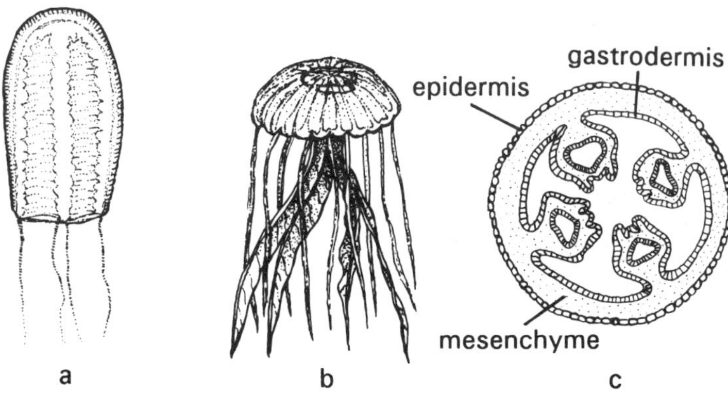

Figure 13.8 (a) Kim (*Kimberella*). (b) Modern jellyfish. (c) Cutaway of bell showing two tissue layers. (a. After Wade; b. After Hyman).

Jellyfish have existed for more than 680 million years, represented in numerous deposits as fossil imprints in which sand has been deposited within the bell and hardened to retain the form outline of the animal. Such unique organisms with radial symmetry, appearing so early in evolution, support the theory of reorganization of DNA in salmutations. There were no known organisms which could be considered, in a Darwinian projection, to be directly ancestral to jellyfish.

However, comparison of DNA of various existing worms with that of existing jellyfish may lead to evidence of possible evolutionary salmutations. The fossil imprints in the rocks of the Precambrian appear to be similar to living species, thus only moderate numbers of jellyfish mutations and salmutations followed their first appearance.

The *Cothonia* genus (Figure 13.9) was one of the first known Rugosa corals, found among the coelenterates of the Ediacara fauna of Australia. It has a curved, conical skeletal form. Cothonians appeared in North America only in the mid Ordovician, then they evolved rapidly in a series of five major salmutation episodes. The last of this genus of Rugosa corals existed in the Permian.

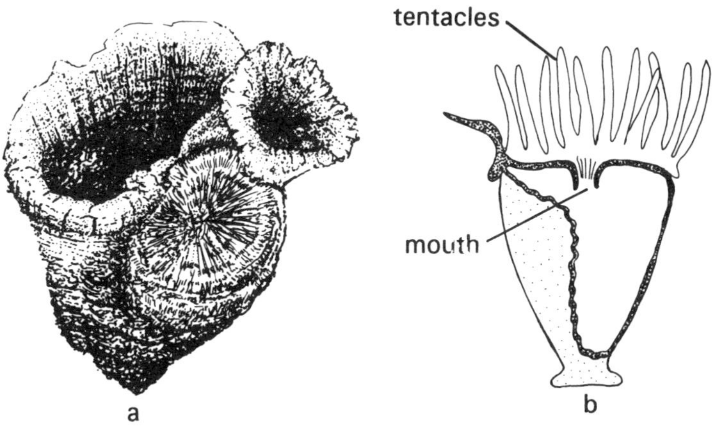

Figure 13.9 (a) Coral (*Cothonia*). (b) Cutaway of modern coral. (a. After Jell and Jell; b. After Clarkson).

Scientists have postulated that the shell form of corals is so highly variable that the many independent stocks represent repeated convergent evolution. This concept is a rejection of Darwinian sequences. The morphology of corals can be explained more satisfactorily by postulating DNA mutations and salmutations occurring throughout coral evolution. I believe perceived convergences represent occurrences of similar DNA configurations in different lines of ascent.

E. N. K. Clarkson (1986: 98) has suggested that an evolutionary principle governing the origin of new Rugosa corals involved a possible sequence from ancestral juveniles to descendent adults (paedomorphy). This concept neglects the reality that both juveniles and adults result from the same DNA production code which turns genes off and on at different times. Therefore, it becomes necessary to study the relationship of DNA nucleotide patterns to varying phyletic patterns of both polyps and adults — to understand the embryological and evolutionary development of these corals.

Clarkson (1986: 99) wrote in relation to the *Amplexizaphrentis delanouei* group of Rugosa corals:

> So many factors seem to have been involved in the evolution of the *A. delanouei* group — including trending, response to local environmental conditions, paedomorphosis, allopatry and polyphyly — that the evolutionary pattern now seems to be so complex as to elude analysis until better data are available.

In my opinion, it is not just better data that are needed, but also new logic to consider the interrelationships of these numerous factors. It is essential to study the DNA patterning of the microorganisms and viruses found within similar lifeforms — to assess the possible routes of transmission of interacting DNA. The field of comparative DNA patterning will open vast new vistas of early evolutionary history.

Division of Pangea

During the Precambrian, Pangea fractured and divided as the tectonic plates moved with gigantic force to split the continent into sections. There is little evidence, so far, that the tectonic movements of this era caused significant extinctions among the organisms in the seas. However, such evidence may have been lost in this, and in subsequent tectonic movements.

14

Paleozoic Era — Cambrian Period Proliferation of New Genera in New Phyla

The Paleozoic Era has been defined as the second major division of geologic time, characterized by massive radiation of new genera. The era opened with the Cambrian period, 570 million years before present (570 MBP). Pangea divided into four continents, creating generally: North America, Europe, Asia and the southern hemispheric continent, Gondwana (Chapter 12, Plate 12.2 #2). The seas adjacent to the continents continued to be warm for most of the period. The year was 428 days long, indicating a more slowly revolving earth than today (Runcorn, 1978: 70).

There appear to be no transitional forms from some of the organisms of the Precambrian to those of Cambrian strata, and no sequential transitional forms between the unicellular and multicellular organisms found together in the Cambrian. Single-celled creatures and complex, fully developed, multicelled animals such as trilobites arose together in the fossil record.

Thus the theory of sequential development must be supplemented by a theory of salmutation.

Many animals of the period developed hard skeletal parts, secreted as calcium carbonate (calcarecious) and silicon (silicious) tests and shells. Sponge-like creatures deposited their spicules of numerous and varied geometric patterns on the sea floor (Figure 14.1 a).

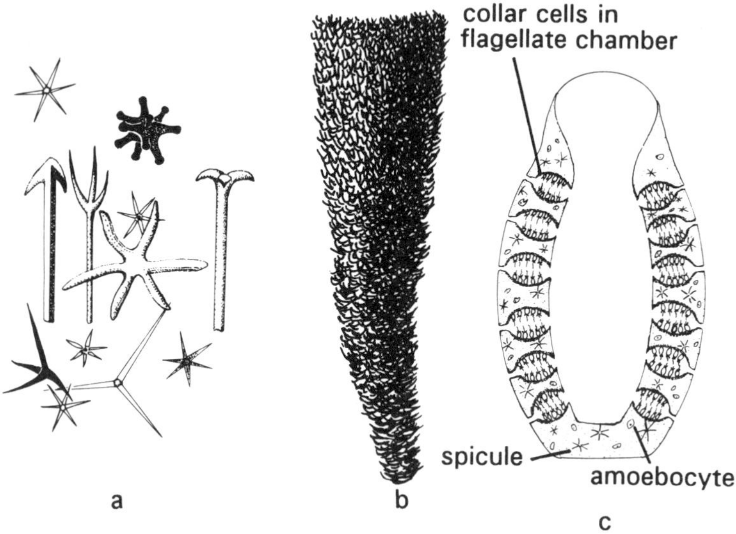

Figure 14.1 (a) Sponge Spicules. (b) Chancel (*Chancelloria eros*, Hyalospongea). (c) Gelatinosan sponge chambers (Demospongea). (a, b, After Hinde; c. After Marshall).

Richard Cowen (1976: 27-28) identified six phyla of lifeforms, found in the earliest Cambrian strata: coelenterates, arthropods, brachiopods and mollusks, echinoderms and chordates. It has been assumed that these developed directly from similar forms of the Precambrian. Except for coelenterates such as jellyfish, there is little fossil evidence to date to verify it. Cowen wrote, "We have to make up a reasonable story for sequence of evolutionary events that would produce the separate phyla by the time we see them with hard parts in the Early Cambrian."

Is this a machination to protect the theory of minor mutations occurring in *sequences*?

The fossils of many kinds of organisms in Cambrian strata give evidence of diverse patterns of evolution in many salmutations. Many different combinations and recombinations of DNA would be required to provide such diversity.

The Wonder of Sponges — Symbiosis and Intelligence

Sponges (*Porifera*) have been considered to be one of the links between single-celled organisms and true metazoans composed of tissues. We have seen that coelenterates with three tissue layers appeared in the Precambrian. Therefore, it is likely that sponges also were Precambrian productions, although their earliest found spicules were in the Cambrian deposits. Sponges have no tissue layers, but rather, appear to be a cell or a congregation of cells.

They are classified by zoologists on the basis of their soft parts. Paleontologists, on the other hand, with evidence only of the hard parts, classify sponge fossils by comparing their spicules with spicules of other fossils and with those of living sponges (Figure 14.1 a).

Chancel (*Chancelloria eros*, Figure 14.1 b) is a thin walled tubular glass sponge, about 3.5 inches in length, from the Cambrian Burgess Shales of British Columbia. It is distinguished by six-rayed individual spicules, held in place by a protoplasmic net. Its class, Hyalospongea, is represented by numerous deep sea fossils from the Cambrian to Recent Period.

The spicules of Chancel were only one of numerous geometric structural patterns represented in sponges which have grown as individuals and as colonies in varied marine and freshwater environments. While Chancel had simple flagellate chambers for passage of seawater through it, later appearing gelatinosan sponges (Demospongea, Figure 14.1c) evolved into species with more complex flagellate chambers.

Sponges have no specialized digestive or nervous system. Flagellated collar cells oscillate to move water currents through the animal, allowing minute food particles to be entrapped on their sticky surfaces for digestion within each collar cell.

A second type of cell lacks a digestive function, receiving nourishment from the collar cells. Some of these form the body wall; some secrete the silicious spicules; a few are male or female reproductive cells. Freely circulating amoebocytes provide transport within the protoplasmic network.

All sponges reproduce sexually. The specialized cells which produce eggs and sperm release them into the water medium. Fertilization results in free swimming larvae which soon settle to the bottom and grow into adult sponges. Sponges reproduce or regenerate asexually as well, by release of buds which grow into new individuals, or by branching of the adults in the colony.

The flagella and amoebocytes within sponges may have evolved from symbiosis of protistan flagellates and amoebae. The presence and organization of these specialized sponge cells, which appear to be identical in structure to protists, lends support to the evolutionary concept of serial symbiosis envisioned by Margulis (1970).

It has been discovered recently that sponges can live both as *single organisms* and as a *colony of communicating cells.* A living sponge has been pulverized in a kitchen blender and sifted through the mesh of a muslin cloth. The cells, so divided, reformed into a living sponge. When two species were blended together, the cells reformed into their own species. There is an organizing intelligence in life which transcends nature's laws as traditionally understood!

Transition of Annelids to Arthropods

The transition from the annelid worms of the Precambrian to a variety of arthropod-like species in the Cambrian was not a smooth evolutionary movement of sequential steps. Rather, each organism found in Cambrian strata around the world seems to be well formed as a functional and functioning system, in tune with its environment, sometimes with characteristics of previous and subsequent lifeforms, and sometimes having little structural similarity to previous or succeeding organisms. It is necessary to introduce some of these organisms and speculate about possible DNA relationships in order to break away from an inadequate Darwinian perspective.

Ayshe (*Aysheaia pedunculata,* Figure 14.2 a) appeared in the Burgess Shale of British Columbia only as impressions in rock. It is one of many Burgess Shale worms, each of such unique form as to defy exact classification. It is thought by some paleontologists to be a connecting link between annelids and arthropods and is classified as Pararthropoda. Although it is worm-like, it has ten pairs of highly developed, claw-bearing, leg appendages which resemble those of Peri (*Peripatus,* Figure 14.2 b) a terrestrial uniramian living today in the damp soil of Brazilian jungles.

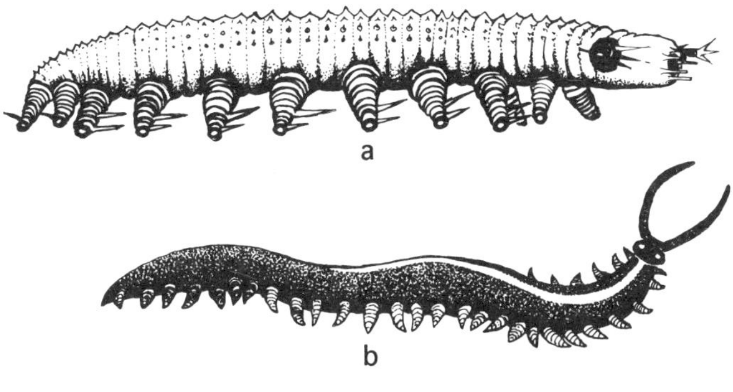

Figure 14.2 (a) Ayshe, (*Aysheaia pedunculata*, Pararthropoda). (b) Peri (*Peripatus*, Uniramia). (a. After Whittington, 1978; b. After Moore et al).

Ayshe's leg appendages probably worked as Peri's do, by muscular movements causing coelomic fluids to pass into and out of them. Legs move forward as they become rigid with inflow, and backward as they shorten and relax, with fluid flow from limb to interior. The head has an annelid-like eye, an anterior pair of branched antenna-like appendages and glands capable of secreting a sticky substance to entangle prey. The suctorial mouth, with no mandibles, suggests that Ayshe fed mainly on sponges, found with them as fossils in areas considered to have been estuaries or swamps, preparatory areas for later terrestrial living.

Modern Peri, a terrestrial organism, with a coelom, and capillaries that bring air from the surface to the tissues, is identified as a uniramian arthropod. Surprisingly, it gives birth to living young. Ayshe is considered to be close to an ancestral uniramian in form, perhaps with both annelid and arthropod DNA from which Peri evolved. Did Ayshe also give birth to live young — a characteristic usually identified with later vertebrates? Is the birth of live young really a vertebrate characteristic, or did this DNA configuration evolve in many other creatures throughout evolution?

Of all the arthropods found in the Burgess Shales, Cana (*Canadaspis perfecta,* Figure 14.3) is one of the most fascinating. It is also one of the most common. This benthic feeder is one of the earliest creatures known to possess a distinctive head with two pair of antenna, one with a pair of eyes, and a mouth with mandible and cusps leading to an intestinal tract, culminating in an anus. If mandibles of Cana allowed it to chew, then how can it be said that the jawed fish that evolved from jawless fish in the Ordovician Period were the first species to solve the problem of chewing? Is there a DNA pattern for mandibles which found its way from arthropods to vertebrates?

Cana is segmented, suggesting worm DNA ancestry, but it has a shell-like carapace, covering the anterior body, thus is considered to be the earliest known malacostracan crustacean. Its ten pairs of walking appendages look like miniature worms with an upper structure of butterfly-like pattern (see Chapter 12, Figure 12.2).

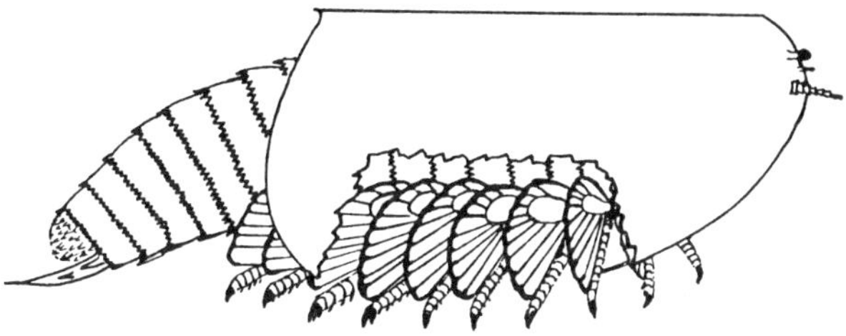

Figure 14.3 Cana (*Canadaspis perfecta*, Arthropoda and/or Crustacea). (After Briggs).

With a little imagination, one can envision Cana having DNA preadaptations for butterflies. Here we have — in one sea bottom creature — legs of worm-like form, like caterpillars in the pupa stage, with wing-like patterning on these appendages and a shell-like carapace — suggesting a chrysalis-like stage. One can imagine that DNA like that of Cana could be recombined to build the DNA structures of the caterpillar-to-butterfly metamorphosis, which appeared 500 million years later in the Eocene.

Since DNA codes form (as well as function) then it seems to me that the DNA patterning of Cana foreshadowed DNA patterning of butterflies. The concept of the butterfly wing pattern as a structure above and perpendicular to the leg appendage in Cana, suggests a preadaptation for a wing which is in a perpendicular position at rest, and moves back and forth to a horizontal position when the butterfly is in flight. A butterfly wing pattern appearing as a perpendicular leg appendage, prior to development of two appendages — legs and wings, suggests the need to understand DNA genes which determine the number and position of different appendages.

The existence of Cana suggests the questions: Although birds and butterflies fly in different ways, is the DNA pattern of butterfly wings, which move from perpendicular to horizontal in relation to the butterfly's body, in any way related to the DNA pattern in birds in which the wings flap up and down from a horizontal plane? Since both birds and butterflies have two X chromosomes in the male and an X and Y chromosome in the female, a reverse configuration from most animals which do not fly, is there any relationship between this chromosome configuration and the fact that birds and butterflies fly? Is there a DNA pattern which is common to birds and butterflies that provides color vision? Is color vision related to flight in these taxa?

The existence of Cana also triggers other preadaptation concepts. It suggests the questions: Is the DNA which gives rise to Cana's appendages in any way related to the DNA for the perpendicular sail of Dime (*Dimetrodon*, Chapter 18, Figure 18.5) and-or of the perpendicular feathered appendage of Longis? (*Longisquama*, Chapter 20, Figure 20.4.) Is the DNA of Cana related to the DNA for the gliding appendages of Kuneo (*Kuehneosaurus*, Chapter 20, Figure 20.5) a lizard of the Triassic? Of course the answers to these questions may never be verifiable, unless the process whereby scientists retrieve and interpret DNA from a hair is also found to apply to retrieval and interpretation of DNA from available unmineralized structures of early lifeforms. Studies of the DNA of living *flying* organisms may assist in interpretation of the complementary nature of structure and function of many animals.

Opa (*Opabinia regalis*, Figure 14.4) an enigmatic animal from the Burgess Shale, seems to be neither annelid nor arthropod. It is believed to be a descendant of the Precambrian ancestral stock of both (Clarkson, 1986: 356). Although some have considered Opa to be an arthropod, it has no legs, a crucial criterion of arthropods. S. M. Manton and D. T. Anderson (1979: 273) wrote that they did not know how Opa should be classified. It contrasted with all other annelids and arthropods. A. G. Sharov (1966) postulated that it represented a hypothetical ancestral proboscidean arthropod line, but H. B. Whittington (1975) rejected that idea.

Opa's 15 segment body contains a head-like area with five mushroom-like eyes, preceded by a long flexible proboscis-like extension with a pincer-like frontal appendage which served to carry food to the ventral mouth. This proboscis was not connected to the gut. Opa had a unique respiratory system; the body dorsal segments each bore a gill-blade overlain by a paddle-like lobe which probably moved up and down, propelling it slowly along the sea floor as it respired. The tail had three dorsal lobes, probably for steering.

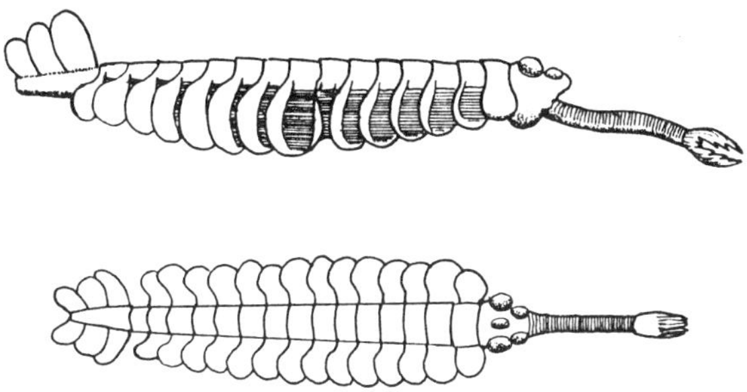

Figure 14.4 Opa (*Opabinia regalis*). (After Whittington, 1975).

The structural pattern of Opa suggests to me a preadaptation for the vertebrae of later appearing vertebrate animals. The DNA of Opa may have coded for the concept of vertebral processes connected to rib processes which first enclosed gills and later, lungs. The elongated neck suggests both worm-like predecessor DNA and a preadaptation for such animals as Tany (Chapter 20, Figure 20.9) Mura (Chapter 21, Figure 21.2) and Diplo (Chapter 21, Figure 21.3). The proboscis suggests preadaptation for an elephant-like appendage. The concept of the pincer suggests a DNA preadaptation for later developing crustacean claws, like those of crayfish and lobsters. Opa is, altogether, an ultimately very satisfactory Cambrian DNA experiment, serving to test many concepts for future development.

Manton and Anderson (1979) believe that the fundamental differences, especially in jaw structures in the various Burgess Shale arthropods indicate that they originated from independent sources of different pre-arthropod stocks.

E. N. K. Clarkson (1986: 357) reported that over 140 species in 119 genera including not only arthropods such as trilobites, but also early brachiopods, mollusks, echinoderms, bryozoans and corals have been found in the Burgess Shales. So diverse are these fauna that some authorities are convinced that metazoans must have arisen independently, many times, in different parts of the world. The polyphyletic origin of various genera and species is becoming a standard belief among paleontologists. Development of new DNA theory will supplant mere restructuring of Darwinian concepts.

Arthropod Trilobites To King Crabs

Cambrian sponges formed an ecosystem with reef building algae, creating shelter for the characteristic lifeform of the period, the arthropod trilobites. There was a massive radiation of trilobites into many very similar organisms with the same habit of dredging the ocean silt and mud for organic nourishment. Trilobites have *not* been classified along a sequence from primitive to more specialized species because their evolving forms have not shown such patterning. Olen (*Olenellus*, Figure 14.5 a, b) has been found near the base of the Cambrian strata in eastern and western United States, Europe, and eastern Asia. It grew to be as long as eight inches.

Trilobites had typical characteristics. The anterior head area (cephalon) of Olen had four pairs of two-branch (biramous) leg appendages, and a pair of attached antennae. The thorax was worm-like, with as many as 45 segments, the most numerous in any known trilobite, suggesting a DNA relationship to annelids. Each segment had a biramous pair of limbs like those of the cephalon. The overlaying dorsal posterior spine (telson) was similar to the spike tail of the living king crab. The cephalon carapace, with ventral mouth and dorsal eyes suggests that trilobites may have had DNA preadaptations for the armored jawless fish of the Ordovician (Chapter 15, Figure 15.5).

Some trilobites could ball like modern pill bugs and mollusk chitons, suggesting DNA relationships between the Cambrian arthropods and mollusks, and between arthropods of the Cambrian and Recent Periods. These DNA relationships do not form continuous sequential patterns, but rather, suggest discontinuities that may be bridged by DNA recombinations.

The trilobites were represented by some species that were blind, and other species with two eyes, which were so complex that their construction was not understood until the 1960s when E.N.K. Clarkson (1966) and Reccardo Levi-Setti described two variant versions of lenses of trilobite eyes capable of seeing in 180 degrees. K. Towe found that the third crystallographic axis of these complex eyes pointed straight so as not to split the image seen by the trilobite. Such a complex structure in the Cambrian does not fit existing theory, but *can* be explained by positing salmutations.

The evolution of trilobite eyes did not show a constant sequential development in size or complexity. Some later trilobites developed larger eyes than their predecessors, while others experienced reduction in eye size. All except one order contained genera with adults having no eyes at all, a seeming evolution toward blindness. Some trilobites had both a pair of simple and a pair of compound eyes. Again, no evolving pattern of complexity is found. Trilobites, rather, represent various complementary patterns of development.

There is no evidence to indicate that trilobites evolved directly from identifiable Precambrian organisms, although the segmentation of both worms and trilobites suggests some common DNA. Even though trilobites lived for at least 300 million years, they were very similar in structure in both Cambrian and Permian deposits (Chapter 12, Figure 12.4, a, b).

On the other hand, there is good evidence that the DNA of trilobites also evolved into new kinds of organisms. The comparison of Olen (*Olenellus*, Figure 14.5 a, b) with Aglas (*Aglaspis*, Figure 14.5 c, d) suggests that Aglas carried DNA similar to that of Olen.

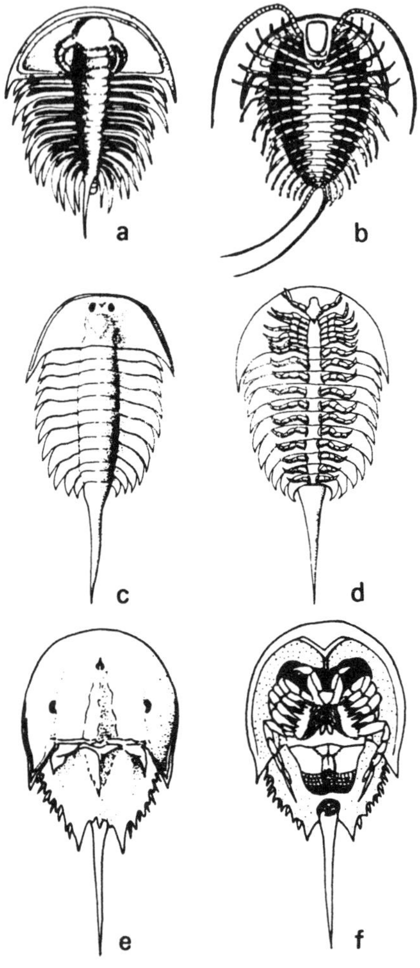

Figure 14.5 (a) Trilobite, Olen (*Olenellus thompson*), dorsal view. (b) Reconstructed ventral view. (c) Aglas (*Aglaspis spinifer* Raosch), dorsal view. (d) Reconstructed ventral view. (e) Living King Crab (*Xiphosura polyphemus*, Chelicerata, Arthropoda) dorsal view. (f) Ventral view. (a, c, d, e, f, After Moore et al).

Aglas, about four inches long (9.7 cm) is characterized by an abdomen of 12 movable segments with six ventral non-branched leg-like appendages, unlike those of trilobites, attached to the cephalothorax formed by fusion of head and thorax segments. The posterior segments ended in a prominent swordlike tail (telson). It had moderately large compound eyes near the front on the dorsal side. It differed from trilobites and crustaceans by having no antennae but rather, one pair of preoral appendages with pincers (chelae) which might be a DNA elaboration of the single pincer on the end of Opa's proboscis.

Aglas lived in the middle Cambrian of Wisconsin and its body structure was similar, in some respects, to living king crabs (Figure 14.5 e, f) found today in abundance along the North American east coast, as well as the in the western Pacific and northern Indian Oceans.

Brachiopods

Lamp Shell Variation of DNA Theme

Brachiopods, known as lamp shells because of their typical Aladdin lamp-like form, have existed from the Lower Cambrian into the present. Bill (*Billingsella,* Figure 14.6 a, b) is a Cambrian fossil of characteristic structure. While there were more than 3,000 fossil genera of brachiopods, less than 100 living genera have beeen identified.

Brachiopods are marine bivalves which have two shell halves of unequal size, contrasted with mollusk bivalves which have two shell halves of equal size. Their valves open by operation of deductor muscles and close by adductor muscles. Most brachiopods live attached to the sea floor by a stem (pedicle) and are characterized by a ciliated food gathering apparatus called a lophophore (Figure 14.6 c).

The calcareous (or phosphatic) four-layer shell is secreted by the mantle, enclosing the lophophore which exhibits a variety of different feathery patterns in different genera. Lophophores are believed by recent experts to function as respiratory organs as well as for food gathering (Clarkson, 1986: 131) although earlier paleontologists did not distinguish such a respiratory function (Moore, 1952: 201). If lophophores are respiratory, perhaps brachiopods carried precursor DNA for the gills of other invertebrates and early vertebrates. On the other hand, lophophores and gills may be variations on a common DNA theme.

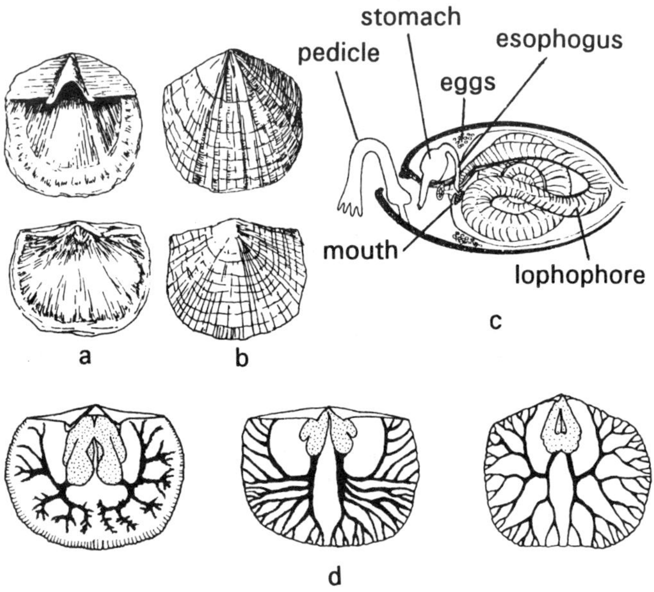

Figure 14.6 (a) Bill, (*Billingsella perfecta*, interior). (b) Exterior. (c) Living brachiopod internal structure. (d) Pallial markings inside brachiopod shells. (a, b, d, After Moore et al; c. After Clarkson).

The lophophore cilia vibrate to generate specific sea water circulation into and out of the mantle, guiding food particles along a mucus belt filament into the centrally located mouth. The mouth leads to esophagus, stomach and intestine. Excreta is expelled into the outgoing sea water flow from the mantle cavity.

Pallial markings (circulation patterns) impressed on the inner surface of one of the valves, are essentially constant in each genus, but vary so widely among different genera as to be useful in classification. Figure 14.6 d shows pallial markings of several different brachiopod shells. Muscle attachment areas are stippled. Does the DNA pattern governing the construction of these shells not only direct the flow of water, but also create beautiful patterns for those with eyes to see them and wonder about their purpose?

Brachiopods are of two sexes. Eggs may be fertilized in the open sea or within the female mantle cavity. In some genera, larvae develop in female brood pouches within the shell.

The earliest appearance of brachiopods in the Lower Cambrian strata showed great diversity of both the inarticulate class (without hinge teeth) and the articulate class (usually with hinge teeth in one valve and sockets in the other). Bill was of the articulate class.

Although paleontologists have traced various lines of similarity in brachiopod organisms from the Cambrian to Recent times, the overwhelming impression from the great variety of patterns of shell structure, lophophores and pallial markings, is that brachiopods represent variations on a DNA theme, more than they represent sequential lines.

Many New Forms of Mollusks

The Mollusca Phylum contains a phenomenal variety of forms — more than 60,000 living species ranging in length from snails of less the .5 mm to giant squid of 16 m (53 feet) the largest invertebrates known. The fossil record of Cambrian mollusks includes not only primitive univalve snails, but also a host of varied clams and many advanced organisms.

Some mollusks drift about passively, some burrow into the ocean floor, some attach or bore into rocks or shells. Some swim actively, but most have a muscular foot and crawl along the ocean floor seeking minute food particles.

Mollusks have been divided (Moore, 1952) into six classes: 1. Monoplacophora, 2. Amphineura, 3. Scaphopoda, 4.

Bivalvia, 5. Cephalopoda, 6. Gastropoda. Five of these classes will be examined here.

From Worms to Univalves

Monophacophorans

The oldest fossil mollusks are considered to be early Cambrian limpet-shaped univalves from Siberia. They are considered by C. R. Stasek (in Runnegar and Pojeta, 1974) to have evolved from free-living flatworms. They are thought to be somewhat similar and ancestral to Cambrian *Pilina* (a Monophacophoran) which has a living relative in *Neopilina* (Figure 14.7 a, b, c). *Neopilina* has a dorsal exoskeleton of excreted calcium plates, a ventral foot, an anterior mouth and an anus on the posterior end of the body. It has been thought that gills developed under the eaves of the shell of *Pilina* because respiration through the epithelial body layer, characteristic of soft worms, was not possible for a shelled animal. Could this represent a salmutation?

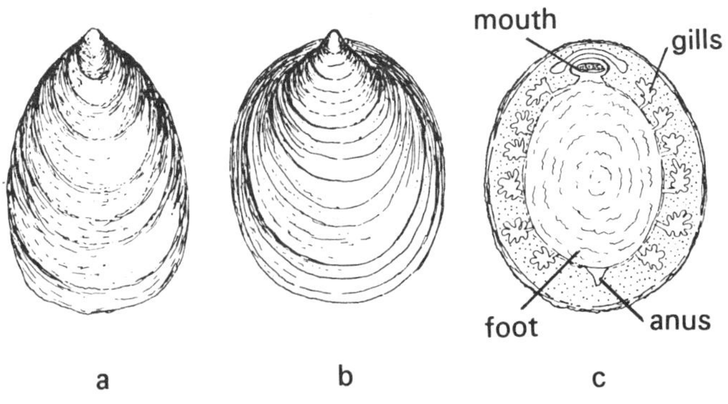

Figure 14.7 (a) *Pilina*, (Cambrian fossil Monophacophora, dorsal). (b) *Neopilina* (living, dorsal). (c) Ventral. (a. After Runnegar and Pojeta; b, c, After Lemche).

Primitive Mollusks

Chitons with Teeth

Although often considered to be the most primitive mollusks, and thus ancestral to other mollusks, the chitons (Amphineura) were first found in Upper Cambrian strata (Figure 14.8). These marine animals have a bilaterally symmetrical shell with seven or eight calcium plates. They have a ventral, anterior mouth with teeth (radula). The mantle cavity contains gills and anus. Under a Darwinian hypothesis, it is necessary to continue to search for earlier chiton ancestor fossils.

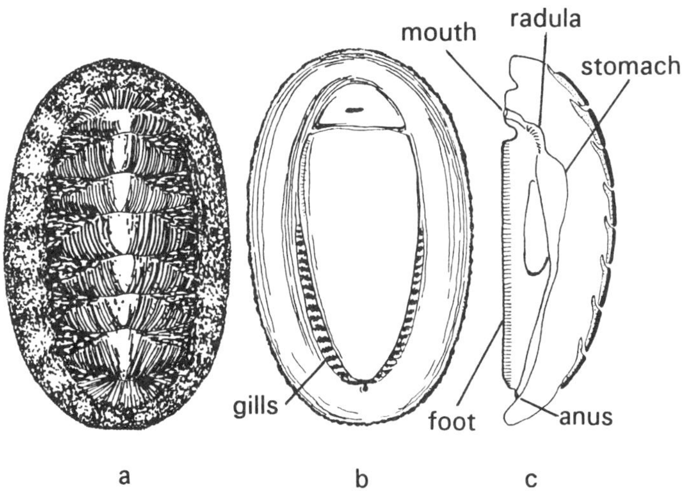

Figure 14.8 (a) *Chiton*, dorsal. (b) Ventral. (c) Side view. (a, b, c, After Moore et al).

Under a DNA hypothesis, the DNA of Cambrian *Pilina* could have been recombined into the DNA of chitons, even though the chitons appear to some scientists to be more primitive. A DNA recombination hypothesis may explain why a large number of primitive species occur later in the fossil record than some of the more advanced species of the same phylum.

Oldest Clams

Bivalves with Primitive Nerve Cords

Ford (*Fordilla,* Figure 14.9 a) is the oldest known Cambrian clam, a bivalve (pelecypod) with a simple ligament connecting the two valves (Pojeta, Runnegar, and Kriz, 1973). It is considered to have evolved from univalves. Clam valves open by dorsal hinge ligament pressure and close by interior muscle action. Later clams developed toothed hinges and are considered, by some scientists, to be descendants of Ford.

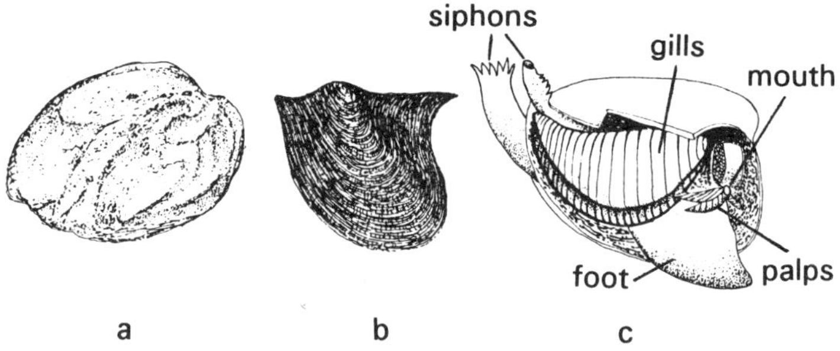

Figure 14.9 (a) Ford (*Fordilla troyensis* Barrande, pelecypod) Cambrian. (b) Teri (*Pterinea*, Pectinacea) a living scallop. (c) Interior of generalized clam. (a. After Pojeta, Runnegar, and Kriz; b, c, After Moore et al).

A scallop, Teri (*Pterinea,* Figure 14.9 b) is a living member of this bivalve class. Teri has not changed much in form from its ancestors, first known in the Ordovician. It is characterized by a hinge on one side, indicating that it lies on its side. Although most bivalves exhibit bilateral symmetry, Teri has a convex upper valve and a flat lower (right) valve. The body of the scallop is enclosed by a membranous mantle which secretes the shell.

Although scallops usually live a sedentary life on the ocean floor, they can clap their valves to swim quickly away from predators. Many clams burrow into the ocean floor or move on a muscular organ called a foot. Such clams, modern counterparts of Ford are illustrated in Figure 14.9 c. The foot protrudes from the shell. It is extended by the force of blood into it, and retracted by muscular movement to evacuate the blood from it. Adductor muscles contract to close the valve, which is slightly open during feeding.

The mouth is above the foot in the anterior region of the clam. Food of microscopic size is brought in through the lower posterior siphon and is conveyed by palps into the mouth. The mouth is connected to a convoluted alimentary tract tube extending to the anus. Waste material empties into the water current expelled through the upper posterior tube. Large gills hang down from the hinge area. A digestive gland and kidney function together with a three chambered heart which pumps blood to the mantle and foot and to the gills for aeration.

The nervous system consists of a primitive looped nerve cord connecting the mouth, foot and viscera. It provides sensitivity to light and touch, while specialized cells react to changes in physical and chemical conditions of the water. Clams do not have heads, although some have eyespots along the mantle edge.

The evolutionary pathway from flat worms to one-valve mollusks to two-valve mollusks is not a smooth Darwinian sequence. Salmutation of DNA is a concept more appropriate to explain these transitions in the fossil record.

First Snails

Gastropods Evolved True Heads

Alda (*Aldanella,* Figure 14.10 a) is a small coiled and ribbed marine snail, about two mm in diameter, with an asymmetric shell, considered by Runnegar and Pojeta (1974) to be one of the first gastropods.

Gastropods, exemplified by living *Buccinum* (Figure 14.10 b) evolved a true head with eyes, tentacles, and a rasping lower jaw with tiny teeth (radula) that scrape against an upper plate to shred food. They shred such hard prey as bivalve shells, in order to reach the inner flesh.

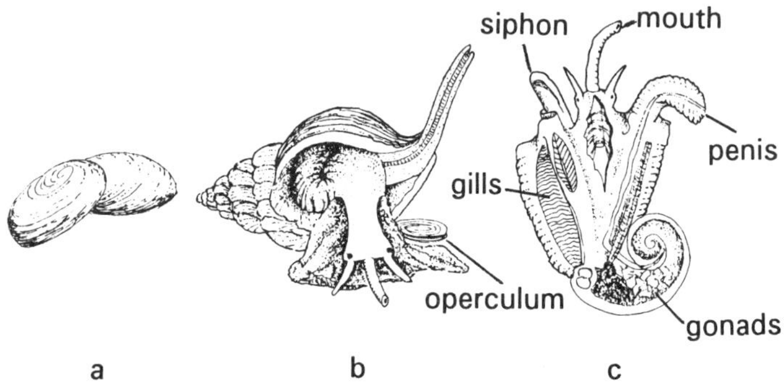

Figure 14.10 (a) Alda (*Aldanella vostokova*) Cambrian. (b) *Buccinum*, living, dorsal. (c) Ventral. (a. After Runnegar and Pojeta; b. After Cox).

The head of *Buccinum* is part of an elongated body with a flat ventral foot which moves by waves of muscular contractions. The viscera is largely within the shell. The body is retracted by muscles attached to the shell, which is closed off by a plate-like trapdoor (operculum). Within the mantle cavity of the anterior shell lie the gills for respiration, specialized organs to sample water entering the cavity, mucous glands to moisten the foot as it moves, and the anus, positioned over the mouth.

Snails, like annelid worms, are hermaphroditic, two sexes in one, a possible indicator of an evolutionary DNA relationship. Gastropods have complex larval stages. The survival advantage of withdrawing the foot-like velum of the second stage larva into its shell has been thought to be the reason that gastropods exemplify peculiar, closely positioned structures of mouth and anus. This relationship, known as torsion, and thought to demonstrate a 180 degree twisting of the viscera from a stage one to a stage two larva. It is in contradistinction to the relationship of mouth and anus at opposite body poles in both earlier lifeforms and in later, more highly evolved creatures.

Torsion has been considered to be support for the idea of survival of the fittest. It is very complex and its adaptive value is not well understood. However, torsion could be a DNA specialization experiment to adapt some of the gastropods for eating feces and detritus as a recycler species, exemplified in those gastropods which cluster around exhalent spires of other organisms. The significance of torsion in gastropods may be clarified by future DNA analysis.

One of the most interesting comparisons of DNA should be that of the common garden snail and the common garden slug. Their soft bodies look identical. The snail has a shell and the slug has a small vestigial shell. What does their DNA show about this relationship?

Fast Swimmers — Cephalopods

Cephalopods were the first creatures to acquire the ability to move very rapidly in their marine environment. They have the most highly developed sense organs and eyes among the invertebrates, and most living species are active predators.

Plectron (*Plectronoceras*, Figure 14.11 a, b) was a small Upper Cambrian cephalopod, about 6-7 mm in length, with a curved shell connected to a tube (siphuncle) for passage of sea water into and out from the interior. A gas chamber was located at the apex, with soft body parts near the aperture of the shell. Different cephalopod species at different times, and in different manner, shifted the centers of buoyancy and weight concentration in conjunction with dramatic shell pattern changes.

Although it is thought that many descendant species of Plectron radiated into the Ordovician (Cyclo, Chapter 15, Figure 15.1) the only cepholopods which evolved into the present are modern nautiloids such as *Nautilus*, squid and octopus.

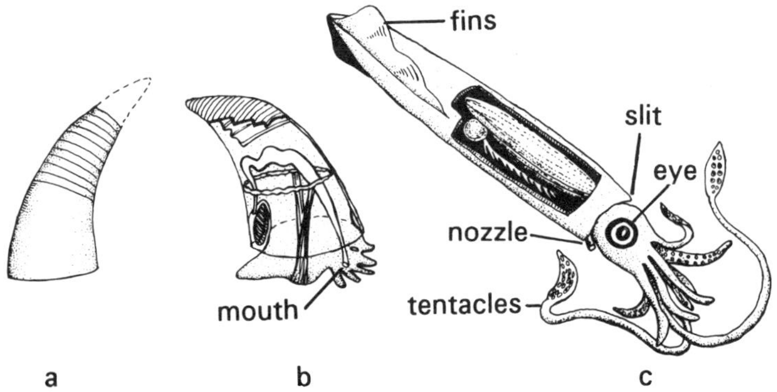

Figure 14.11 (a) Plectron (*Plectronoceras cambria* Walcott). (b) Postulated interior structure. (c) *Loligo*, living squid. (a, c, After Moore et al; b. After Yochelson, Flower, and Webers).

It was formerly assumed that the fossil record showed a gradual increase in coiling from straight-shell species to fully coiled-shell species. Then Clarkson wrote (1986: 200):

> This, however is an incorrect view, for each shell form, even if of unusual appearance, is fully adapted to its own particular mode of life, and buoyancy adaptations in particular have been a primary evolutionary control.

I believe this concept supports the idea of DNA recombination in cephalopods initiating both *sequential and discontinuous* shell shape development. Although evolutionary trends from straight to coiled shell, from small to large size, and from simple to complex sutures (junction of inner sepia and shell wall) have been posited, Moore (1952: 386) records reversals of suture evolutionary trends, and examples of large and small, coiled and straight, simple and complex suture forms, all appearing together in deposits of various periods.

It is quite impossible to detect clear sequential developments among fossil and living cephalopods. It is clear that many structural characteristics have appeared in similar lifeforms in successive strata, and it is also evident that some of these same characteristics appear in species of different phyla or classes. The similarities and differences between the nautiloids (Chapter 15, Figure 15.1) and ammonites (Chapter 17, Figure 17.6) for example, support a theory of interacting DNA.

It has been postulated that in advanced cephalopod species, the cumbersome shell armor surrounding the air chamber was progressively reduced until the apex region remained only a remnant. The squid, *Loligo*, is an example of a living soft-body cephalopod with many advanced systems (Figure 14.11 c). Modern *Loligo*, averaging about one foot in length, has been used as a basis for understanding fossil cephalopods. It has a round head with two large, well developed eyes. It has ten tentacles, two of which are longer than the others, with suckers concentrated in pads used to grasp prey and hold it while it is killed and chewed in the beaklike jaws.

Behind the head is a sleek body with the mantle cavity containing the viscera, including a pair of gills, digestive tract and heart. The mantle cavity also provides the jet-propelled locomotion system. Water enters a slit behind the head, aerating the gills. The slit closes and the mantle then contracts, ejecting water through a movable nozzle under the head area. The nozzle positioning allows the animal to jet foreword or backward, achieving maximum speed in the backward direction. The tail area fins allow slow speeds, fine adjustments and hovering.

Moore (1952: 337) noted, "Like many other cephalopods, *Loligo* has the ability to change color and pattern kaleidoscopically, before the eyes of an astonished observer." Is there a common configuration of DNA which provides such a wonder in *Loligo* and also in animals like chameleons? It seems to me that this characteristic of cephalopods and chameleons can be understood only by positing a system of DNA transfer and-or a system of DNA recombination among various animals.

Echinoderms — Ancestors of Chordates

Echinoderms have been classified (Clarkson, 1986: 223-224) in four subphyla and nine classes ranging from primitive Cambrian eocrinoids to classes of sea lilies (stalked crinoids) sand dollars and sea urchins (echinoids) sea cucumbers (holothuroids) starfish (asteroids) brittlestars (ophiuroids) and a class of rare organisms (homalozoa) recently considered for reclassification as a separate chordate subphylum, Calcichordata (Figure 14.14). This reclassification indicates a belief that an evolutionary pathway occurred from Cambrian eocrinoids to a primitive class of chordates.

Echinoderms all have characteristic skeletal microstructures of calcite arranged in honeycomb patterns with minute passageways that are not only definitive of the phylum, but also very beautiful in their crystallographic units.

The mouth is centrally located in most classes of echinoderms and a long coiled intestine extends to an anus. The body cavity contains the nervous and reproductive systems. A water circulation system may be internal or may be associated with ventral tube feet in some genera. There is no heart.

Echinoderms were represented in Middle Cambrian rock strata by the class Eocrinoidea. Eocrinozoans were primitively stalked marine animals having a sac-shaped body (theca) covered by geometrical plates (with pores for water circulation), a stem, and long arm-like tentacles. It was postulated by Moore (1952: 583) that this class may have been ancestral to cystoids and crinoids as well as other echinoderms.

Although the class Eocrinoidea is believed by some to contain the first fossilized ancestral genera in the Cambrian, it also has been postulated that echinoderms may have originated in the Precambrian. There was a marked proliferation of a variety of echinoderm classes following the Cambrian, but the Cambrian genera all became extinct. Together with other ancient sea creatures, their remains have formed the limestone of Earth.

Macrocy (*Macrocystella*, Figure 14.12 a) was a Cambrian eocrinozoan. Its theca was composed of four circles of five plates of solid crystalline calcite. Attached to the base of the theca was a tapering stem which anchored Macrocy to the ocean floor. The mode of life of this animal can only be known by comparison with living echinoderms. Perhaps its life cycle began with a larval form like that of crinoids living today (Figure 14.12 b). Similar larvae developed into both crinoids and starfish (Figure 14.12 c, d).

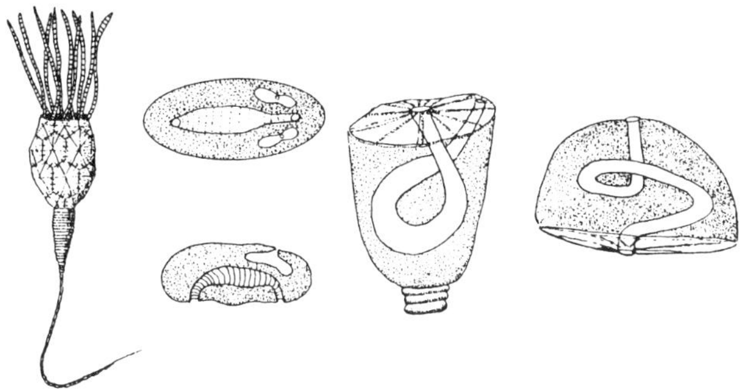

a b c d

Figure 14.12 (a) Macrocy (*Macrocystella*, Eocrinozoa). (b) Crinoid larva. (c) Stem-bearing echinoderm alimentary canal. (d) Star fish alimentary canal. (a, b, c, d, After Moore et al).

The larva of a crinoid is a strangely advanced form, most closely similar to that of primitive hemichordates, the classification immediately preceding chordates and vertebrates. It is an interesting subject for DNA research, for it is very unusual. The crinoid egg, which generally was released and fertilized in the sea, developed in radially symmetrical cell divisions through the hollow sphere (blastula) and two-layer (gastrula) stages to a bilaterally symmetrical first stage larva. It had a soft body, with a cavity connecting mouth, straight digestive tube and anus (on a horizontal plane). During development, the alimentary canal shifted from a horizontal to a vertical position.

It has been assumed that the larva of Macrocy, like that of living crinoids, attached itself to some sea floor object, elongated into a stalk with the mouth migrating upward as it developed into an adult (Figure 14.12 c). In similar fashion, the alimentary tract of a star fish moves, in the second larval stage, from a bisymmetrical configuration to the radially symmetrical adult stage (Figure 14.12 d).

Echinoderms were the first animals in the Cambrian record, above protozoan size, to have skeletonized bodies rather than soft bodies or confining shells. It has been thought by some scientists, that echinoderms were on a sequential evolutionary pathway following soft worms and jellyfish, to be followed by bony jawless fish. This seemed possible because the larval stage of echinoderms exemplified bisymmetrical patterning even though the adult stage, like jellyfish, had radial symmetry.

A radial body plan such as that of jellyfish and corals has been thought to be more primitive than a bisymmetrical plan. Echinoderms, with radial symmetry in adults and bisymmetrical larvae, are much advanced in a number of features, over both other radially symmetrical and also over bisymmetrical invertebrates.

How can it be thought that radial symmetry is more primitive than bisymmetry, when both appear in complementary fashion in the structure of one organism, and may be alternated in different stages of development? It seems to me that DNA is capable of structuring innumerable complementary combinations in each organism. It is capable, also, of structuring combinations which change in radical ways in short periods of time.

The processes of larval development, metamorphosis, and embryological unfolding in different organisms, in relation to evolution of adult structure, have often defied understanding. In spite of these problems, scientists are convinced that DNA codes not only the structural form that an adult animal will assume and its functional parameters, but also the program of its development from egg to adult. What a phenomenal intelligence is exhibited by DNA!

The concept of salmutations seems to be required for understanding of evolution from worms and jellyfish to early echinoderms, because each of these taxa has unique characteristics which do not follow a sequential pathway. Although echinoderm evolution is complex, perhaps the DNA configuration of jellyfish such as Brook (*Brooksella,* Figure 14.13 a) and of annelids such as Worth (*Worthanella,* Figure 14.13 b) contributed to the DNA configuration of early eocrinoids and then of starfish and brittlestars. Starfish have an important characteristic (DNA configuration?) like that of worms. Both worms and starfish are able to regenerate extensive parts of their bodies, when destroyed.

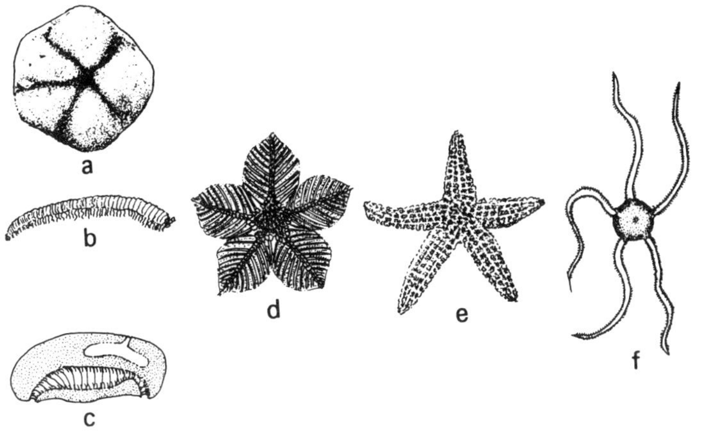

Figure 14.13 Evolution of Starfish DNA? (a) Brook (*Brooksella*, jellyfish). (b) Worth (*Worthanella*, annelid worm). c) Echinoderm larva. (d) Ville (*Villebrunaster*, somasteroid, ventral view). (e) Starfish (asteroid, dorsal view). (f) Brittlestar, (ophiuroid, dorsal view). (a. After Wolcott; b, c, d, f, After Moore et al; e. After Johnson).

Starfish (Figure 14.13 e) do not fossilize readily because they break up easily and their parts are scattered. It is not known exactly when they first appeared on Earth, but if they are more primitive than Ordovician brittlestars (Figure 14.13 f) as postulated by some scientists, they may have occurred in the Cambrian. Fossils of Cambrian Ville (*Villebrunaster*, Figure 14.13 d) have been classified as Somasteroidea by some scientists, indicating ancestry to the asteroid starfish.

Starfish (asteroids) usually have five arms which contain a series of flexible plates within an epidermal layer. Inner coelom extensions of the animal extrude between the plates on the dorsal surface for respiration, indicating a possible DNA affinity with jellyfish which have soft, flexible coeloms through which they respire. Large plates on the ventral side of starfish arms are penetrated by tube-feet. The mouth is located in the center of the ventral surface.

Although Cambrian Ville may have fed on detritus, as some scientists believe, the staple diet for starfish for eons of time, at least since the Devonian, has been clams, including scallops such as Teri (Figure 14.9 b). The starfish uses the suckers on its tube feet to pull the two valves in opposite directions until the clam's adductor muscles relax and the shell opens slightly.

The starfish then extrudes its stomach through its mouth and squeezes it inside the clam shell. It digests the clam flesh in this position and afterwards returns its stomach into its own body. The clams are not without defensive mechanisms, but this predator-prey relationship has been verified to have existed 400 million years ago, and may have originated in the Cambrian, more than 500 million years ago. The DNA development demonstrated by predators and prey in interacting systems in each period of geologic time warrants careful examination. It is not explained by survival of the fittest. Predators and prey survive together.

How is the DNA of starfish related to that of snails? The larval torsion development of the alimentary canal in gastropods and echinoderms must be patterned in the DNA codes of the two phyla. What are these codes?

The evolutionary genesis of echinoderms is murky at best. Although similarities with earlier organisms exist, echinoderms seemed to arise fully formed as many unique organisms. Starfish have evolved into the present, appearing in forms similar to those of the Cambrian.

Transition from Echinoderms To Chordates

Although Richard Cowen (1976: 27-28) noted that chordates appeared in the Cambrian, evidence had been meager and generally unknown to the public until 1989, when Stephen Jay Gould published *Wonderful Life; The Burgess Shale and the Nature of History*. Before his discussion of the earliest known chordate, scientists had presented examples of protochordates and living animals, as the type of animals postulated to be similar to the earliest chordates.

Raymond Moore (1952: 718) classified in the Protochordata Phylum, Cambrian animals which possessed a notochord (a primitive stiffening pre-vertebral rod structure) during at least part of their lives. The graptolite class of these Cambrian organisms consisted of planktonic colonial branched tube animals with chitinous skeletons covering the soft inner parts. Although these organisms were originally classified as coelenterates or bryozoans, it was discovered that they were strikingly similar to modern pterobranchs, which are now considered to be in the Phylum Hemichordata (half-chordate).

A modern pterobrach, Rhab (*Rhabdopleura*, Figure 14.14 a) is a colonial animal consisting of both budding individuals, and feeding, sexually reproducing individuals. They attach to a submerged rock and grow upward in a contractible stalk tube structure with a head of plumes, proboscis and mouth above the body mass, looking rather like a marine flower.

In addition to these animals, hemichordates include the living genus, Cephalo (*Cephalodiscus*, Figure 14.14 b). This adult animal, seen in a cross section, suggests the embryonic stage of vertebrates. It lives in tubes on the sea floor. The head area contains a heart above the mouth. The tentacled arms, with food grooves similar to crinoid echinoderms, appear in an area of the animal which is above and to the rear of the mouth. The Rhab, Chephalo brachial slit for respiration is a hemichordate characteristic. The stomach takes up a large part of the body cavity and curves around into a rectum leading to the anus below the tentacles.

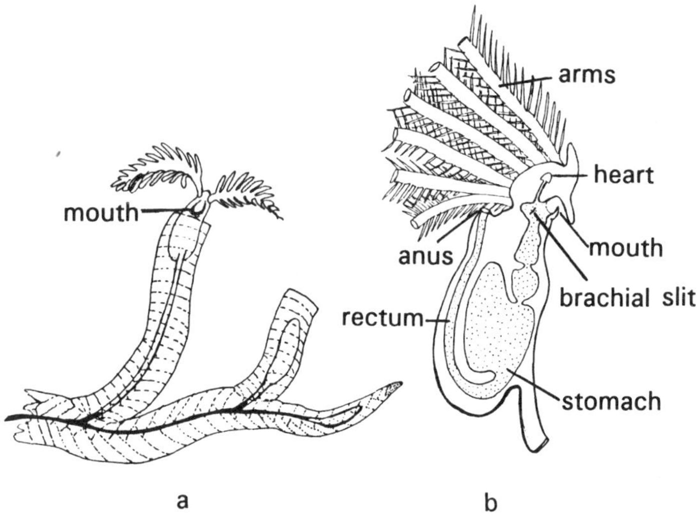

Figure 14.14 (a) Rhab (*Rhabdopleura*, living Calcichordata). (b) Chephalo, (*Cephalodiscus*, living). (a. After Moore et al; b. After Jefferies).

R. P. S. Jefferies (1979: 462) has identified a group of fossil organisms from the Cambrian as calcichordates, which may have evolved from echinoderms and Cephalo-like hemichordate ancestors.

It is difficult to imagine that such organisms might represent ancestors of chordates, because humans have tended to identify with the vertebrae-developed branch of the chordate family, rather than with such primitive organisms. Yet it has been envisioned that pterobranchs and calcichordates might have been ancestors of Amphio (*Amphioxus,* Figure 14.15 a) a living chordate.

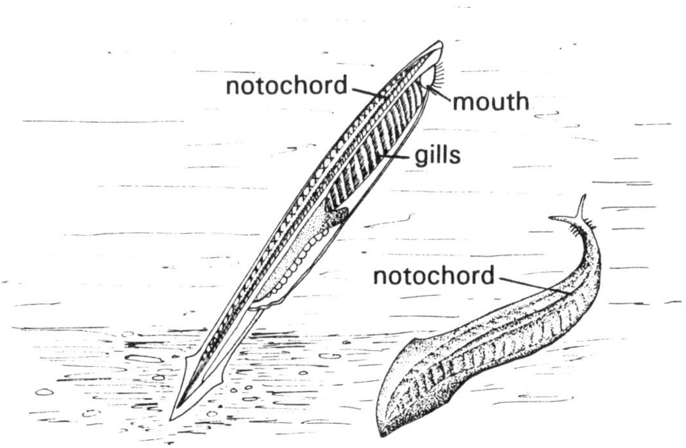

Figure 14.15 (a) Amphio (*Amphioxus,* living, Chordata, Cephalochordata). (b) Pika (*Pikaia gracilens,* Chordata). (a. After Gregory, 1951; b. After Gould, 1989).

Scientists said that if Amphio had not existed, someone would have had to invent it. They considered it to exemplify the prototype for vertebrates; it has been that basic to explanation of vertebrate evolution. Although Amphio was known only from living species, Jefferies (1979) envisioned that it had evolved from ancestors like the calcichordates of the Cambrian.

Amphio is a small, living, fish-like marine animal which floats near the sea floor, often with its tail buried in mud. It is a filter feeder with a distinct notochord and a well developed nerve cord. Water enters the tentacle surrounded mouth, passing through pharyngeal gill slits into an organ called an atrium. The water exits through a ventral pore while food particles are absorbed in the intestine. Wastes passes out the anus. The liver and gonads of each of the two sexes are beneath the intestine.

Amphio was the basis for the scientific theory of chordate origins when Stephen Jay Gould (1989: 321-323) enlightened the public about Pika (*Pikaia gracilens,* Figure 14.15 b). He gave credit to Simon Conway Morris and H. B. Whittington for rediscovering (in the Cambrian Burgess Shales) this little creature originally named by Charles Doolittle Wolcott in 1911. Walcott considered Pika to be an annelid worm. But Morris reached a firm conclusion from study of about 30 specimens that Pika was a chordate, the first recorded member of our own phylum.

Pika could easily be considered to be ancestral to Amphio, as well as being ancestral to the first vertebrate fish of the Ordovician, and thus ancestral to a whole line of vertebrate evolution.

The confusion of Walcott resulted from misidentification of the bands of muscle which have the zigzag bend of chordates, not the segmentation of annelids. Pika is a laterally compressed, ribbon-shaped aquatic swimmer, about two inches long. It has a notochord, the feature which gives it its identity and its classification.

The next step in understanding the place of early chordates is identification of the DNA patterns which produce them. Perhaps Amphio may be useful again, in providing the DNA to coordinate DNA patterns of early chordates.

During the Cambrian Period almost every important phylum (except subphylum Vertebrata) arose. It is significant that the Cambrian organisms exemplified innumerable DNA configurations which would become apparent in the vertebrates which began to arise in the Ordovician Period.

The end of the Cambrian brought massive extinctions noted in the disappearance of algae reefs, the sponge-like reef animals, and two thirds of the families of trilobites, which apparently could no longer survive in the mud deposits of their vanishing ancestral habitats. About 90% of marine species have lived in water less than 660 feet deep, thus changes in land and sea formations drastically affected the ecosystems of marine fauna.

This first recorded massive extinction of species seems to be associated with the great tectonic earth movements of the Cambrian and accompanying climatic changes. It was only in the 1960s that earth scientists recognized the relationships of tectonic movements, climatic changes and extinctions. Much work still needs to be done to reconstruct these patterns. A new field, paleoecology, has been organized to study the relationships between environments and ancient organisms.

15

Paleozoic Era — Ordovician Period First Vertebrates

The earliest known vertebrates, jawless fish, appeared in the Ordovician period beginning about 500 million years before present (500 MBP). The climate tended toward moderation and stabilization, as the four major continents divided the seas into numerous and varied water environments. The vanished ecosystem of algae reefs and sponge-like reef dwellers was renewed when blue green algae stromatolites began to harbor animal newcomers including stony sponges and simple corals. Reef builders today are confined to within 30 degrees north and south of the equator, so since the coral fossils were found in New York State, USA, it is estimated that the state was much farther south in the Ordovician than it is today.

The period was characterized by a diversity of green, brown and red algae. Sea plants had life cycles of sexual reproduction. The sea was dominated by filtering suspension feeders such as brachiopods, and characterized by invertebrate mud dwellers such as trilobites and starfish.

Exploding Variety of Cephalopods

Cephalopods like Cyclo, looking rather like today's Pearly Nautilus, exploded in variety in the early Ordovician (Figure 15.1). Some cephalopods were sizable animals which swam freely, escaping the life of the Cambrian mud feeders. One nautiloid carnivore reached 13 feet in length. Nautiloid designs showed various solutions to problems of depth control, buoyancy and orientation of shell. Did the cephalopods of the Ordovician ascend from Cambrian species like Plectron? (Chapter 14, Figure 14.11 a.) What can we learn from them about the changes in DNA patterns through time?

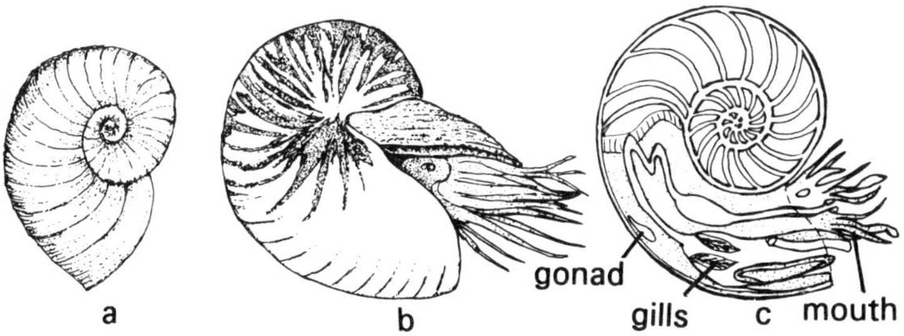

Figure 15.1 (a) Cyclo (*Cycloplectoceras miseri*, Ordovician nautiloid cephalopod). (b) *Nautilus*, a living cephalopod, exterior. (c) Cutaway. (a. After Moore et al; b. After Trauth; c. After Clarkson).

First Bryozoan Colonies

The Phylum Bryozoa (Ectoprocta) was first known in the Ordovician in four orders of the Class Trepostomata (Moore, 1952: 161). The remains of bryozoans are so abundant that they form a large part of the rock of Middle and Upper Ordovician strata. Bryozoans are tiny aquatic animals, generally less than 1 mm in length, which usually grow in calcareous skeletonized colonies, secreted by individual zooids. The colony may spread to as much as 60 cm (2 ft.) in diameter. *Bugula*, a modern genera which shows one form of growth, has chitinous, rather than calcarecious hard parts (Figure 15.2).

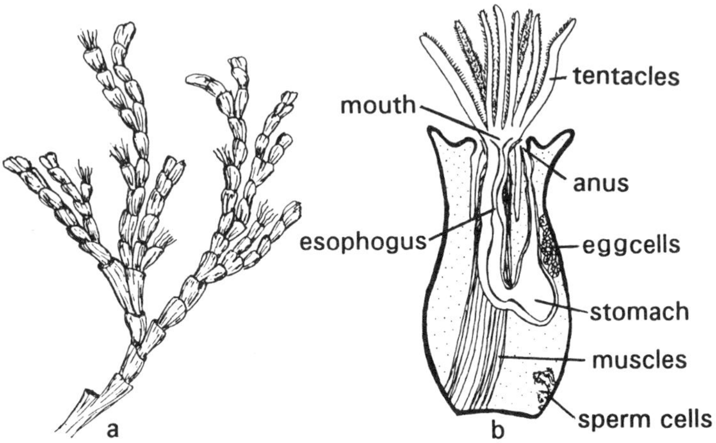

Figure 15.2 (a) Bryozoan colony. *Floridinella.* (b) *Bugula*, living zooid, internal structure. (a, b After Moore et al).

Individual bryozoan zooids are considerably more advanced in structure than coelenterates, although both have tentacles and saclike bodies. Bryozoans are distinguished by a U-shape digestive tract, which leads from the mouth to esophagus, stomach, intestine and anus. A nerve ganglion connects mouth and anal area with nerves running to the rest of the body. Retractor muscles can pull the animal, tentacles and all, within the body wall.

Most bryozoans are hermaphroditic, the egg cells being located above the sperm cells. The eggs, after being fertilized in the ocean, grow into larvae, which swim about or are carried by currents before they grow to adulthood and settle on the ocean floor, each capable of beginning a new colony. Asexual budding then creates a whole colony from this one adult. Bryozoans have no gills, but probably breathe by means of their lophophore-tentacles, similarly to brachiopods. They have no heart nor blood vessels, but do have numerous white blood cells, which flow within the body cavity.

The evolution of bryozoans has not been well studied, even though 3,500 living and 15,000 fossil species are known. Perhaps consideration of their DNA will assist in learning more about the relationship between their genetic and phylogenetic patterns and their relationships to other organisms.

The Largest Arthropod

The Ordovician strata contains evidence of the family of the largest arthropod ever known, Terago (*Pteragotus*, Figure 15.3). It grew to be almost 2 m (6 ft.) in length. Its gills were adapted to remain moist so Terago may have been an early land-living arthropod. It was a fascinating creature in its own right with many characteristics suggesting evolution from DNA of trilobites and their successors (Chapter 14, Figure 14.5). Why did such a large arthropod exist so early in history? How did its DNA evolve?

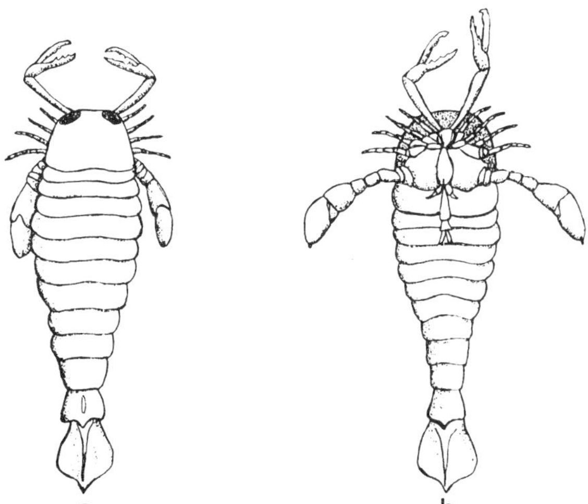

Figure 15.3 (a) Terago (*Pteragotus*, Eurypterida, dorsal. (b) Ventral. (a. After Moore et al; b. After Snodgrass).

The Mystery of the Conodonts

The Ordovician is considered by some scientists to be the beginning of the mystery of the conodonts (Figure 15.4). Others believe conodonts arose in Precambrian times. Conodonts are microscopic fossils of calcium phosphate — with wide stratigraphic and geographic distribution — commonly occurring in sandstones, limestones and shales from at least the Ordovician to the Triassic. They were especially abundant in black shales which lack association with any known marine invertebrates.

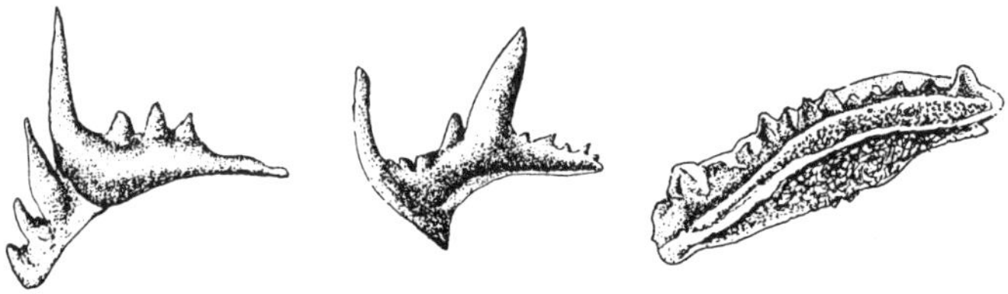

Figure 15.4 Conodonts. (After Clark).

Various experts have considered them to be fossil parts of worms, crustaceans, gastropods, cephalopods or fish. Because of their chemical composition, because they were frequently attached to bone-like material and looked like teeth, and because they have been found in the Harding sandstone formation of Colorado, they may have been the teeth of fish.

However, since no fossil fish with teeth were known to exist as early as the Ordovician, and toothed fish in this time frame would not fit into classical evolution theory, conodonts have been neglected in many modern biology textbooks.

It is possible that the DNA configuration for toothed fish could have occurred much earlier than previously recognized. Complementarity dynamics logic allows the concept that toothed fish could have evolved from DNA combinations originating in the Precambrian and Cambrian.

Toothless, jawless fish, usually represented as ancestral to jawed and toothed fish of the Devonian, could have arisen at the same time as the jawed fish. Jawless fish could have arisen later than the fish represented only by Ordovician conodonts. At least the minds of scientists should be freed from the Darwinian legacy that prevents exploration of the full ramifications of the evidence.

Earliest Known Vertebrates

The Subphylum Vertebrata has been thought by some to begin with members of the class Cyclostomata, the classification for the earliest known jawless fish, which are represented today only by lampreys and hagfish.

The first fossil evidence of vertebrates (apart from conodonts), consists of fragments of dermal armor found near Saint Petersburg, Russia, and in the Harding sandstone of central Colorado, USA. These fossils are considered to be from ostracoderms of a kind that were common later in the Silurian and early Devonian deposits in which better preserved fossils were found.

The only cyclostome fossils, from which the internal anatomy of the head could be reconstructed, were a few of the ninety species of cephalaspids identified from the Old Red Sandstone formation of Great Britain.

Hemi (*Hemicyclaspis*, Figure 15.5) is an example of such an early jawless fish, believed to be a fresh water genus. In addition to bony armor-plated and flattened head, Hemi had an ossified inner cranial skeleton fused with gill structures. The brain, nerves, blood vessels and two semicircular canals of the ears were similar to those of current day lampreys. The head had two closely spaced eyes directed upward, a central pineal eye, one nostril slit, and a soft ventral mouth in front of moveable plates with openings for gill pouches. Thus Hemi was probably a bottom dwelling species, which strained its food as it passed water through its gill chambers. Hemi was able to eat only small organisms, as its restricted mouth operated like a siphon.

Figure 15.5 Hemi (*Hemicyclaspis*, Cephalaspida, ostracoderm cyclostome) an early jawless fish. (After Stensio).

The lower ridge line of the head structure is believed to be either an electric or vibrational sensory organ. Head shields of cephalaspids varied in shape. Some were 30 cm or more in breadth, but most were in the range of 1-4 cm. Cephalaspid bodies ranged in length from a few inches to two feet. Posterior to the bony head shield was a scale covered trunk, flattened ventrally with paired pectoral fins of unknown internal design, and a heterocercal tail.

Jawless fish were believed to be true vertebrates, with primitive cartilaginous vertebral structure. The shape of the head shield suggests that DNA of Hemi may have salmutated from DNA of trilobites or other early arthropods. The evolution from invertebrates to jawless fish involved complex salmutations with development of three complex tissue layers in embryos and adults, as well as new spinal structures.

There is evidence of extinctions of numerous taxa at the end of the Ordovician, including many kinds of jawless fish, as the climate turned relatively cold for a time, with sea ice forming in the Antarctic, southern Africa and Australia.

16

Paleozoic Era — Silurian Period First Land Plants

The Silurian period of 450 million years before present (450 MBP) began with a configuration of four continents as the proto-Atlantic ocean closed in tectonic movement, locking North America and Europe together, building the mountains of Scandinavia and Appalachia and the hills of Scotland and New England. The climate again warmed during this time. The first land plants arose — lower vascular plants. Although classical theory is strangely silent regarding the evolution of early land plants, salmutations from a variety of green algae appear to be their source.

The midwestern United States was covered by an epicontinental sea. The Silurian seas were characterized by plant plankton eaten by animal plankton, eaten by suspension feeders with scavengers such as snails and sea urchins roaming the sea floor. Larger carnivores preyed on smaller ones. Sponges ate the organic material decomposed by bacteria.

Early Vertebrates — Jawless Fish

Although Hemi-type fossil bone fragments were evident in Ordovician deposits, fragments of heterostracan cyclostomes, presumed ancestors of Tera (*Pteraspida,* Figure 16.1) have been dated from the late Cambrian of America, as well as from the Ordovician of Europe and Australia. Thus, according to some paleontologists, heterostracans rather than cephalaspids were the oldest known vertebrates.

Figure 16.1 Tera (*Pteraspis*, heterostracan cyclostome). (After White).

Tera (*Pteraspida,* Figure 16.1) is thought to have been present in significant numbers in the Silurian, and became common in lower Devonian deposits. Tera did not have a single head plate like Hemi, but rather — lacking bone cells — had individual carapace plates, which formed in the young and grew by successive additions. There was a dorsal spine on the cephalon. The eyes were on the sides of the head. Tera's head had no opening for a pineal eye like Hemi's. The ventral mouth slit was followed by flexible plates and paired nasal sacs. There were gill outlets on the sides. The rounded front (rostral) plate varied in different species, as did the ventral and dorsal plate discs.

Jawless fish radiated into a great variety of species which swam in open water, sucked small particles through their tube mouths and scooped plankton from surface waters. Complex ecosystems were evident.

Vertebrates — Earliest Known Jawed Fish

Authorities have postulated that jawed fish evolved from jawless fish in the Silurian. Jawed fish were the first vertebrates to prey on quarry larger than microorganisms. Their strong jaws projected a new way of life in the seas, with development of ecosystems of prey and predators of large size as well as small. The increasing complexities of evolution are evident in the increased diversity in reef ecosystems which developed with the advent of the jawed fish.

Alfred Romer (1966: 316) believed that jawed fish may have arisen in inland streams, but left few fossils. Some catfish-like fossils have been found in fresh water deposits of this time when mountains rose and water flowed down from higher elevations. Although it has been presumed that jawed fish evolved from jawless fish, Richard Cowen (1976: 51) observed that all living jawless fish have gills inside the gill arches while all jawed fish have gills outside the gill arches. There seems to be no intermediate gill pattern in the fossil record and the two existing patterns each solved the breathing problem of the particular species. Salmutations could explain such evidence.

Erik Jarvik (1980) supported classification of early jawed fish, known only by such fossils as scales and jaw bones, as acanthodians of the Upper Silurian. He believed they were related to sharks. Romer (1966) questioned this, though he reported that acanthodians were of shark-like proportions and had the same shape tail fin. Ossified endoskeletons found in later Devonian acanthodian fossils, were weakly developed.

Restorations of fossil evidence of Clima (*Climatius*, Figure 16.2) a fresh water genus from Scotland, show details of exterior structure. Clima had an exoskeletal shoulder girdle with a series of bones leading to a pectoral pair of dermal spine supported fins. It was considered by some experts to have had a primitive form of ventrally paired fins, arising from the body wall, rather than from the endoskeleton.

Figure 16.2 Clima (*Climatius*, acanthodian). (After D. M. S. Watson).

Jarvik (1980: 328) maintained that bony structures such as spines may have developed independently of fins, in median fin folds. This developmental feature showed similarity between acanthodians and certain later teleost fish. Such a characteristic, considered by paleontologists to show structural relationships between species, must also be significant in considering DNA configurations which may have been common to both.

Tooth whorls of acanthodians found in Upper Silurian deposits of Germany (Figure 16.3 a) are strangely reminiscent of smaller conodonts (Chapter 15, Figure 15.4) as are acathodian jawbones of Lower Devonian deposits (Figure 16.3 b). Could DNA of acanthodians and conodonts be similar? Could acanthodians have evolved from ancestors with conodont dentition?

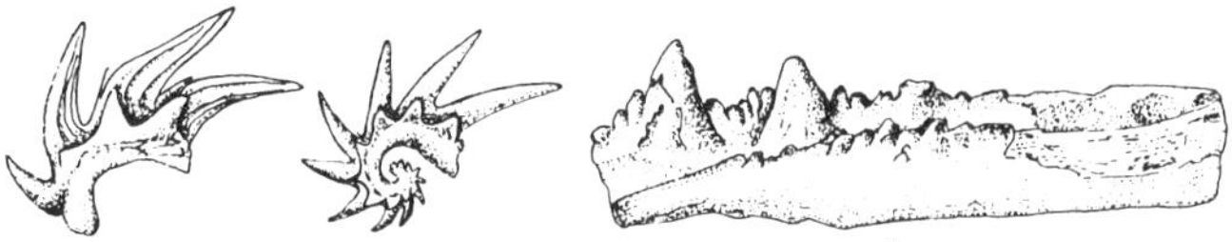

Figure 16.3 (a) Acathodian tooth whorls. (b) Acathodian jawbones. (a, b, After Jarvik).

Scientists believe that as some Silurian fish invaded oxygen poor waters clogged with decaying plant life, they developed lungs, an innovation which improved upon oxygen processing by gills. These lungs appeared also to be preadaptations for the lungs of future lifeforms which were to live on land. Such preadaptations present a problem for classical evolution theory which maintains that variations in lifeforms are random. The salmutation theory embraces the concept of purposeful preadaptations.

The Silurian closed with extinctions of many sponges, fish, trilobites, brachiopods and echinoderms. However, the extinctions of the Ordovician and Silurian were minor compared with the extinctions yet to come in the Permian and Cretaceous.

17

Paleozoic Era — Devonian Period "Age of Fish"

In the Devonian period, 400 million years before present (400 MBP) three continents existed: Euroamerica, Asia and Gondwana (Chapter 12, Plate 12.2 # 3). The year was 400 days long. The climate remained warm. Swampy areas, developed from the shallow seas, were the life beds for primitive vascular plants and club mosses. Liverworts and cycads arose in the late Devonian. Although swamps were widespread, there were also marked seasonal droughts in some areas.

A class of armored fish, placoderms, the dominant fish of the Devonian "Age of Fish" radiated from the late Silurian. They are considered by some paleontologists to be among the earliest known vertebrates; their fragmentary remains were evident as early as the Cambrian and Ordovician. E. A. Stensio, a Swedish paleontologist, was the first to clarify the classification of placoderms into arthrodires and antiarchs.

When placoderms appeared in the fossil record of the Devonian, they were divided into numerous distinct groups of highly specialized individuals, not closely related to later sharks nor to lungfish. Although early placoderms were not known beyond the Devonian, Erik Jarvik wrote that living rays and chimaeras could justly be considered to be living placoderms. Are rays and chimaeras evidence of DNA restructuring from the DNA patterns of early placoderm ancestors?

Arthrodire placoderms were mainly marine fish with well developed jaws and an ossified exterior and interior skeleton. Most had a head shield and armor plates around the thorax, connected by ball and socket joints. In the middle Devonian, arthrodires two feet in length were common. In the late Devonian, there appeared Dinic (*Dinichtys* or *Dunkleosteus*, Figure 17.1) a similarly structured giant fish, 30 feet in length.

Figure 17.1 Dinic (*Dinichthys*, arthrodire placoderm). (After Heintz).

Its body was broad and flat, with skin believed to be smooth and unscaled, posterior to the series of bony armor head plates. Dinic had a pineal eye area and large paired eyes, similar in structure to those of higher bony fish and tetrapods. The unique jaws consisted of an upper jaw of two pairs of bony parts: the front pair with sharp tusk-like projections, the posterior with a shearing edge. The similar lower jaw was a single structure. The projections were not teeth, but rather hard bone, connected to cartilaginous jaw bars, unlike the jaws and teeth of other fish.

The side head shield covered gill chambers. Head movements, on ball and socket joints, were thought to facilitate gill breathing. In contrast to the arthrodire Dinic, certain upper Devonian antiarchs, although classified as placoderms, looked rather like Tera, the heterostracan (Chapter 16, Figure 16.1) illustrating the difficulty in establishing strict sequential classifications.

Alfred Romer (1966) believed that sharks may have evolved independently of placoderms, although they were somewhat related. It seems to me that relatedness, in independent evolutionary pathways, could only occur through a process such as salmutation.

Sharks, with jaws in the earliest known species, were noted for cartilaginous endoskeletons, and with some exceptions, internal fertilization of large, shell-covered eggs. They had no lung structure, in contrast with many higher bony fish, being exclusively gill breathing ocean dwellers from their middle Devonian appearance.

The shark Clado (*Cladoselache*, Figure 17.2) was found in the Cleveland Shales of the late Devonian as fossil teeth and calcified cartilage skeleton with impressions of soft tissue, skin and fins in body outline.

Clado had well developed spineless fins: two dorsal fins, large paired pectoral fins and small paired pelvic fins. There was a pair of horizontal keels at the base of the heterocercal tail fin. Could these keels, absent in later sharks, represent production of prototype DNA for subsequent development of the horizontal tail fluke of whales (though whales are considered to be mammals, not fish, and have been thought to have evolved from land animals)?

Did the lack of claspers, characteristic of later sharks, mean that Clado did not fertilize internally, that the feature of internal fertilization appeared as a DNA mutation or salmutation in later shark species, perhaps as a DNA concept carried from invertebrates such as ancestors of *Buccinum*? (Chapter 14, Figure 14.10 b.)

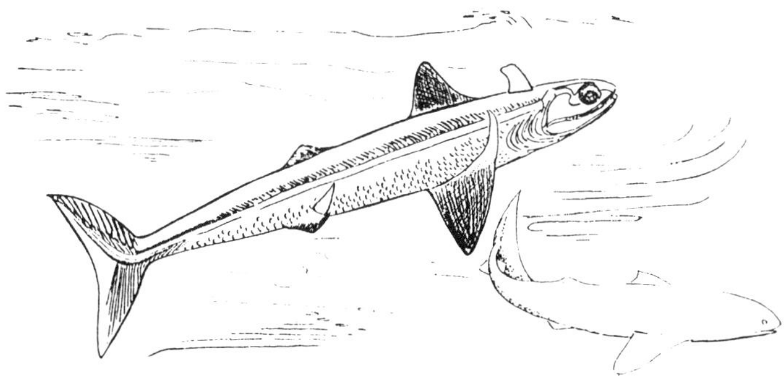

Figure 17.2 Clado (*Cladoselache*, Elasmobranchii, shark). (After Dean).

Advanced bony fish appeared suddenly in the early Devonian. Authorities continue to search, without success, for a common ancestor in the Silurian. They have postulated that these bony fish arose separately from archaic stock, since the known acanthodians are considered to be too specialized to be ancestral to them. Scientists have projected backward to search for the unknown ancestor of both placoderms and bony fish, but they have recognized this as pure speculation, without fossil evidence. Is there DNA evidence in modern fish to assist the speculation?

In the middle Devonian, three kinds of bony fish appeared, each fully developed, diversified and dominant: ray-fin fish, lungfish and lobe-fin fish. They all had functional lungs as well as gills. Of living fish, only lungfish still breathe with lungs; the ray-fins have evolved a buoyancy-related swim blatter in lieu of the lungs evident in their predecessors, and the lobe-fins are mostly extinct except for the coelacanths that also have a swim blatter.

Lungfish (Dipnoi) showed certain resemblances to the lobe-fins, although they were strikingly different in skulls and dentition. Both were well adapted to the freshwater environments in which they arose. Early lungfish probably lived with seasonal aridity, with special adaptations for it, as do living lungfish. Fleura (*Fleurantia,* a Devonian lungfish from America, Figure 17.3) had thickened bony mouth rims for cropping food, and a modification of small teeth in the median portion of the mouth, in lieu of the tooth plates characteristic of the group.

Figure 17.3 Fleura (*Fleurantia,* Dipnoi, lungfish). (After Graham-Smith and Westoll).

Scientists have postulated an evolutionary line from Devonian lungfish to living Australian lungfish, although Fleura is considered to be merely an interesting side branch.

Salmutation from Fish to Amphibians

It is believed that a lobe-fin fish, Eusty (*Eusthenopteron,* Figure 17.4) evolved in the Devonian as a fast swimming fresh water genus with a unique set of head joints allowing the snout to be raised and lowered, and the lower jaw to be closed independently. These innovations inaugurated a change from breathing mainly through the gills to air breathing through the lungs. These fish could breathe without lowering the lower jaw, a distinct advantage for swallowing prey at the surface, without drowning. It was also an evolutionary advance toward amphibians, which arose with lungs and four (tetrapod) limbs in the Devonian.

Erik Jarvik (1980: 234) said:

> We cannot explain why the tetrapod limb...and other prerequisites for a terrestrial life arose in the fish ancestors of the tetrapods. Theories about these things found in the literature are pure guesswork.

A significant challenge to Darwinian thought is: Did lobe-fin fish evolve in the Devonian into amphibians, by gradual successive stages of random mutations, solving the problems inherent in the move from a water to an air and land environment? I believe the mutations or salmutations required to make this change could not be random changes, but rather must be DNA interactional responses to the requirements of a different environment. In addition to solving numerous problems of land living, amphibians introduced new processes of metamorphosis and development from the swimming larval stage to a walking adult stage.

Amphi (*Ichthyostega,* Figure 17.5) the first known fossil amphibian, was discovered in Greenland, which was then connected to both North America and Europe. Amphi looks rather like a lobe-fin fish with feet. Amphi could walk on land and swim in the fresh water of streams, or propel itself in shallow water by both tail and feet like a modern crocodile. Yet Amphi is not recognized by all scientists as a link between fish and later land vertebrates, because it was thought to have a weak backbone. and its strength was in its ribcage not its vertebrae. It did not have the vertebral solution of later land animals.

This first, four legged, land walking, air breathing Amphi fossil had no known amphibian-like precursor. Only fish preceded it in the fossil record. It is still the choice of many

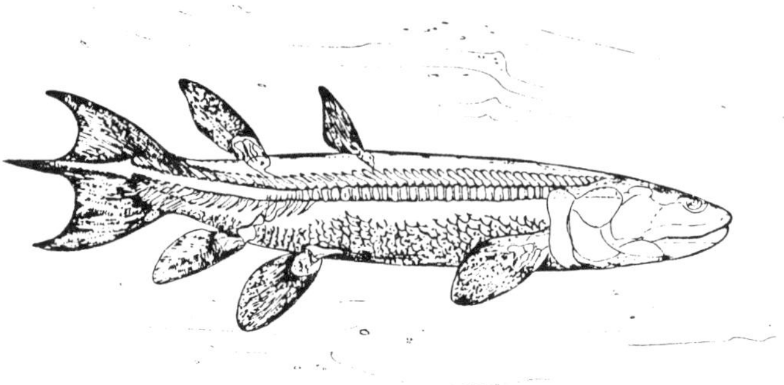

Figure 17.4 Eusty (*Eusthenopteron*, Rhapidistia, lobe-fin fish). (After Gregory, 1951).

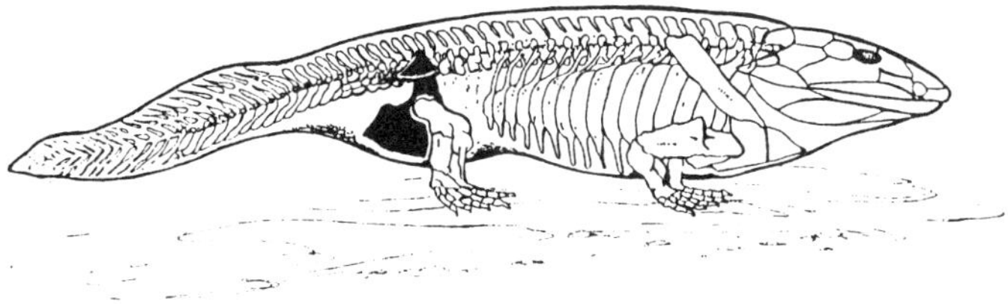

Figure 17.5 Amphi (*Ichthyostega*, Amphibia). (After Jarvik).

scientists as the transition possibility between fish and amphibians, whether or not it is an ancestor of later vertebrates.

J. J. Schmalhausen (1968: 24) found it difficult to assess the relationships of the living taxa of amphibians, let alone the fossil amphibian relationships, saying "...the vertebral evidence offers no suggestion of the relationship of the living amphibian orders."

The lungfish is another proposed link between fish and amphibian, contrasted with the viewpoint that the transitional fish was lobe-fin. The lungfish has fins, gills and a fish-like intestine. It has lungs, heart and larval stage, like those of amphibians. However, the lungfish heart is not intermediate between that of fish and amphibians. There is no transition between organ systems, which are quite unique, and seem rather to illustrate that amphibians are remarkably well defined and isolated from fish.

Examples of possible transitions from fish to amphibians are instructive of the need for complementarity dynamics and the concept of salmutation. Although scientists differ about the sequence from fish to amphibian, for the moment let us assume that the choice of fish representing the transitional position is Eusty, and the amphibian representing the progeny, is Amphi. The transition is a paradox, if considered in sequential logic. One individual must give birth to the next individual, but one individual has never been known to give birth to a different genus.

However, Eusty, the parent genus, and Amphi, the daughter genus, may be considered to represent complementary processes of continuity and discontinuity. It becomes possible to visualize a reorganization of genetic material to form a new genus of lobe-fins if viruses inserted new DNA into their eggs. I believe that the fossil record of radiation of various amphibian populations in the Devonian must have been a time of accelerated viral activity.

In the transition between genera, it is necessary to consider not just an individual birth as suggested by a single cause and effect chain in sequential logic, but to consider a population of lobe-fins about to parent a new genera. Consider that this population spawned and their fertilized eggs floated in the shallows of a stagnant stream. If the same kind of viruses inserted their DNA into a population of lobe-fin eggs, concurrent reorganization of DNA could be effected — the lobe-fin eggs hatching as amphibians. The change from fish to amphibians required massive changes in structure and function.

It becomes possible to envision insertion of foreign DNA into lobe-fin or lungfish eggs. In development of the larval stage of fish and amphibians, reorganization of DNA may be demonstrated not only by cell DNA but also by the protein products of DNA. The Darwinian concept of vertical sequences of germ lines is not an absolute principle. In addition, eggs of individuals of one genus may salmutate into individuals of a different genus by means of horizontal gene transfer. Admittedly, it is difficult to envision a natural process which may not occur during the short lifetimes of humans. One must be able to contemplate instantaneous events, in reciprocal moments, during a time span of thousands or even millions of years.

Evolutionary movement of vertebrates from a water environment to an air and land environment involved design and development hurtles, in addition to the problem of continuous breathing in the new medium of air. There was the difficulty of physical support, requiring new skeletal features — organisms weigh more in air than in water. Walking appendages were a necessary development. Water contains a great variety of nutrients which fish eat, so a new problem of food supply arose with land living, and changes in the digestive system were required. The chemical makeup of the body actually changed.

Because an air environment is very drying, it required a new kind of epidermis, and with the rise of reptiles, a new kind of egg. Air living required changes in eyes and ears. A different system for elimination of toxic carbonic acid was necessary.

New DNA of amphibians must have provided the blueprint for those systems. The modification of the DNA of fish to create the DNA of amphibians was a masterpiece of intelligence and control, a cybernetic revolution.

First Insects — Food for Amphibians

On diagrams of the geologic periods and trees of life, and from information in many resources, it appears that insects arose in the Devonian without direct evidence of possible progenitors. They seemed to arise fully formed, unique and functional. Such a development requires a concept of DNA transfer from simpler lifeforms into insects — major salmutations. It demonstrates purposeful salmutations; insects were needed to provide food for amphibians as they crawled onto the land for the first time.

Found in the Devonian strata, were the first wingless insects: spiders and scorpions. A recent discovery suggests that an insect like a silverfish, a brissletail, may have been the first insect, perhaps as early as 390 million years ago. Were wingless insects preadaptations for the later appearing winged insects of the Carboniferous Period? Winged insects were a necessary complement in preparation for insect-plant symbiosis. A second element of the coevolutionary complement, flowering plants (angiosperms) arose millions of years later in the Cretaceous.

Beautiful Ammonites

Beautiful ammonites arose in the Devonian (Figure 17.6) presumably from nautiloids (Chapter 15, Figure 15.1) and radiated in various sea habitats. Unlike nautiloids, ammonites appeared in two sexes of different size. If ammonites evolved from DNA of nautiloids, does this development require consideration of both Mendelian principles and the salmutation principle?

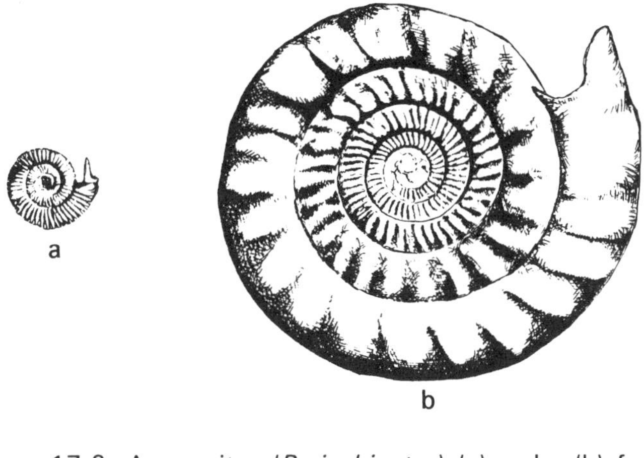

Figure 17.6 Ammonites (*Perisphinctes*) (a) male. (b) female. (After Callomon).

End of the Period Extinctions

After the aggregation of land masses, the Devonian closed with massive extinctions of additional families of trilobites, of lung fish, and ray-fin fish. Romer (1966: 317) said that almost every ancient fish type present in the period was exterminated, except for a few of the acanthodians. This exterminiation included numerous taxa of ostracoderms, antiarchs and arthrodires. The massive tectonic movements and resulting climatic changes are presumed to be one cause of these extinctions.

18

Paleozoic Era — Carboniferous Period Survival of the Fewest

The Carboniferous period of 340 million years before present (340 MBP) included two divisions of land masses: Euramerica-Gondwana and Asia. The seas impinged upon areas now known as dry land. Much of middle North America was an inland sea.

This time period was divided into two sub-periods. The earlier Mississippian was predominately represented by fossils of marine life in series of limestones, during continuation of a warm trend. The later Pennsylvanian was characterized as a colder period with shallow swamp water deposits and coal seams of continental origin in Europe and North America. These coal deposits reveal the fossils of vegetative masses, early mosses and cordaites (ancestral conifers). Tree fern and scale tree forests were laid down in swampy areas.

In the strata represented by freshwater fish, the proportions of the three predominate types of fish which survived the Devonian were reversed. In the Carboniferous, the lobe-fins survived in reduced numbers, the lungfish were few. Ray-fin fish radiated in large numbers in lakes and rivers. Such reversal of evolutionary direction by survival of the fewest was not envisioned by Darwin. In the marine environment, sharks proliferated, as did their prey.

Amphibians and Aquatic Reptiles

Various species of amphibians radiated profusely in the swamps. Several kinds of amphibians developed different reptile-like skull characteristics, perhaps indicating that reptiles were evolving from several different DNA sources, to the surprise of Darwinians and as evidence of salmutations. Erik Jarvik (1980: 267) noted that in trying to unravel the interrelationships of reptiles, scientists encountered the same dilemmas as those faced in understanding the early fish. He said,

> Like [early fish] the various reptile groups appear suddenly in the fossil record and several of them (mesosaurs, ichthyosaurs, chelonians) show the same reluctance to change as many of the fish groups.

In constructing a family tree of reptiles, "avoiding preconceived opinions and arbitrary arrangements" and keeping strictly to anatomical facts, he showed that the evolutionary lines do not branch; they run parallel, as do the lines in general vertebrate evolution. I identify these parallel lines of ascent evident in the fossil record, as indicative of salmutations occurring at the beginning of new lines.

Meso (*Mesosaurus,* Figure 18.1) was an early aquatic reptile about 16 inches in length, found in fresh water strata of South Africa and South America, evidence of conjunction of the two continents in the Carboniferous Period. Some Mesos grew to be three feet in length.

The tail and webbed feet were adapted for swimming. The ribs were rather like those of later marine sirenians. The shoulder girdle scapula was similar to that of other aquatic forms. The delicate marginal teeth suggest the crustacean straining structure which was perfected in the DNA of later appearing baleen whales.

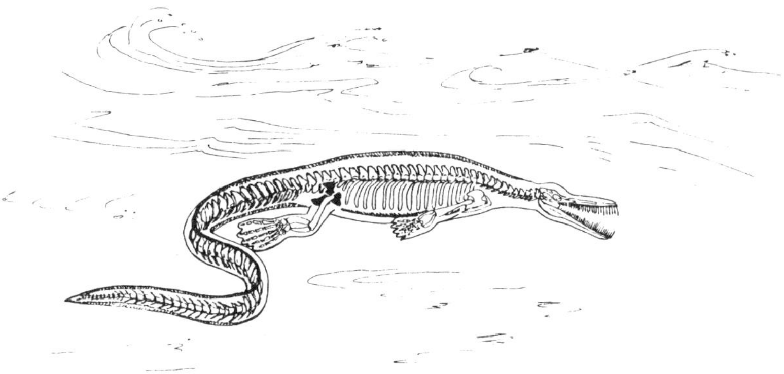

Figure 18.1 Meso (*Mesosaurus*, Reptilia). (After McGregor, 1908).

Meso has been considered to be a descendent of terrestrial stem reptiles (Figure 18.4 Hylon). Early aquatic reptiles, both freshwater and marine varieties, may have evolved by salmutations from land-living species, or, aquatic reptiles may have evolved by salmutations from earlier aquatic taxa. It is difficult to conceive of land-adapted Hylon and other stem reptiles as possible predecessors of Meso. This little animal carries some characteristics similar to land animals but more characteristics similar to aquatic animals. More must be learned about the organization of DNA which produces these kinds of adaptations before we can identify Meso's progenitors.

The use of complementarity dynamics frees the mind to reevaluate the concept that Meso evolved from one maternal terrestrial vertebrate, and to consider, in addition, combinations of DNA patterns that might have structured it.

Micro (*Microbrachis,* Figure 18.2) has been classified with Ophi (Figure 18.6) as a lepospondyl reptile, although it has more affinities with labyrinthodont amphibians. This little five-inch-long animal had a rib structure and head suggestive of both reptiles and amphibians, and it had the webbed feet of amphibians. It thus exemplified recombinations of DNA of two major phyla, support for the concept of salmutation. It appeared in the late Pennsylvanian but did not survive the Permian Period.

Figure 18.2 Micro (*Microbrachis*, Lepospondyli, Amphibia). (After Steen).

I believe it is possible that early aquatic reptiles did not evolve from terrestrial reptiles, but rather, that both amphibians and early aquatic reptiles salmutated from DNA of earlier sea creatures. Ichthy (*Icthyosaur*, Figure 18.3) also thought to be an early aquatic reptile, may have evolved from some early shark-like ancestors. It seems logical to assume that Ichthy, a giant swimming reptile, must have had some previous aquatic ancestors before its sudden appearance in the Carboniferous Period marine deposits.

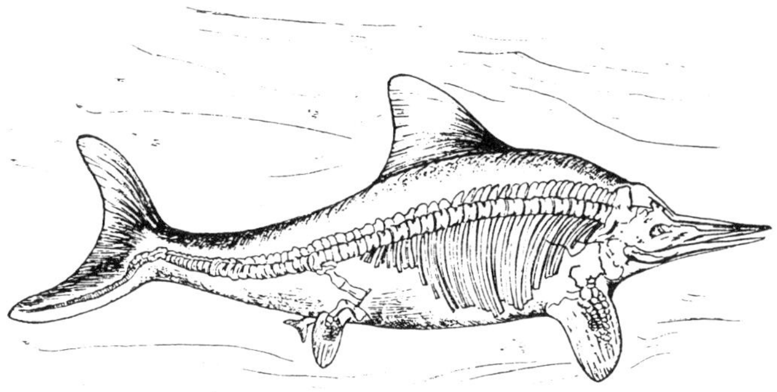

Figure 18.3 Ichthy (*Icthyosaur*, Euryapsida, Reptilia) giving birth. (After Merriam).

Darwin seemed to have had little or no evidence of these sea-going reptiles when he developed his theory. Great gaps in theory must be filled to account for those taxa which were discovered beginning in 1810 when 11 year old Mary Anning dug from sea cliffs in the south of England, the first *Ichthyosaur* fossil.

Theory must account for the fact that those fossils have been found containing skeletons of Ichthy babies. Therefore, animals considered to be early reptiles gave birth to living young, rather than passing through an amphibian-type egg-laying stage — a difficult concept for strict Darwinians — but understandable as a salmutation from some shark-like ancestor known for internal fertilization and live births.

First Land Reptiles

Lower Pennsylvanian tree stumps contained fossils of various kinds of reptiles. Therefore, it would appear that reptiles radiated rather quickly on land in the Carboniferous. An early Pennsylvanian, Hylon (*Hylonomus*, Figure 18.4) about two feet long, found in tree stumps of Nova Scotia, is thought by some paleontologists to be the first reptile. Fossil eggs have not been located with Hylon to verify its mode of reproduction, although some believe that it may have produced the more complex amniotic eggs of reptiles rather than the simpler eggs of amphibians (Chapter 20, Figure 20.1).

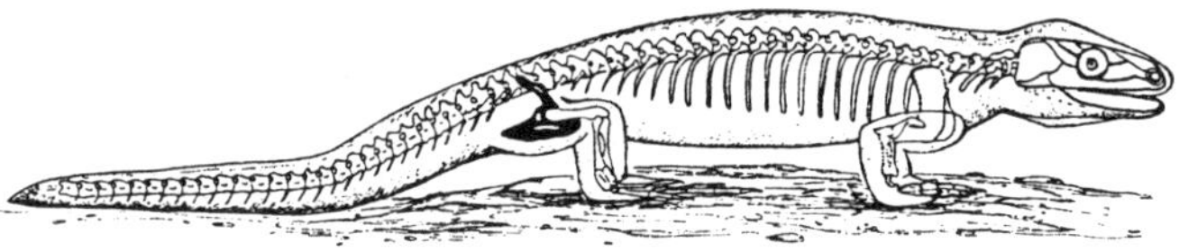

Figure 18.4 Hylon (*Hylonomus*, Cotylosauria, Reptilia). (After Carroll).

Stem reptiles such as Hylon are distinguished from amphibians mainly by their definitive jaw structure and cheek region of the skull. Later reptiles are classified mainly by the configuration of temporal openings of the skull. Reptilian skulls tend to be higher and narrower than the flattened skulls of amphibians. A real neck is developed in most reptiles, and the attached neck ribs shorten or disappear.

Some reptiles developed underslung abdominal ribs such as those of turtles (Chapter 20, Figure 20.7) and Moren (Chapter 21, Figure 21.2). The pelvic region became more marked in reptiles, without a change in the number of pelvic bones. Pelvic bones of the animals illustrated in this work have been highlighted to call attention to the variety and unique adaptations of this pivotal structure. Comparison of pelvic bones also leads to the conclusion that there is no strict sequential development of this important structure from amphibian to reptile to mammal.

The limb bones of reptiles, generally became slimmer and longer, although some developed massively. Hylon carried DNA combinations which probably developed into a long and distinguished group of fascinating reptiles, although its genus became extinct in the Permian.

It is interesting to note that the first pelycosaurs also appeared in the Pennsylvanian: Eda (*Edaphasaurus*), and Dime (*Dimetrodon,* Figure 18.5) the reptile which is credited with giving rise to the mammal-like reptiles of the Permian.

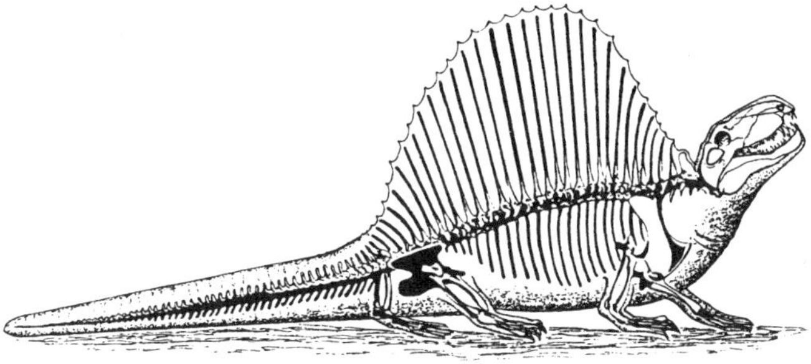

Figure 18.5 Dime (*Dimetrodon*, Pelycosauria, Synapsida, Reptilia). (After skeleton, Natural History Museum of Los Angeles County).

Dime (*Dimetrodon,* Figure 18.5) a most unusual-looking creature from the redbeds of Texas, USA, was a voracious predator, ranging to 11 feet in length. Dime is notable because of the controversy over the nature of its strange "sail" a perpendicular extension of the vertebrae which is thought by some to be a temperature control mechanism. A herbivore of the period, Eda, had a similar sail, studded with crossbars connected by membranes. It may have been prey for Dime. It is thought that these pelycosaurs arose from cotylosaurs such as Hylon.

Because both these pelycosaurs lived in an area of sedimentary formation, it would not be totally unwarranted to assume that they spent time in water as well as on land. Perhaps the "sail" *did serve as a sail,* as some experts posited. The sail may have warmed them and sped them on their way through their watery habitat. Their strong sprawling legs could provide speed on land. The delicate nature of Eda's sail membrane would have prohibited land-living near those plant species which could have snagged and damaged it.

Reptiles were the first animals which could live their whole life spans on land. They are reported to be cold-blooded animals, requiring the heat of the sun or the captured heat in their surroundings to raise their temperatures to a functional level. Although snakes have been said to have arisen only in the Cretaceous Period, Ophi (*Ophiderpeton*, Figure 18.6) looking very much like a snake, was found in the Mississippian deposits of Nova Scotia. Did Ophi's DNA salmutate into snakes?

Figure 18.6 Ophi (*Ophiderpeton*, Lepospondyli, Amphibia). (After Fritsch).

First Flying Insects

The first air-born insects, including huge dragonflies (*Meganeura*, Figure 18.7) with a wingspan of 75 cm — the largest insects known — appeared in the Carboniferous, apparently without known flying insect antecedents. Did winged insects evolve in the Mississippian? No such fossils have been discovered, but in the Pennsylvanian, there were at least a dozen orders of winged insects, many similar to living representatives. Did they salmutate from wingless insects by recombinations of DNA in a manner suggested by the metamorphosis of butterflies from caterpillars?

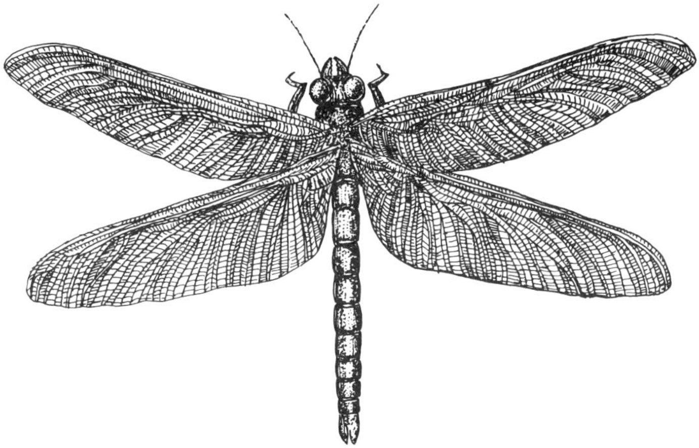

Figure 18.7 Dragonfly (*Meganeura*, Insecta, Arthropoda). (After Moore et al).

End of the Period

Toward the middle of the Pennsylvanian, there was a crash of amphibians, and trilobites all but disappeared. During this period, much of Gondwana was covered by glaciers, and ice caps were prevalent in that part of Gondwana consisting of South Africa, South America, India, Australia and Antarctica. The two continents of Euramerica and Asia were gradually joining as the Carboniferous Period ended.

19

Paleozoic Era — Permian Period Rise of Mammal-like Reptiles

There were 390 days in a year when the Permian began 280 million years before present (280 MBP). Pangea was again configured as one continent in the late Permian, 240 million years ago. The two major continents of the Carboniferous had, by then, collided and become one. The red formations of the Alps, the Rockies and the Appalachians had risen into giant ridges (Chapter 12, Plate 12.2 # 4).

In the early Permian the weather was as cold as it was during part of the Precambrian. The icy winds even brought cooling to the inland equatorial regions. An ice age with glaciation extended to sea-level elevations. However, the conjunction of the continents also originated an equatorial current which brought warm water and a relatively uniform environment to easterly areas of islands and barrier reefs in the equatorial region. This provided for some of the richest reef faunas ever found. The assemblage of reefs and reef dwellers was evidenced by fossils of new kinds of green algae columns, bryozoans, brachiopods, rugose corals and new types of sponges and sea lilies.

It is postulated that the conditions in those regions changed rather abruptly during the Permian, possibly by retreat of the sea from the Pangean continent. The rich life of the reefs was then exterminated, leaving fossil reefs in areas now known as Texas, USA and the Mediterranean. A warming trend began after mid Permian until a warm, tropical climate prevailed in much of southern Laurasia (mid Pangea) and the southern hemispheric area.

Amphibians and Stem Reptiles

Common Ancestry of Predators and Prey

The Permian supported numerous amphibians. *Eryops* (Figure 19.1) about five feet in length, was one of the best known amphibians from the redbeds of Texas, USA. It lived a truly amphibious life from the late Carboniferous through the early Permian. It was noted for a strong spinal column and well developed limbs. Its ribs and shoulder girdle appeared as though they could have evolved from Amphi (Chapter 17, Figure 17.5). Its pelvic structure was very different. Was the pelvis from a different genetic source?

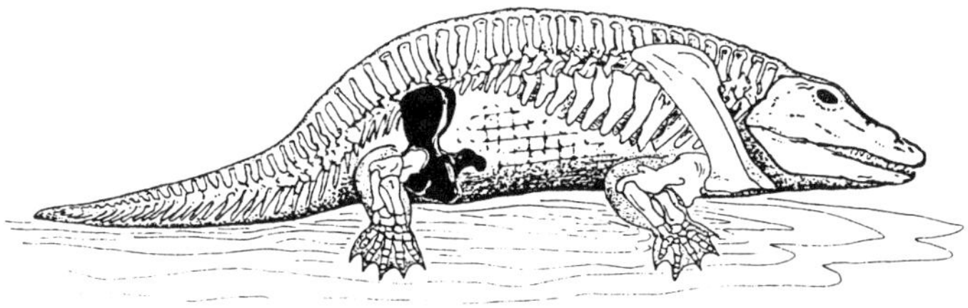

Figure 19.1 *Eryops*, (Rhachitomi, Temnospondyli, Amphibia). (After Gregory, 1951).

Although *Eryops* looked like a crocodile, it has not been considered by all scientists to be a direct ancestor of crocodiles. Crocodiles, rather, are thought by some scientists to be derived from phytosaurs like Mystrio (Chapter 20, Figure 20.8) or thecodonts like Salty of the Triassic (Chapter 20, Figure 20.2).

It seems more likely that crocodiles and alligators, and perhaps Mystrio too, replicated DNA patterns derived from *Eryops* and its kind.

Seymouria (Figure 19.2) about two feet long, was an early Permian amphibian tetrapod (or cotylosaurian reptile) from the redbeds of Texas, USA. Scientists see its skeletal structure near the dividing line between amphibians and reptiles. It had an amphibian-like skull and a reptilian skeleton, rather like the stem reptiles, cotylosaurs (Hylon, Chapter 18, Figure 18.4). *Seymouria* was believed to be amphibian in its egg structure. The existence of several animals which are thought to be on a dividing line between phyla, sugggests the principle of salmutation.

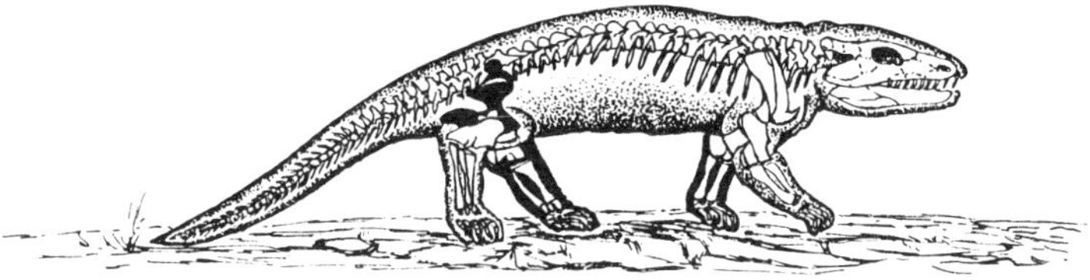

Figure 19.2 *Seymouria* (Anthracosauria, Amphibia, or Cotylosauria, Reptilia). (After Gregory, 1951).

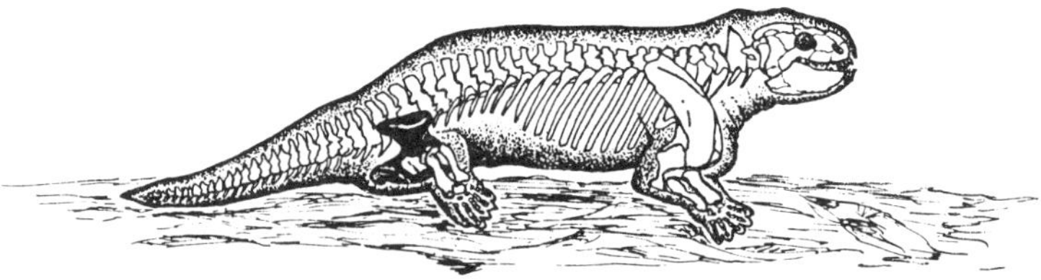

Figure 19.3 *Diadectes* (Anthracosauria, Amphibia, or Cotylosauria, Reptilia). (After Gregory, 1951).

Diadectes (Figure 19.3) about 10 feet long, another early Permian anthracosaurian amphibian (or a cotylosaurian reptile) is considered to be the first land herbivore. Most of the early tetrapods appear to have been predacious, possibly an anomaly of fossil preservation rather than an indication of true incidence. The concept of both predators and prey arising from the same stem animals supports the theory of salmutation. Common ancestry of predators and prey at various stages of evolution, is not a Darwinian concept.

A growing number of reptiles, 15 to 20 orders, have been identified in Permian deposits. Early reptiles exhibited three skull types: anapsid, a primitive roofed skull; diapsid, a two arched type; and synapsid, a fused arch type. The latter synapsid type represented by *Diadectes* seemed to be leading toward the therapsid, mammal-like reptiles of this period.

Darwinian theory has been called upon to explain, with basic sequential logic, the evolution from early stem reptiles of the Paleozoic to the dinosaurs of the Mesozoic Era. Although this pattern has not been greatly disputed, this theory has not provided a ready answer to the question: From whence came the earliest Mesozoic mammals?

Mammal-like Reptiles or Early Mammals?

The origination of the mammal line is frequently depicted in the scientific literature as beginning with small, nocturnal, mouse-like creatures, hiding from the giant dinosaurs in underbrush, during the "Age of Reptiles" (Chapter 20, Figure 20.10, Eo). These early mammals are known only from scattered teeth, bone fragments and a few skulls. They were thought to arise from small reptilian ancestors, but fossil evidence for this is limited.

Profound evidence in the fossil record, suggesting the ancestry of mammals, has come to light only recently in excavations of the Karroo beds of South Africa. This evidence does not support the theory that small mammals arose from small reptilian ancestors. On the contrary, the Permian Period Karroo formations revealed mammal-like animals (therapsids) of various sizes ranging from Thrin, 18 inches long (Figure 19.6) to *Moschops* (Figure 19.8) eight feet long, and *Jonkeria*, 12 feet long, not counting the tail. These animals did not give rise to mouse-like mammals. They carried DNA which appears to represent preadaptations for a large array of different kinds of mammals.

It is not too surprising that Darwinian theory is inadequate to explain the origination of the mammal-like reptiles of the Permian-Triassic boundary, although it has been posited that they evolved from pelycosaurs like Dime (Chapter 18, Figure 18.5). It was only in the early 1960s that the eminent paleontologist Edwin H. Colbert (1973) journeyed to the Karroo beds, north of Johannesburg, South Africa, to seek out Lystro (*Lystrosaurus*, Figure 19.4) a mammal-like reptile of the therapsid group known as a dicynodont (two dog-like-teeth). It was in 1968 that Colbert identified a fossil bone fragment of the same genus from Antarctica. Two years later he found a full skeleton of Lystro in Antarctica.

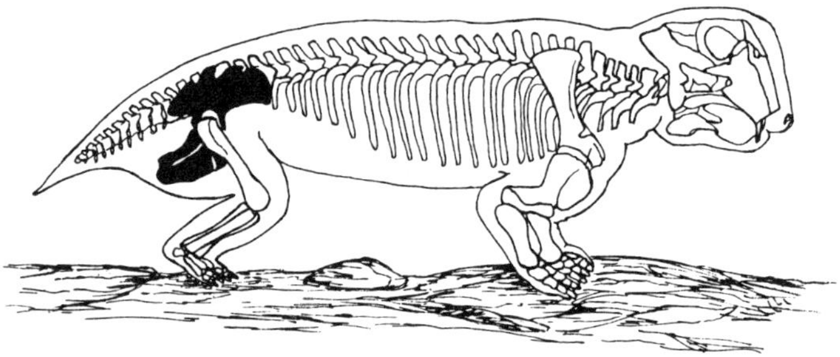

Figure 19.4 Lystro (*Lystrosaurus*, Dicynodontia, Therapsida). (After D. M. S. Watson, 1914).

This was the first fossil evidence to substantiate not only the existence of Lystro in two locations of similar age, Antarctica and Africa, but also strong support for the theory of continental drift, first proposed in 1912 as a detailed hypothesis by the German, Alfred Wegener.

Lystro, about five feet in length and somewhat similar in appearance to Kanne (*Kannemeyeria*, Figure 19.5) lived during a period of dicynodont proliferation; its kind constituted 90% of the late Permian fossil skulls found in South Africa. From 250 to 200 million years ago, in tropical or semitropical conditions, herds of these herbivores lived rather like hippopotami, grazing on the flora of marshes and lakes in the Southern Hemisphere.

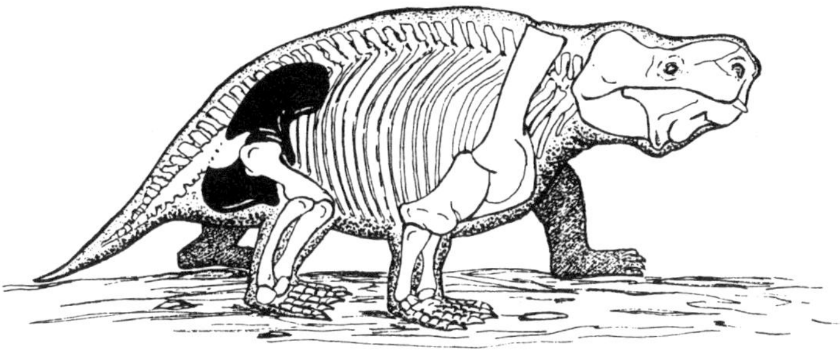

Figure 19.5 Kanne (*Kannemeyeria*, Dicynodontia, Therapsida). (After Broom).

Kanne, a wider ranging genus of dicynodont than Lystro, has been found in South America and China, leading some scientists to believe that China may also have been in the Southern Hemisphere in the Permian. Kanne also ranged into the western United States, indicating a vast, connected, watery, rangeland for mighty herds of dicynodonts. It is strange to contemplate that Kanne had no teeth, apart from the two tusks of its name, but rather had bony pads to masticate its food. Yet Kanne, with a skull nearly two feet in length, became very common in the early Triassic. In the later Triassic, other genera similar to Kanne grew skulls up to a yard in length!

Found with Lystro in both South African and Antarctic fossil deposits was the 18 inch cynodont (dog-teeth) Thrin (*Thrinaxodon,* Figure 19.6) a typical carnivorous theriodont of the period. Thrin had developed a secondary palate in the roof of its mouth, a mammalian characteristic which assists breathing during mastication and suckling. Suckling of the young may have been present in this advanced therapsid. Thrin may even have had hair, and was probably warm blooded; however these assumptions cannot be verified by fossil bones.

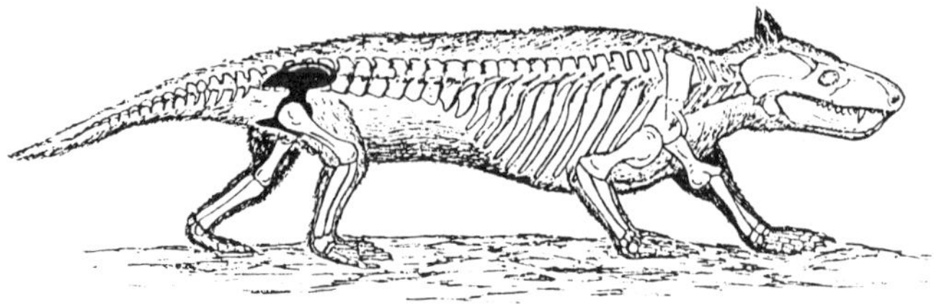

Figure 19.6 Thrin (*Thrinaxodon*, Cynodontia, Theriodontia). (After Brink).

Thrin's dentition included front nipping teeth similar to human incisors, upper canines, and cheek teeth with cusps. The backbone was more mammal-like than Dime's (Chapter 18, Figure 18.5) with a more distinct neck region, shorter ribs and fused pelvis and vertebrae, leading toward mammalian lumbar development. The front leg elbows bent backward, the back knees bent forward, with legs underneath the body, instead of sprawling like those of reptiles. The pelvic ilium extended forward, unlike that of many therapsids. The brain case was small in relation to Thrin's size.

Thrin appears to be so dog-like, that it necessitates the concept of similar DNA configurations in the Permian and the present, with all that that implies! The combination of the mammal-like animals: Lystro, Kanne the toothless herbivore, and Thrin with truly mammal-like dentition, suggests the concept of a lush marshy ecosystem, extending from Antarctica and Africa to South America and China, roamed by predators and prey as in the African veldt of today.

In addition to theriodonts like Thrin, the Permian gave rise to the Anomodontia, a suborder of the order Synapsida. Early anomodonts are dinocephalians of the mid Permian of Russia and South America.

Titan (*Titanophoneus*, Figure 19.7) found in Russia, was an enigma. Classified from its chisel-shaped incisors characteristic of herbivores, it also had powerful canine tusks like a carnivore. Titan was five feet long in head and trunk;

with tail, it was six and a half feet long. Could Titan represent an early omnivore ancestral DNA experiment?

Figure 19.7 Titan (*Titanophoneus*, Dinocephalia, Anomodontia). (After Orlof).

Moschops (Figure 19.8) another giant mammal-like reptile, eight feet in length, was definitely herbivorous, with chisel-shaped cropping teeth. How lush the vegetation of the world must have been in the Permian to support such huge herbivores! Moschops' skull suggests an eerie foreshadowing of that of sheep and goats of the present. Is its DNA skull pattern recreated in animals of today?

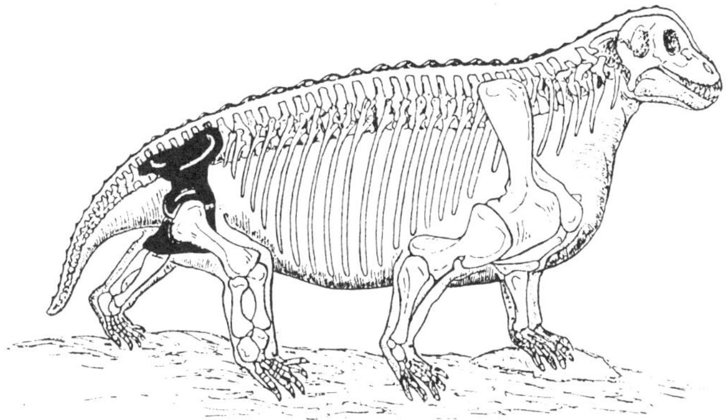

Figure 19.8 *Moschops* (Dinocephalia, Anomodontia, Synapsida). (After skeleton, American Museum of Natural History).

An advanced anomodont, *Galepus* (Figure 19.9) was a tiny animal with a skull only one and half inches in length. A characteristic of Galepus, apparent in both dinocephalians and dicynodonts, was the jaw articulation below the level of the tooth row. The most amazing attribute of *Galepus*, however, is the skull configuration, which is strikingly similar to later appearing camels, thus the name, dromasauria (camel-like reptile). What better evidence is there not only of DNA preadaptation, but also of unusual DNA combinations? This little animal simply cannot be a Darwinian chance variation in a sequential pattern of lifeforms.

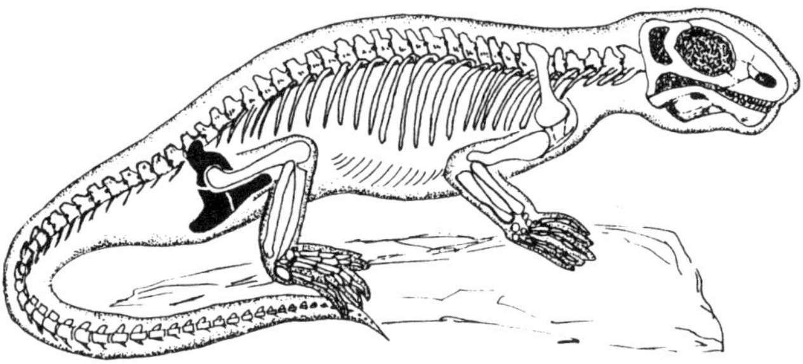

Figure 19.9 *Galepus* (Dromosauria, Dinocephalia, Anomodontia). (After Broom).

Although found in late Triassic strata of Europe, Oli (*Oligokyphus*, Figure 19.10) about 20 inches long, must be considered with the Permian-Triassic boundary mammal-like reptiles. Similar animal fossils have been found in South Africa, China and North America. They are thought to be derived from cynodonts.

Oli's classification is based on its dentition. The front of the mouth has large incisors. There is a characteristic gap and a final battery of highly developed cheek teeth, with longitudinal rows of cusps. The tooth structure looks to be ancestral to that of later Mesozoic herbivorous marsupials, multituberculates and rodents. Oli's skull and skeleton were very similar to that of early mammals (Eo, Chapter 20, Figure 20.10).

Romer (1966: 184) noted, "... in one group or another of the more advanced therapsids almost every diagnostic feature of mammals has been attained." He didn't explain how these features could have been transmitted into the later-appearing mammals. The concept of salmutation is needed.

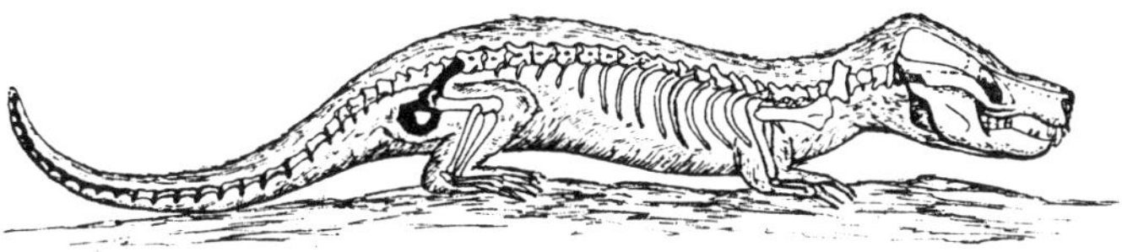

Figure 19.10 Oli (*Oligokyphus*, Tritylodontia, Therapsida of the Triassic). (After Kuhne).

End of the Period Extinctions

By the late Triassic, all the therapsids were extinct except one last tritylodont found in the middle Jurassic strata in England. Within five million years of the late Permian, much of Permian life had been wiped out. The fusulinids, complex protozoans which had lived for 80 million years, creating vast deposits of limestone from their shells, utterly vanished. The suspension feeders were almost wiped out. Trilobites, after proliferating for about 300 million years, became extinct. Gone were shark-like fish, pavement toothed sharks and many primitive fish. Of amphibians, few other than primitive tetrapod coal swamp types survived. The ranks were thinned of dominant lifeforms such as paired-fin fish, stem reptiles, aggressive carnivorous reptiles such as Dime, and the mammal-like reptiles.

One half of all known families and 24 orders and superfamilies of marine and land animals disappeared. It is estimated that 80% to 85% of the species then living vanished, including 75% of the amphibians and more than 80% of the reptiles. It appeared that survival at the end of the Permian was not related to a creature's stage of evolutionary progress as expected by classical evolution theory. It seems more like a massive housecleaning of the natural world. The reasons for the Permian extinctions are unknown, but various speculations cite the extinctions as resulting from tectonic colliding of the two major continents, and the explosive volcanism and weather changes associated with them.

Analysis of Paleozoic Evolution

This brief suggestive analysis of the evolution of animals during the Paleozoic Era leads to the conclusion that the unknowns remain enormous! Scientists began their studies of the fossil record by describing those fossils found in identified geologic strata. They compared and contrasted identified fossils with those that preceded and followed them, looking for possible evolutionary lines. They have compared fossils with living organisms to envision their internal systems and describe possible similarities and differences.

Each scientist or group, in each defined area of expertise, has assumed responsibility for designating which animal characteristics were to be used for classification and evolution analysis. This has led to very different emphases in selection of the traits studied in different animals.

Some of the major characteristics which have been selected for analysis include: spicules of sponges; tissue structure of coelenterates; segmentation in annelids and in arthropods; shell structure and muscle attachment sites in mollusks, configuration of cephalon plates of jawless fish; configuration of jaws in jawed fish; identification of notochord in chordates. The parts of vertebrates selected for emphasis include not only jaws and vertebrae, but especially the skull configuration and tooth structure.

Although the pelvic structure has been emphasized in the classification and study of evolutionary pathways of dinosaurs of the Mesozoic, it has not received emphasis in analysis of many other vertebrates, including mammal-like animals of the Paleozoic and the mammals of later periods. I believe that if the pelvic region is especially important for classification and evolution analysis of dinosaurs because it is the pivotal region of these vertebrate skeletons — then it is also significant for consideration of other vertebrate quadrupeds. If one alters perspective from skull and tooth to added pelvic emphasis, in the mammal-like animals, for example, one is struck not only by the uniqueness and contrast between each animal's skull and pelvic bones, but also by the similarities of both skull and pelvis when these animals are compared with later appearing mammals.

Such an expansion of emphasis is encouraged by complementarity dynamics. It becomes possible to consider not only immediate predecessors and successors of the particular fossil, but also encourages comparison of pivotal bones in fossils of other time periods, and in living animals. This additional pelvic emphasis in the illustrations of the vertebrates of the Paleozoic as well as the Mesozoic and Cenozoic, is not done here to support establishment of one main definitive vertebrate characteristic, but rather to establish the importance of encouraging different perspectives in the development of scientific concepts.

It also suggests the question: Can DNA analysis show a connection between DNA configuration and skull and pelvic configurations across long time spans? Perhaps the emphasis on pelvic bones, illustrated in this work, will encourage new concepts to be developed for evolution analyses.

20

Mesozoic Era — Triassic Period First Dinosaurs

The Paleozoic (old) Era was at an end and the Mesozoic (middle) Era, the "Age of Reptiles," began with the Triassic Period, 230 million years before present (230 MBP). The continent of Pangea split as the Tethys Sea encroachment toward the west created the two continents of Laurasia to the north and Gondwana to the south, completing the division about 190 million years ago (Chapter 12, Plate 12.2 # 5). The moderate climate continued, then warmed to unprecedented heights.

Reef life seemed again to be absent from the Earth for the first 10 million years of the Triassic Period. Then in the Mediterranean part of the expanded Tethys Sea, a reef community revitalized green algae, sponges and a new group of corals. Six families of coral fossils were found in Germany, Sicily and the Southern Alps. Also dwelling in the reef area were foraminifera, sea urchins and mollusks. Red algae began to flourish.

The warming climate was accompanied by radiating diversity of many species. The period was dominated by new kinds of ocean deposit feeders: bivalve mollusks and burrowing echinoids. Marine reptiles reappeared in the fossil record. Land reptiles like those in Carboniferous formations, radiated in great profusion in the Triassic. They lived in swamps and forests of cycads, tree ferns and conifers.

Reptile Eggs and Salmutations

There seems to be no transition in the fossil record from amphibian eggs to unique and complex amniotic eggs of reptiles (Figure 20.1). How could this be? The first reptile fossil amniotic eggs, totally different in structure and environmental complementation from the eggs of fish or amphibians, were found fully developed in the Triassic, without indication of a precursor. I believe amniotic eggs represented salmutations, reorganization of DNA for numerous new genera.

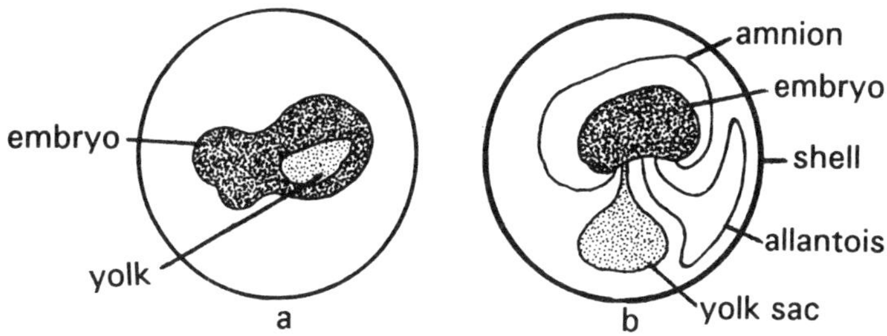

Figure 20.1 (a) Amphibian egg; (b) Reptile amniotic egg. (After Denton).

Conventional evolution theory, positing a series of individual mutations in one species line — as the basis for a new species line — neglects the fact that such mutations probably would make the DNA of the mutated individual incompatible with the DNA of a potential mate which had not mutated. Unless male and female mutated in a compatible way, in those cases in which the chromosome number is changed, they would fail to produce viable offspring.

No. The beginning of a new taxon line, a new genera, is not from the mutation of a single individual. The beginning of a new genera requires the same salmutation throughout an interbreeding colony — an insertion, something like an infection, changing the DNA of breeding stock all at once. This condition could be met by a viral invasion of eggs in a specific location, with the same DNA reorganization of many or all individuals, like influenza carried into a biological community.

This concept is mind-boggling, yet, it is not more difficult to envision than the caterpillar metamophosis into butterfly. Cladists have found patterns of regularity in various genera DNA configurations. Although the DNA of one genus may not appear to be ancestral to — or progeny of — another, there are intricate relationships of form that have not yet been fully explained. I believe salmutation theory will assist in these deliberations.

The cladism of W. Hennig did not originally concentrate on DNA, but continued a conventional emphasis, reconstructing organism relationships from homologous bones or other structures. Hennig searched for sister groups with homologous features in common. R. P. S. Jefferies (1979: 457) accepted Hennig's perception and supported it:

> This method is more powerful than searching for ancestors because ancestor-descendant relationships will exist within a group of known organisms only by good luck, can exist between contemporaneous organisms only by grace of metaphor, and are logically impossible to prove, by purely morphological methods, between any two known species.

Jefferies may have rung a death knell for the Darwinian gradual-sequential ancestor-descendant succession concept, without actually proclaiming it. But his support for Hennig's cladistic "sister groups" classifications, based on recency of common descent, merely sidesteps the ancestor-descendant logic problem.

I believe DNA analysis, based upon salmutation theory, must supplement cladistic identification and analysis of closely related, so called "sister" species. Because DNA is not often available from fossils, analysis of living taxa DNA may be considered in relation to analysis of living taxa physical features to provide clues to the pattern of DNA evolution.

Triassic Diversity

As the Triassic opened, both large and small thecodont reptiles were radiating. They were thought by some scientists to stem from cotylosaurs like Hylon (Chapter 18, Figure 18.4). Salty (*Saltoposuchis*, Figure 20.2) was a small bipedal reptile, just under four feet long, whose kind has been considered by some experts to be a progenitor of flying reptiles, birds and dinosaurs. Its front limbs were tiny in relationship to the hind limbs. Perhaps it had DNA precursor patterning for the relatively tiny front limbs of the giant, *Tyrannosaurus*, of the Jurassic. Salty had a double row of armor plates down its back, perhaps indicating a DNA preadaptation for the armored dinosaurs of the Cretaceous.

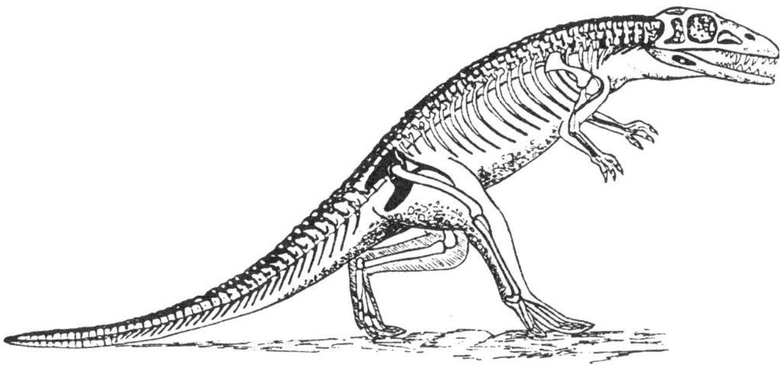

Figure 20.2 Salty (*Saltoposuchus*, Pseudosuchia, Thecodontia, Diapsida). (After Huene).

The Triassic is noted for the first true dinosaurs, thought to have arisen from thecodonts such as Salty. During this period, dinosaurs replaced more primitive reptiles and amphibians. They appeared in North America, Europe, South Africa and South America. They varied considerably in length, from three feet to 21 feet. Some smaller dinosaurs, of fowl size, appeared later in the Jurassic.

Scientists have hypothesized that the first dinosaurs were bipedal and carnivorous, with some of them reverting later to a quadrupedal gait, and becoming herbivorous. Dinosaurs have been divided into two orders, the Saurischia, with triangular (triradiate) pelvic structure, first identified as carnivores, and the Ornithiscia, dinosaurs with parallel pelvic bones, (tetraradiate — bird hipped) first identified as herbivores.

Celo (*Coelophysis,* Figure 20.3) considered a small carnivore, six to eight feet in length, was common in the late Triassic of North America. It was one of many similar genera extending into the Cretaceous Period. Its kind has been thought to be ancestral to ornithiscian dinosaurs and birds. Perhaps it had DNA which appeared in a recombined form as the DNA of Archy (Chapter 21, Figure 21.7) considered to be the first true bird of the Jurassic.

Figure 20.3 Celo (*Coelophysis*, Coelurosauria, Saurischia, Diapsida). (After Colbert, 1964).

Longis (*Longisquama,* Figure 20.4) was found in the lake beds of Turkestan (north of Afganistan). Although it is shown here, on a tree limb as a terrestrial reptile, capable of "parachuting," I think it is possible that Longis was at least partially aquatic. Because of its lake bed location and its keeled scales, the perpendicular appendage would be limited in its advantage as a "parachute" but might be useful in sailing.

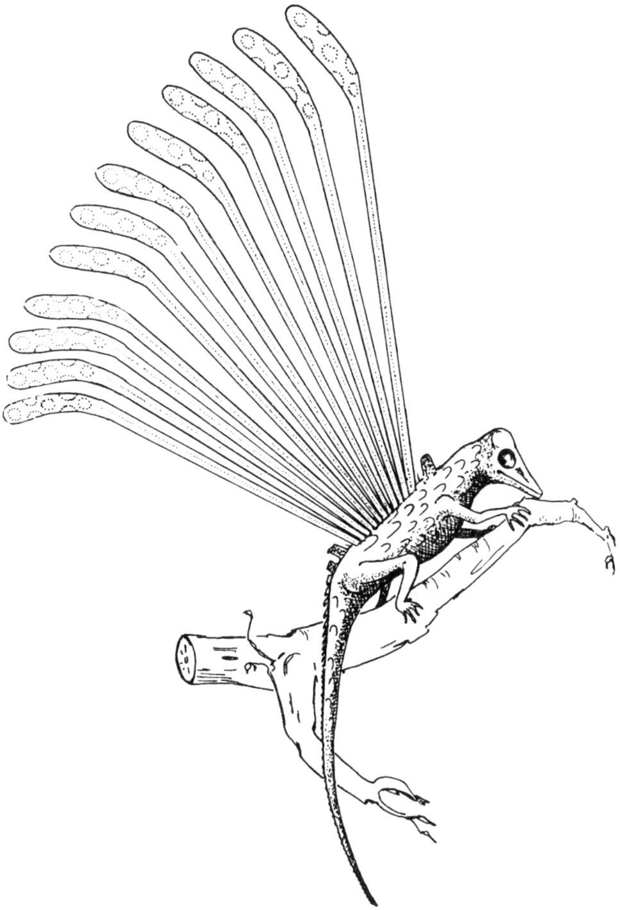

Figure 20.4 Longis (*Longisquama,* Thecodontia, Diapsia, Reptilia). (After Sharov).

This perpendicular "sail" and keeled scales were perhaps the first indication of feather-like structures in the fossil record, and may have been adaptations for water transport, and DNA preadaptations for the feathers of aquatic birds. It is possible that sea birds may have preceded land birds of the Jurassic. Unusual animals like Longis may be keys to DNA concepts that otherwise, might be neglected.

Pelycosaurs like Dime (Chapter 18, Figure 18.5) which had preceded the mammal-like reptiles of the Permian were present in the Triassic in smaller numbers than the thecodonts. Thecodonts continued to appear, along with dinosaurs.

Kuneo (*Kuehneosaurus*, Figure 20.5) about two feet long, has been considered to be an early lizard, a descendent of eosuchian ancestors that had primitive two-arch skulls. Kuneo had specialized ribs projected from the vertebrae to form a gliding "sail" similar to the modern lizard, *Draco*. How interesting it will be to learn the DNA relationships of lizard rib sails and the sails and wings of other lifeforms! The DNA concepts for sailing, gliding, and flying must have some logical connections, waiting to be discovered! The pelvic area of Kuneo seemed to be similar to that of Homoee (Chapter 21, Figure 21.8 a). What was their DNA relationship?

The parallelism of species discussed by Edwin Colbert (1965: 91) was exemplified in the early sea turtle progenitor *Henodus* , a placodont (Figure 20.6) about four feet long, and Triasso, an early dry land turtle (*Triassochelys*, Figure 20.7). Such similar concepts in totally different environments, each arising from unknown (but probably different) sources, triggers the question: How is DNA patterning communicated to provide for such parallel developments? There is still mystery remaining at the core of DNA functioning.

Mystrio (*Mystriosuchus*, Figure 20.8) over 11 feet long, presents another mystery. Is such an animal to be considered near the base of crocodilian life, or is it a parallel development? Is *Eryops* (Chapter 19, Figure 19.1) which lived in the Permian, long before Mystrio, a crocodile ancestor, though there is a long gap in the fossil record? Surely, if there is validity to

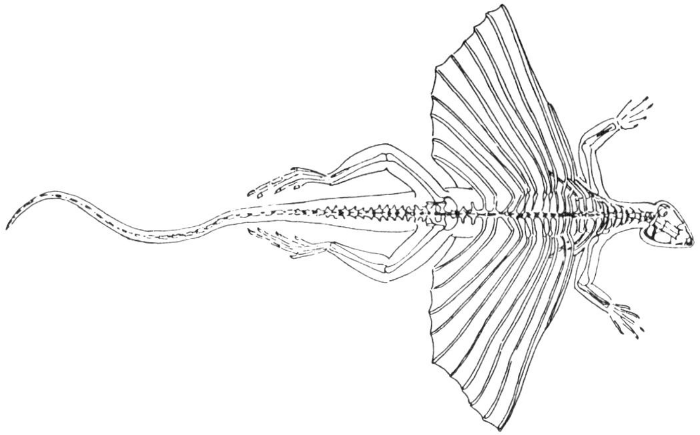

Figure 20.5 Kuneo (*Kuehneosaurus*, Lacertilia, Squamata, Diapsida). (After Robinson).

Figure 20.6 *Henodus*, aquatic placodont. (After Huene).

Figure 20.7 Triasso (*Triassochelys*, Chelonia, Anapsida, dry land turtle). (After Jackel).

consideration of homologies, the similarities of *Eryops* and early crocodiles cannot be neglected. Are the pelvic bones of modern crocodiles more like *Eryops* or like Mystrio? Are they a valid clue to an *Eryops* — crocodile relationship?

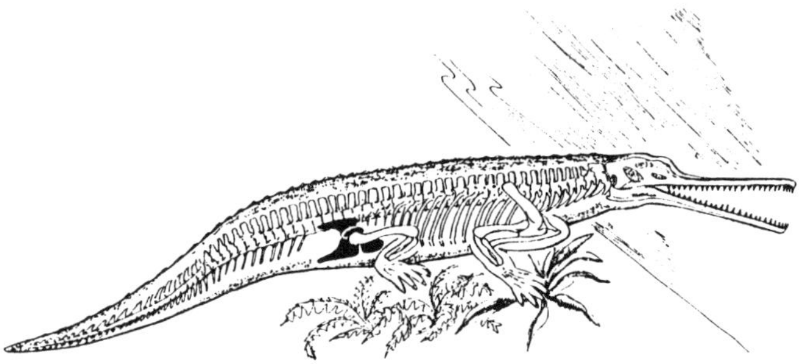

Figure 20.8 Mystrio (*Mystriosuchus*, Phytosauria, Thecodontia). (After McGregor, 1906).

The DNA of such creatures as *Eryops* and Mystrio may have been combined in salmutations to create the radiating variety of crocodiles found from the Triassic to the present. Progress in science and philosophy is just as dependent upon discovery of the mysteries remaining, as on explainations of the fossil record.

Tany (*Tanystropheus*, Figure 20.9) about two and a half feet in length, was almost as long in neck as in the rest of its body. The neck bones were elongate and dissimilar to those of most other reptiles, while its skull was similar to other euryapsid reptiles. It is interesting that similar skulls often are mounted on very dissimilar necks, a noteworthy characteristic of euryapsids.

Did the long neck concept arise from the DNA of a snake-like ancestor like Ophi? (Chapter 18, Figure 18.6.) Can the DNA concept of the long neck of aquatic-living Tany be a preadaptation for the long necks of the Jurassic herbivorous dinosaurs and the sea going plesiosaurs? Fossils of Tany raise the question: Why do animals with long necks have such a tremendous variety of different neck bones? It appears that the concept of a long neck is not directly associated with a single form of neck bone structure. DNA seems to move and-or be moved in interchangeable combinations of patterns among diverse animals.

Figure 20.9 Tany (*Tanystropheus*, Euryapsida, Reptilia). (After Peyer).

Mammal Development

Fossil teeth, thought by some scientists to be those of the earliest mammals, appeared in Triassic deposits of 200 million years ago. Although paleontologists in the 1940s to 1960s identified mammal-like reptiles of the Permian and Triassic, there has been confusion in determining times of transition or identifying the progression of animals from reptiles to mammals.

Colbert (March, 1949: 42) wrote that the reptile became a mammal when it had a constant body temperature, had hair, and when reproduction had reached the stage in which the female lactated and suckled its offspring. "Unfortunately we can obtain no direct evidence on these changes, for temperature controls, hair, mammary glands and other soft parts of the anatomy are not preserved as fossils."

Also, of course, there is no information about the ability of similar ancient animals to interbreed. Thus there is no foundation for distinguishing separate species, based on the modern definition of species. Lacking the information to determine temperature control, hair, lactation, and the ability for mammal-like reptiles and early mammals to interbreed, paleontologists have ignored the problem, and based classification upon tooth and bone structure. The case of reptile or mammal and their relationship hinges on the idea that the two bones of the reptile jaw became a middle ear bone, and the three bones of the mammal inner ear appeared to be similar in shape to those reptile bones.

The identification of *Ictidosaur,* considered a reptile, as a postulated transitional animal to mammals, must be based, according to Colbert (March, 1949: 43):

> ...on this difference between a single ear ossicle and three ear ossicles. Though seemingly insignificant, the difference is so constant that it serves admirably as a reference point at which to draw the very fine line between reptiles and mammals.

P. D. Gingerich (1977: 470) suggested that the first known mammals were prototherians such as Eo (*Eozostrodon,* Figure 20.10) whose fossil teeth have been found on three continents. A late Triassic animal, known as *Kuehneotherium,* with significantly advanced cheek teeth, and *Megazostrodon,* a tiny shrew-like triconodont, were also named as early mammals.

The problem, again, in attempting to determine a transition from one taxon line to another is a logic problem. Either/or logic requires a choice between two alternatives. Either an animal is a reptile or it is not a reptile. Either an animal is a mammal or it is not a mammal. An animal may be a reptile with mammal-like characteristics or it may be a mammal with reptile-like characteristics. However, as soon as an animal has been placed in a category, certain characteristics do not fit. No matter how these animals are classified, paradox remains because the modern test of the basic category *species* (viable interbreeding possibility), cannot be applied to the fossil record.

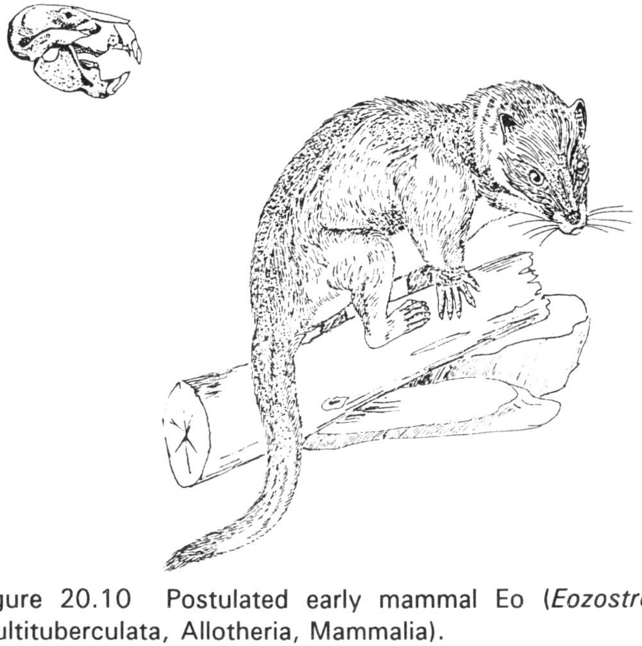

Figure 20.10 Postulated early mammal Eo (*Eozostrodon*, Multituberculata, Allotheria, Mammalia).

In the future, DNA configurations for numerous mammal characteristics, including especially skulls and pelvises, may supplement consideration of ear ossicles and cheek teeth as criteria for classifying both reptiles and mammals.

Some scientists, who have ventured the opinion that mammals evolved from the mammal-like reptiles of the Permian and Triassic, believe that mammals arose polyphyletically from a number of sources. However they have not proclaimed that this belief requires a new theory of evolution and new illustrations of mammalian development. I believe it is necessary to consider that the mammal-like animals of the Permian had some DNA configurations which are also found in animals we call mammals.

This must change the belief that mammals arose by means of random mutations selected by competitive processes. There was a majesty of infinite intelligence in processes of development of new lifeforms by means of recombinations of DNA. The animals we call mammals began their rise to dominance from the DNA patterns of a number of Permian and Triassic progenitors.

Triassic Extinctions

As the two major continents split into several new continents and seas began to swamp land areas, the extinctions were almost as great as those at the end of the Permian. Despite the richness of the radiations in the Triassic, with most modern types of reptiles appearing, along with early dinosaurs and mammals, the period ended in massive extinctions on land and sea.

Twenty-four of the 25 families of ammonites disappeared. The bryozoans and many kinds of fish were gone. There was near extinction of sea lilies, brachiopods and many corals. Coal swamp amphibians vanished. The dinosaurs' ancestral thecodonts and some of the mammal-like reptiles became extinct. By the end of the Triassic, there were about 17 kinds of mammal-like reptiles and mammals, and 83 kinds of other reptiles remaining.

21

Mesozoic Era — Jurassic Period Giant Dinosaurs

The Jurassic began about 180 million years before present (180 MBP) — continents continuing the process of massive splitting, creating new ocean areas. The year comprised 380 days, as the Earth's rotation continued to slow. The North Atlantic ocean formed as Europe split from North America. Gondwana began to break into three continents: Antarctica, Africa and South America (Chapter 12, Plate 12.2 # 6).

Most of the fossil discoveries of this period are in marine deposits. Continental strata increased near the close of the period. Typical European fossils are almost all marine. Areas of mid-America, England, Germany, Normandy, southern India and Tanganyika provide about the only fossils of land animals.

In the Jurassic, a resurgence began of cartilaginous fish, including sharks, skates and rays. Bony fish were almost absent from fresh water, but swarmed in the seas. The earliest sturgeon appeared. There were swarms of primitive ray-finned fish, seemingly ancestral to modern ray-finned types.

Pros (*Proscinetes*, Figure 21.1) five inches in length, was a common Jurassic fish. Pros had a very complex and unusual bone structure. Its upper rib bones seemed to be patterned like the lower rib bones of earlier amphibian vertebrates like Amphy (Chapter 17, Figure 17.5). It is out of Darwinian order to contemplate a Jurassic fish bone structure evolving from DNA similar to that of a Devonian amphibian, but a DNA theorist may be permitted such a concept. DNA patterns will not be found if they are not sought.

Figure 21.1 Pros (*Proscinetes*, Pycnodontiformes, Osteichthyes, bony fish). (After Smith-Woodward).

Amphibians were almost absent from the fossil record of the Jurassic. However, there was a great radiation of large and small marine reptiles. Were the marine reptiles developed from combinations of fish and amphibian DNA?

The Jurassic seas swarmed with the aquatic equivalent of dinosaurs, whale-size short-necked plesiosaurs, and long-necked plesiosaurs like Moren (*Morenosaurus*, Figure 21.2). Fossil bones of Moren, 21 feet long, are on display in the Museum of Natural History in Los Angeles, California, USA, but originally it lived in the Proto-Atlantic Ocean. Moren's bony undercarriage like that of a turtle, evoked assumption that it evolved from a turtle-like ancestor. However, Moren has other very unturtle-like characteristics. That long neck is surely not from turtle DNA. I postulate that such a neck may have originated from DNA inherent in a creature like Tany (Chapter 20, Figure 20.9) although the neck bones are entirely different. Salmutation theory is needed to contemplate Moren's interesting combination of characteristics.

Figure 21.2 Moren (Morenosaurus, Plesiosauria, Euryapsida). (After Natural History Museum of Los Angeles County).

Spectacular Radiation of Dinosaurs

There are no clear lines of evolution of the two main classifications of dinosaurs, based upon similar characteristics. The evolution of both saurischians and ornithiscians, of large and small dinosaurs, of herbivores and carnivores, of bipeds and quadrupeds — apparently from similar stem reptiles — challenges classical evolution theory.

The Morrison formations of Utah, Colorado, Wyoming and Montana, USA, reveal that dinosaurs radiated in spectacular fashion in the Jurassic, some of small size and some as huge as

Allosaurus. Dinosaurs found in the United States and in many other parts of the world are classified in two orders — saurischians and ornithiscians — based on pelvic structure. They do not fit into Darwinian patterns of evolution.

Although conventional wisdom might suggest that since early saurischians were carnivorous, that line of dinosaurs would continue to be carnivorous, this was clearly not the case. There was no clear-cut, continuing, sequential pattern of relationships among pelvic structure, bipedal or quadrupedal gait, and dentition. DNA of dinosaurs, must have been a profusion of varying interesting patterns.

The longest quadrupedal dinosaur known prior to the 1980s, Diplo (*Diplodocus*, Figure 21.3) was an herbivorous saurischian, 87 feet long and more than 14 feet high at the hips. It is difficult to imagine that Diplo could have evolved from a saurischian like Celo (Chapter 20, Figure 20.3). Scientists may say that Diplo *reverted* from a carnivorous biped to a herbivorous quadruped, but such a statement gives no real meaning to the concept of evolution from a six-foot long, rather bird-like dinosaur to a huge 87 foot long, lumbering monster.

Figure 21.3 Diplo (*Diplodocus*, Saurischia). (After Hatcher).

Darwinians have continued to blame the deficiencies of the fossil record as cause of failed hypotheses and confused classifications, but science must make sense of the millions of fossils which have been found. Diplo needs more than Triassic saurischians to explain its existence.

Not all herbivores were quadrupeds. Many bipeds, such as Campto (*Camptosaurus*, Figure 21.4) varying in length from six to 20 feet, were herbivorous. Campto combined the bird-hip characteristic (thought to originate in a small carnivore) with herbivorous dentition and way of life. Duck billed herbivores of the Cretaceous, with ornithiscian pelvises (Chapter 22, Figure 22.2) seemed equally competent on two legs or four, much to the confusion of Darwinian thought.

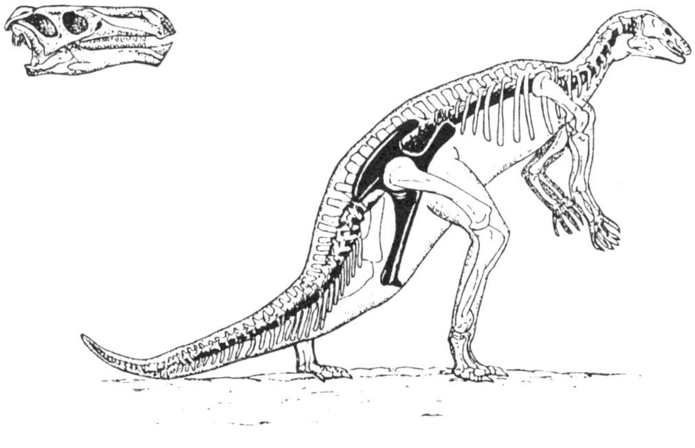

Figure 21.4 Campto (*Camptosaurus*, Ornithiscia). (After Gilmore, 1909).

Campto's skull had many characteristics in common with Diplo and Stego (*Stegosaurus*, another herbivorous ornithiscian quadruped, Figure 21.5) which grew to 30 feet in length. Stego is noted for its double row of dorsal plates and the sharp bones on the end of its tail. The dorsal plates carried blood within them, thus were a form of radiator for the animal, an innovation with unknown precedents. Possibly, the DNA for these plates is related to the DNA for the sail of Dime.

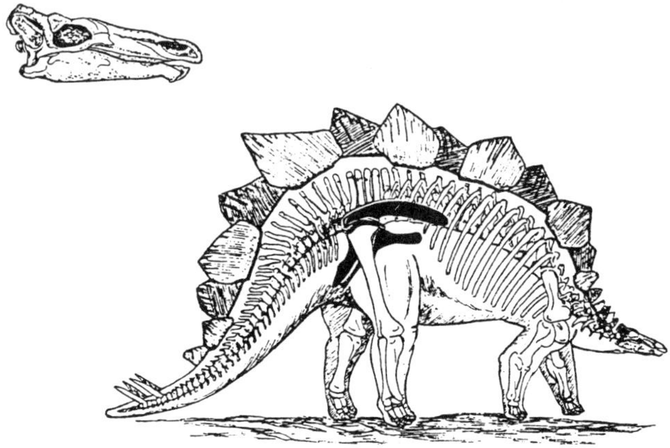

Figure 21.5 Stego (*Stegosaurus*, Ornithiscia). (After Gilmore, 1914).

The hands and feet of dinosaurs did not fit into sequential patterns of increasing or decreasing numbers of digits. Rather the number of digits varied among the dinosaurs, in some cases even differing in front and rear limbs. Celo had large clawed hands used for grasping, with three long middle digits and two short outside ones. The feet had three toes, with a forth small inner toe above the others. Later coelurosaurs had fewer toes and fingers.

Diplo had five-toed, elephant-like feet. Campto had five fingers, the first three were large with hoof-like nails and the next two were tiny. The rear feet had four toes with tiny hooves. Stego had hoof-like nails on its five-toed front feet and four-toed hind feet. The 27 genera of duckbilled dinosaurs had a great variety of hands and feet. One, *Saurolophus* had four front webbed fingers of which two also had hooves. Its rear feet had three hoofed toes. Proto (*Protoceratops*, Chapter 22, Figure 22.3) had four toes on each of his front and rear feet, while Mono (*Monoclinus*, Chapter 22, Figure 22.4) also a ceratopsian had five digits on the front limbs and four on the rear limbs. Digit numbers seem to vary without direct correlation with digit similarities, and without expected correlations to skull and-or pelvic structures. DNA moved and-or was moved among dinosaurs in marvelous combinations of patterns.

One fascinating feature common to Diplo and *Allosaurus* is the similar double openings on the rear dorsal surface of their skulls, which have some similarities to a horse's skull. Could this be a location for ears? Or what is connected here? Scientists should ask such questions. The fixation on dinosaurs as reptiles without projecting ears, should not restrict the questions which are posed.

Recently, scientists have postulated that the large size of many of the dinosaurs gave them equivalent adaptation to warm-blooded animals, and thus they were better able to endure changes in temperature than previously thought. Or perhaps, large dinosaurs were warm-blooded animals. Were the small ones warm blooded also? If small land birds such as Archy (Figure 21.7) salmutated from dinosaur DNA, perhaps the higher temperature of birds, necessary for flight, evolved from an earlier salmutation to warm bloodedness.

DNA code combinations for a variety of characteristics must eventually be considered a basis for classification of dinosaurs, as of other lifeforms. This can be approached only after DNA research on modern animals has progressed to the point that the "rules of the game" of DNA recombinations are well known. It is important to to identify possible DNA characteristics which show repeat appearances in different animals.

Preadaptive DNA?

Marine turtles were in evidence in the Jurassic as were early frogs. Lizards were rare. Primitive crocodiles were abundant. Flying reptiles like Ramphor (*Rhamphorhynchus*, Figure 21.6) with leathery wings similar to those of Eocene bats (Chapter 24, Figure 24.7) appeared in long tail and short tail forms and apparently flew with little wing flapping, gliding well, over long distances.

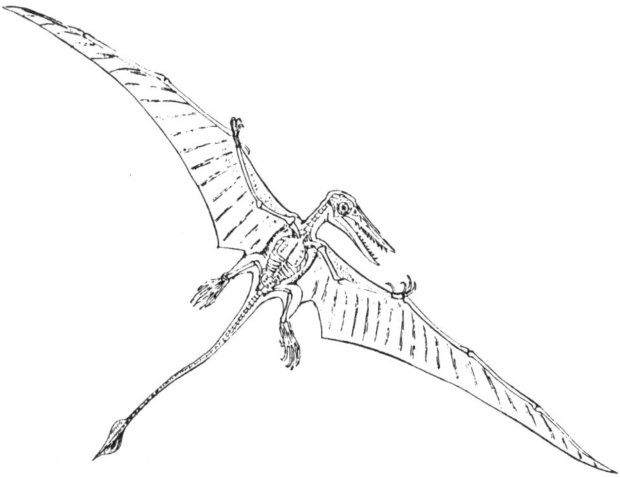

Figure 21.6 Ramphor (*Rhamphorhynchus*, Pterosauria). (After Seeley).

Five specimens of the first known bird, Archy (*Archaeopteryx*, Figure 21.7) were found in Germany. Archy was different from its progenitors of scale-covered dinosaurs. It had true feathers. It probably could not fly well due to its weight and lack of bone structure which could support strong muscles. John Ostrom, of Yale University, USA, believed that Archy lived on the ground and made flying leaps after insect prey. For 50 million years after Archy, no land birds appeared in the fossil record. Nature seemed to pause to considered how to combine the gliding advances of pterasaurs and the feathered wings of Archy to create modern birds.

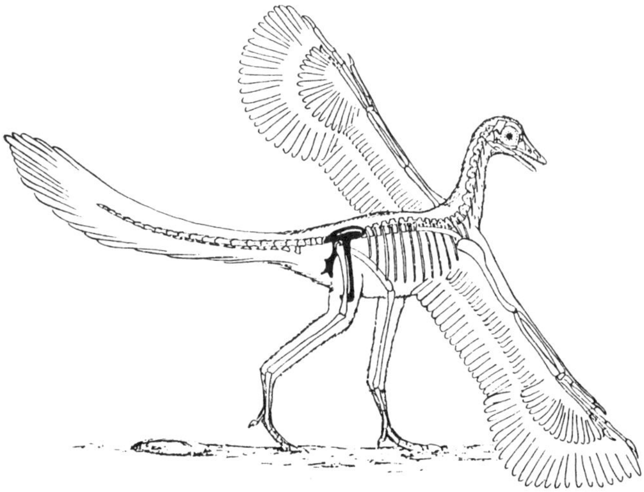

Figure 21.7 Archy (*Archaeopteryx*, Aves). (After Heilman).

Homoee Speculation

A very interesting animal of the Jurassic, Homoee (*Homoesaurus,* Figure 21.8 a) about seven inches long, was named in a rather whimsical fashion. The Greek word "homo" is defined as "same" as in the word, homology. Saurus means lizard, so *Homoesaurus* means "like a lizard." However, the word "homo" is identified very closely with the designation for human, *Homo sapiens,* and therefore, perhaps inadvertently, suggested to me a relationship of this animal to a human.

As I turned the drawing of Homoee to a vertical position, it suggested the imaginary concept of Homoeewhatsit? (Figure 21.8 b.) (This flight of imagination, not conventional for scientists, is certainly allowed a philosopher.) What is this creature? Where is its place in nature?

With a little adjustment of position of arms and legs, and removal of Homoee's tail, Homoeewhatsit? suggested to me, a concept of a human preadaptation. In a lizard-like reptile? How could that be? Humans are thought to have evolved from apes, not from lizards. But if DNA does not always pattern in sequential lines, could there be preadaptive DNA here? DNA is made up of chemicals, after all. Why not consider the possibility that preadaptive concepts appeared in the fossil record in many forms?

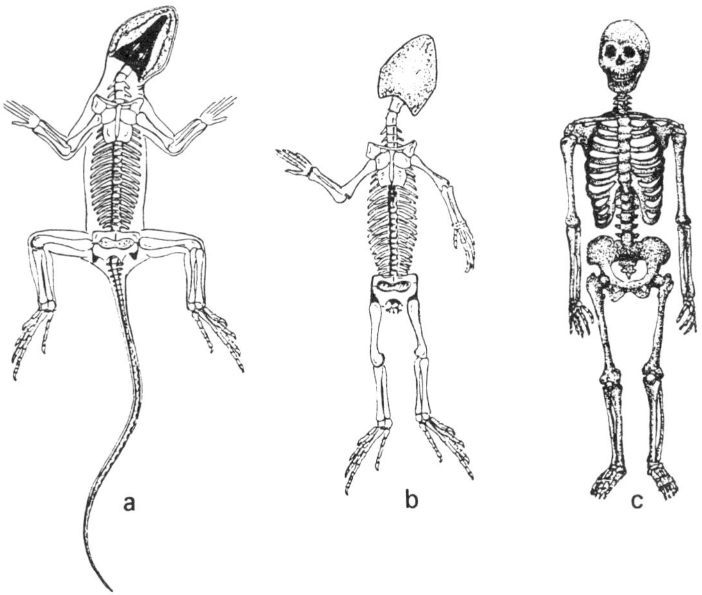

Figure 21.8 (a) Homoee (*Homoesaurus*, Rhynchocephalia, Diapsida, ventral view). (b) Homoewhatsit? imaginary animal. (c) Woman, of 9,000 years ago). (a. After Williston; c. After skeleton, George C. Page Museum).

For comparison to Homoee, the human skeleton in Figure 21.8 c is a woman of 9,000 years ago, from the La Brea Tar Pits, Los Angeles, California, USA.

There is something else about Homoee that warrants consideration. This ryncocephalian reptile has a bone growth pattern which is unlike other reptiles. Most reptile bones grow from their ends, thus reptiles may continue to grow during their entire lifetimes, some reaching huge size. On the other hand, mammal bones grow from a point within the bone, not from the end points. This point of growth is called the epiphysis, a part of bone separated from the main shaft and from the ends, by a layer of cartilage which unites with the bone through ossification, reaching completion at maturity. Thus mammals grow to adulthood, then stop growing.

Homoee bones grew like mammal bones, not like reptile bones. Homoee was restricted in size like mammals. How did Homoee acquire a DNA bone-producing pattern like that of mammals? Homoee lends support to the idea that certain elements of DNA leap and jump, (actually, certain DNA "transposable elements" provide a regulator — operator system of mutation and control of gene action). The original discovery of "jumping genes" is credited to Barbara McClintock (1981) for her studies of maize, beginning in 1931 and continuing through the 1980s.

I am not alone in visualizing human preadaptations in early vertebrates. Erik Jarvik (1980: V 2, vi) whose life-long study of the Eusty-Amphi, fish-amphibian transition brought many new insights, found similarities of the pectoral and pelvic fins of Eusty (Chapter 17, Figure 17.4) and the embryonic human hand and foot. He also found similarities in their ear ossicles. He considered that the majority of tetrapods, including man evolved from the Osteolopiformes, of which Eusty was representative. To explain his findings, he presented "a new theory of the origin of mammalian ear ossicles based on a comparison of *Eusthenopteron* and man..."

As scientists continue to find skeletal similarities which cannot be explained by positing minor mutations and sequential births, new theories arise. If DNA codes — and-or is coded — for all characteristics of organisms, then it would seem that observed preadaptations must be coded. Coded preadaptations would only be meaningful if there were some expectation of the kind of conditions under which a new taxon were to occur. For instance, a preadaptation for air breathing and land living presupposes information that there will be a suitable air and land environment. This could occur only if information about the future reached backwards into the past. This idea requires the concept of reciprocal time. In the case of Homoee, the bone growth pattern strongly suggests a preadaptation toward mammalian, and ultimately human skeletons — lifeforms destined to live in the far distant future.

As we move into the 21st Century, perhaps we will become more open-minded in considering the possible philosophical implications of scientific work. Perhaps we will fulfill Einstein's hope that we can change our mode of thinking.

As the Jurassic closed, mammalian life was about to burgeon into significance. Were mammals preadapted to radiate after the demise of the dinosaurs?

22

Mesozoic Era — Cretaceous Period Burgeoning of New Life

About 130 million years before present (130 MBP) the Cretaceous began and lasted for about 65 million years. By the end of the period, the Earth was configured as six continents, with Australia remaining joined to Antarctica — and India, an island. Africa, including Arabia, also remained an island evolving its own characteristic fauna, different from the other continents, until it was joined to Eurasia by tectonic movements in the Miocene (Chapter 12, Figure 12.2, # 7).

The Earth in the Cretaceous, was beginning to show modern continental divisions with low-lying lands and widespread seas. As Gondwana continued to split and the Atlantic Ocean opened completely, the continental shelves experienced great flooding. The ocean currents altered and the climate became relatively stable, and (on average) very warm — perhaps the warmest it had ever been.

During the middle of the Cretaceous, oceans covered about one-third of the continents; the midsection of the United

States was an inland sea. This flooding was accompanied by the buildup of the mid-oceanic ridges and accelerated sea-floor spreading. Niche and area diversity increased. It was a time of optimum conditions for major radiations of many groups in the sea: ammonites, reef corals and many varieties of plankton. Primitive fish were replaced by more advanced teleosts. Snake-like amphibians appeared. Marine lizards and turtles arose but not one *Ichthyosaur* remained in the rich North American fossil beds. Marine crocodiles disappeared while fresh water crocodiles experienced progressive changes, some growing to gigantic size. Phobo (*Phobosuchus*, Figure 22.1 a) is known by its huge fossil skull, indicating the crocodile was at least 50 feet in length, a monster capable of eating dinosaurs! Note the comparison to the size of a modern crocodile skull (Figure 22.1 b).

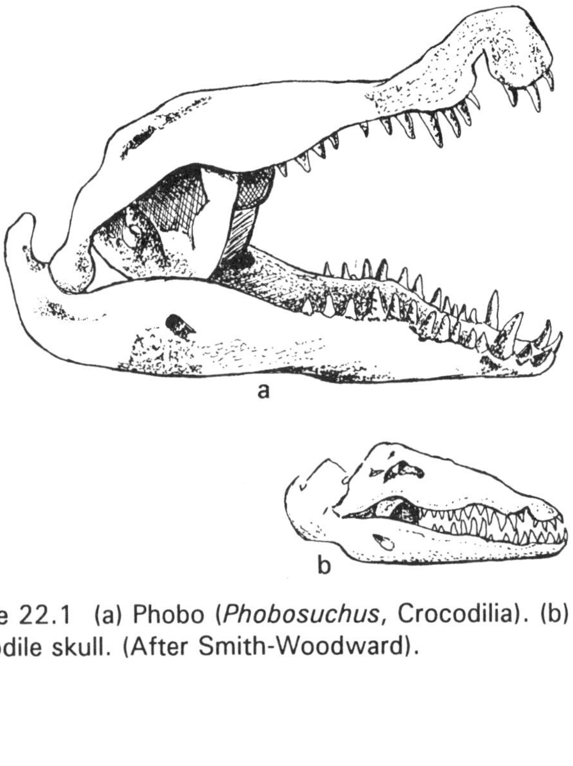

Figure 22.1 (a) Phobo (*Phobosuchus*, Crocodilia). (b) Modern crocodile skull. (After Smith-Woodward).

Flying reptiles, pterosaurs, reached their largest size. The giant *Pteranodon* had a wing span of over 50 feet and, it is estimated, could soar up to 30 miles per hour.

First Flowering Plants

The first flowering plants, the Angiosperms, pollinated by winged insects, arose in spectacular array and radiated on earth. Such coevolution is inexplicable in Darwinian terms. The evolutionary interaction of flowers and pollinators, of predators and prey, is too much to ask of a random process. Such evident coordination requires intelligence — the intelligence of and-or within organisms and the infinite intelligence exemplified in environmentally responsive, progressive and complex salmutations.

Evidence of DNA Recombination

in Dinosaurs, other Reptiles and Birds

The ability of DNA to recombine and-or to be recombined requires that all possible systems of recombination be explored, including movement of DNA patterns not passed directly by an animal to its offspring. To consider the relatedness of characteristics which do not fit the Darwinian pattern is a philosophical necessity in a world in which scientists are beginning DNA recombination experiments on a broad scale.

Dinosaurs radiated into armored ankylosaurs and duckbills (*Anatosaurus*, Figure 22.2 a). The duckbilled dinosaurs arose in the Cretaceous in at least 27 genera, mainly noted for a variety of very different head crests. These dinosaurs were the first known creatures to have duckbills. Was the DNA for this shape of bill carried into the DNA for the platypus bill configuration and that of ducks? (Figure 22.2.) Is shape of features significant apart from chemical composition of those features? The leathery composition of the platypus bill might be more comparable to a beaver's tail than to the bony dinosaur bill. Is the leathery composition of beaver bill and tail a similar DNA relationship?

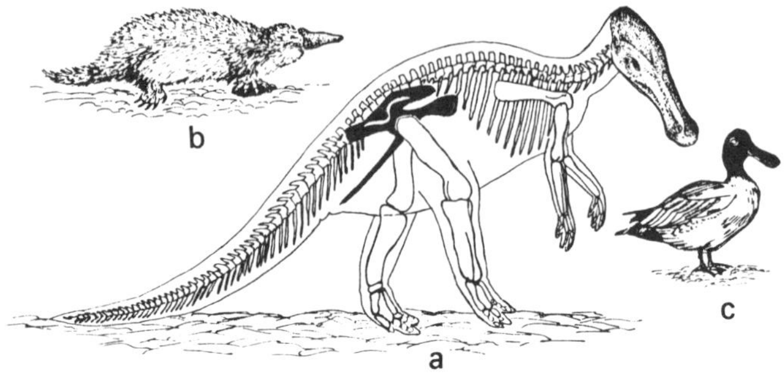

Figure 22.2 (a) Duckbill dinosaur (*Anatosaurus*, Ornithiscia). (b) Living *Platypus* (Monotremata, Mammalia). (c) Living duck (Anseriformes, Avis).

Scientists steeped in Darwinian theory are prevented from considering the relationship between the duckbill of a dinosaur, a platypus and a duck, because there is no obvious successional relationship, no close connection in time nor in total body configuration of these animals. To consider the relatedness of duck-like bills in such diverse animals is scientific heresy. However, if the universe is one, and everything integrates within that wholeness, it only remains for humans to learn how similar patterns are related. The complementary nature of DNA and its functioning allows the concept that similar patterns in lifeforms are created by similar patterns of DNA.

Many kinds of horned dinosaur fossils have been found in western North America; Alberta, Canada; Mongolia and China. Proto (*Protoceratops*, from Mongolia, Figure 22.3 a) was the first dinosaur found with a nest of eggs, thus verifying dinosaur mode of reproduction. Proto was thought to be ancestral to many genera of later ceratopsians including Mono (*Monoclinus*, Figure 22.4) of rhino proportions. Mono was one of many ceratopsians which lived only in the Cretaceous Period, becoming extinct with all the other dinosaurs at the end of this period.

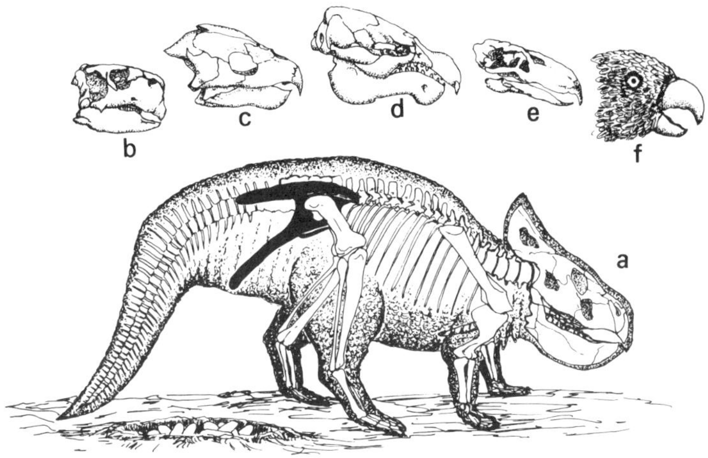

Figure 22.3 (a) Proto (*Protoceratops*, Cretaceous Ceratopsia, Ornithiscia). (b) Scaphon (*Scaphonyrx*, Triassic Rynchosauria). (c) *Archelon* (Cretaceous Chelonia, turtle). (d) *Protosiren* (Eocene Sirenia). (e) *Phororhacos* (Miocene Aves). (f) Parrot (Aves, living). (a. After Brown, Barnum, and Schlaikjer; b. After Huene; c. After Hay; d. After Sickenberg.

Proto and Mono had prominent beaks, but they were not the first animals, nor the last, to be so endowed (Figure 22.3). The Triassic rynchosaur, *Scaphonyrx* also had a beaked skull. The Cretaceous turtle, *Archelon; Protosiren* of the Eocene; *Phororhacos*, a Miocene bird; and many living birds, including parrots, have beaks. Are they all recipients of a similar DNA configuration?

Tylo (*Tylosaurus*, Figure 22.5) from the Cretaceous inland sea deposits of Kansas was 26 feet long, a giant sea serpent which has been thought to be ancestral to living land reptiles such as 10 feet long Dragon Lizards of Komodo Island, Indonesia (*Varanus*, Figure 22.6).

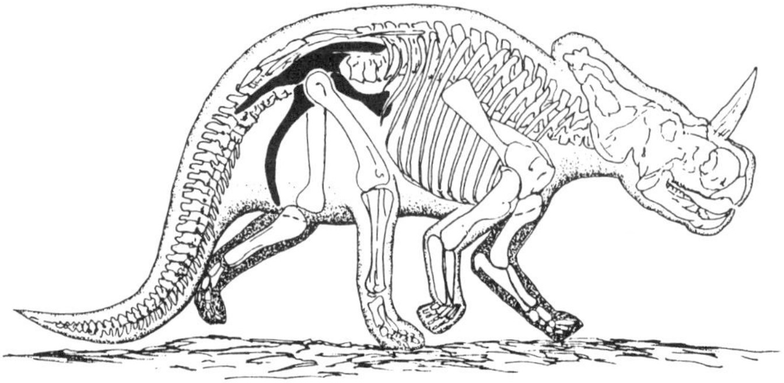

Figure 22.4 Mono (*Monoclinus*, Ceratopsia Ornithiscia). (After Brown, Barnum, and Schlaikjer).

It is very difficult to envision evolution from Tylo, a sea creature, into land living Komodo Dragons, requiring totally different environmental adaptations. It is easier to envision that DNA similar to that of Tylo or Ichthy (Chapter 18, Figure 18.3) might salmutate into modern sea mammals, and that small lizard DNA might salmutate into Komodo Dragons.

Tylo (Figure 22.5) has also been thought to be ancestral to the sea bird Hesper (*Hesperornis*, Figure 22.7) because of homologous jaw bone structure. It is easier to envision DNA salmutations from small bird-like dinosaurs to explain the existence of Hesper, than it is to envision a sequential evolutionary line from a huge sea reptile to a small sea bird.

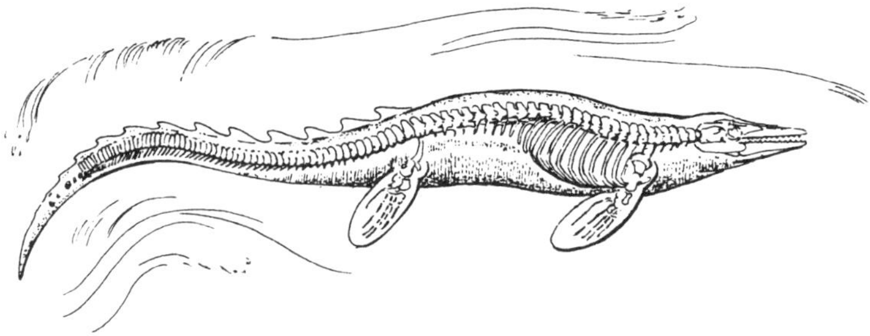

Figure 22.5 Tylo (*Tylosaurus*, Mososauria, Reptilia). (After Osborn, 1899).

Figure 22.6 Komodo dragon, monitor lizard (*Varanus,* Reptilia, living)

How do scientists decide what represents a homologous (legitimate) similarity, analogous (illegitimate) similarity, and superficial resemblance (a similarity to be ignored or called parallel evolution, outside the range of sequential theory)? I believe *nothing* can be ignored. Everything known about the subject must fit into an adequate theory.

The Null Hypothesis

Darwinians have been slow to acknowledge the significance of Hesper (*Hesperornis,* Figure 22.7) a sea bird with feathers but no wings, which apparently lived in the period between Archy (*Archaeopteryx,* Chapter 21, Figure 21.7) and modern birds. Hesper appears to be a fish eater, well adjusted to a life in the sea.

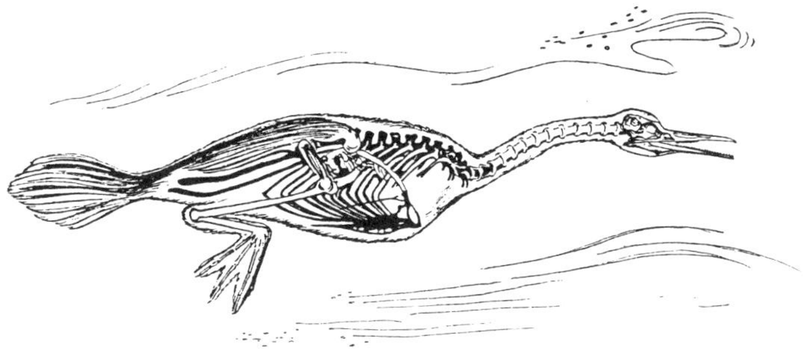

Figure 22.7 Hesper (*Hesperornis,* Neornithes, Aves). (After Marsh).

A bird without wings? The threat of the null hypothesis has been realized! The basis for disproving the major premise in Chapter Two, that "All birds have wings" (in the example of deductive logic) is upon us. Why question that "All birds have wings" before knowledge of Hesper? Deductive logicians must indeed face the fact of the null hypothesis in this case!

Was Hesper really a bird? It does appear so. It looked like a bird. It had feathers like a bird, head and feet like a bird, but it apparently lacked the DNA (genes) for developmment of wings. If it is considered a sea bird and a progenitor of later sea birds, it would be strong support for the salmutation theory because it challenges land-based Archy as the prototype for sea birds. It suggests that a wingless bird could give rise to sea birds with wings (with a little recombination of DNA) — polyphyletic bird origins.

The concept of polyphyletic origins, appearing again and again in modern biological literature, requires new theory, the development of salmutation theory. Hesper also is evidence to support the use of both old and new logics to discover new truths. It is necessary to remember that an objective in the use of deductive logic is to uncover the facts which do *not* fit within an all-inclusive category, as well as the facts which do.

Until recently it was thought that 50 million years intervened between fossil evidence of Archy and the rise of modern birds. It is now thought that birds such as ducks and geese, a few with teeth, may have appeared earlier than other modern birds, some as early as the Cretaceous. How is the DNA of ducks and geese related to the DNA of Hesper?

Enigma of Marsupials and Placentals

Some scientists have posited that an early interchange of animals along a North and South American island arc occurred in the Cretaceous with movements of aniliid snakes, teiid lizards, hadrosaurian dinosaurs and early marsupials. These scientists believe that marsupials, primitive opossums, arose in North America and migrated to South America, thence to Antarctica and on to Australia along a southern land route.

Since Africa had already separated from the other Gondwana continents, no marsupials entered that continent.

After the late Cretaceous or early Paleocene, tectonic movements separated the South American continent further from North America, and Australia separated from Antarctica becoming an island. These two southern continents then evolved separate unique fauna of marsupial animals.

South America, having a few herbivorous placental taxa at that time, developed a mixed fauna of placentals and marsupials. New Zealand, having separated from Australia before marsupials became established there, remained without any mammals until their later introduction by humans.

Darwin thought that placentals arose from marsupials. Other scientists have thought that both evolved from a common therapsid ancestor. In one hemisphere the young marsupials were born prepared to crawl into a pouch in an almost helpless stage of development, and in the other hemisphere the young remained in the placenta until more fully developed. The first known placentals were discovered in the Northern Hemisphere; shrews and lemurs in North America, moles and hedgehogs in Europe.

Could these animals develop from recombination of DNA of then existing lifeforms? Why did both placentals and marsupials evolve similar higher taxa, with similar solutions to environmental problems in their separate environments? Both marsupials and placentals later appeared as mice-like, rat-like, dog-like and wolf-like animals, saber tooth cats, and a host of other similar types. Marsupial characteristics appeared in a range of kangaroos of many sizes, some quite gigantic, and in a variety of other animals found only on the Australian continent.

Such parallel developments cannot be envisioned without acknowledging the interconnectedness of life, and positing forms of intelligence and-or communication beyond our present understanding. There must be fascinating similarities and differences to be discovered in the DNA of animals that arose in the separate hemispheres.

Early Primates

An early primate fossil, lemur-like *Purgatorius*, with teeth similar to primitive ungulates was found in tropical North America. Was this a beginning of primates leading to man? Were primates contemporaries of other early mammals? Did insectivores arise earlier in Europe and primates arise in the North American Cretaceous? Traditional theory posited that insectivores gave rise to primates, but the fossil record is not clear.

Close of Cretaceous Period

As the Cretaceous drew to a close and the shallow seas withdrew from the continents, another massive extinction occurred. About one fourth of all known animals disappeared. Sixteen superfamilies and orders vanished, including all the existing dinosaurs, the huge marine reptiles, flying reptiles, all the remaining ammonites, some bottom dwelling mollusks, cephalapods and many kinds of plankton. No terrestrial vertebrate weighing more than 55 pounds survived. Among the survivors at the end of this Mesozoic Era were many varieties of plants, many nautiloids, fish, turtles, crocodiles and primitive mammals.

Cretaceous Tectonic Movements

The Cretaceous is the earliest period with excellent geologic evidence of the effects of tectonic plate movements. Much prior evidence was destroyed in continuous plate displacements. As the marine plates rode under the continental plates (subduction) heat was generated, rocks melted and volcanoes were created inland from plate subduction areas. Although it is not clear exactly when and where various plates have converged throughout the Earth's history, the volcanic activity in the Cretaceous provides evidence of devastating effects on land and in the sea. (It may be well for Californians (USA) to ponder that the San Andreas Fault rests upon the convergence of two tectonic plates which are currently slipping along their juncture at an average rate of about two inches per year.)

During the last ten million years of the Cretaceous Period, explosive volcanism was evidenced in at least 200 episodes in India, Japan, China, Alberta Canada, Mexico and Wyoming, Colorado, Utah, Arizona and New Mexico, USA. These volcanoes produced such heavy ashfalls that the blanketing effect is thought by some scientists to have been the initiating cause for the extinction of the dinosaurs and many other lifeforms (Axelrod, 1981).

Another theory is that a giant asteroid impacted on Earth near the end of the Cretaceous and was the cause of the demise of the dinosaurs and other lifeforms. Since the disappearances occurred during a period of several million years, I think it is likely that tectonic events were a significant cause of the extinctions. However, it probably will be found that both hypotheses have validity.

Marine core samples of Cretaceous deposits show that ashfalls exceeded one meter in thickness in some areas and spread over thousands of square kilometers. Volcanic dust 1.5 millimeters deep was recorded in Texas, Oklahoma, Louisiana, Arkansas and southern Illinois, USA. Daniel I. Axelrod (1981: 18) said, "Continuing heavy ashfalls ...probably had a more drastic effect on land and sea life than the gradual retreat of the Cretaceous seas..."

The scenario of the Cretaceous extinctions may have been rather like this: Animals living in a warm and equitable climate were faced with conditions of severe ash pollution of their whole environment. Not only was their breathing choked and their herbaceous food supply destroyed, but they must have become disoriented in the dust clouds of the changing landscape. Carnivores would have collapsed as the herbivores died.

In the sea, the ash clogged fish gills and led to digestive problems; bivalves were smothered by more that 30 to 50 centimeters of sediment. Ashfall killed plankton. Axelrod estimated that the ashfall blocked enough solar radiation on several occasions to reduce the ocean surface temperature by three to 3.5 degrees Celsius and bottom water temperature about five degrees Celsius.

Although the average atmospheric temperatures then (polar 12 degrees C, to equatorial 30 degrees C) were still warmer than today's average temperatures (polar 0 degrees C, to equatorial 30 degrees C) the thermal shocks of sudden cooling on organisms which had evolved in warm isothermal conditions for no less than 100 million years must have been devastating. The combination of massive ash deposition and unseasonal cooling was probably enough change to cause the demise of the lifeforms which are known to humankind only as the magnificent fossils of the Cretaceous.

23

Cenozoic Era — Tertiary Period
Paleocene Epoch
"Age of Mammals"

Claims and counter claims are rampant in paleontology as in many other fields of science. Expert opinion about the evolutionary relationships of Paleocene animals is so diverse that new DNA information will inevitably enlighten the dialogue.

The Cenozoic Era, "Age of Mammals" began about 65 million years before present (65 MBP) and included two designated periods, the Tertiary, lasting about 63 million years and the Quaternary, beginning about two MBP. The Tertiary Period is divided into five epochs, beginning with the Paleocene. The year consisted of 365 days at the beginning of Paleocene.

Sea floor spreading rates declined and more land became exposed. Sea level was almost 100 meters above present levels. The epicontinental sea that had separated eastern and western North America at various times during the Mesozoic Era retreated from the continent, allowing all North America to be configured as one continental land mass, even as the proto-

Atlantic ocean grew and further separated eastern North America from Europe.

The continents approached their present positions, although North and South America were separated by an island arc. India and Africa-Arabia remained islands, with the Tethys sea separating Africa-Arabia from the European continent. Australia had begun to separate from Antarctica, and move northward, rotating on its tectonic plate during the era. Gross movements of the tectonic plates and massive extinctions of lifeforms slowed after the end of the Cretaceous. The geological record reveals many additional Earth movements in the Cenozoic Era but they are generally much less extensive and severe than those of previous eras.

The splitting of continental land masses generally led to warmer and more seasonal climatic patterns in the Paleocene. Seasonal cooling brought droughts in some areas. Increased horizontal and vertical ranges of temperature in both ocean and atmosphere increased circulation and brought increased vigor to the biosphere.

Seas of tropical and temperate latitudes evidenced differences between them in more than 80% of marine species. The seas were dominated by mollusks: clams and snails that were similar to today's species. Many species of bottom dwelling organisms, as well as fish, nautiloids and turtles radiated in the seas.

Flowering plants dominated the landscape in well watered areas. Snakes and crocodiles were active reptilian carnivores, especially in South America which lacked placental carnivores during most of the era.

Proliferation of Early Mammals

Although the Cenozoic is called the "Age of Mammals," the earliest known mammal fossils were found in Triassic deposits. Scientists generally believe that the early mammals of the Triassic, found only as fossil teeth and a few skulls and skeletal parts, represent a variety of shrew-like, mouse to rat-size animals. Those early fossils were known from England,

Europe, China, South Africa and North America (*Eozostrodon*, Chapter 20, Figure 20.10).

In the 1960s to 1980s extensive new fossil evidence of Permian and Triassic mammal-like reptiles in Antarctica, Africa, Eurasia, South America and North America gave support to the idea that different mammals evolved from different therapsid sources, in both northern and southern hemispheric locations.

There continues to be lack of understanding of the evolutionary connections between the therapsids and the early mammals that succeeded them. A question that scientists must answer is: How were evolutionary connections made between the many very large therapsids and the uniformly small mammals thought to have evolved from them?

In any case, the mostly small mammals present by the Cretaceous began to radiate into larger sizes and varieties in the Paleocene, following the extinction of the dinosaurs. The Mesozoic animals which had ranged from western North America to Asia across the Bering connection had represented different evolutionary patterns from the eastern North American-European animals. But as the Cenozoic began, new evolutionary patterns emerged.

The Tertiary fossil record is probably more complete in the Western Hemisphere than in the rest of the world, with the exception of evidence of later primates, which has been developed in various areas, but especially in Africa. Therefore the Tertiary of the Western Hemisphere will be emphasized in this review, turning to world-wide primate history.

Although American scientists believe that the best early Cenozoic mammal fossil record is concentrated in western North America, this may change. Early Cenozoic mammal fossils may in the future be discovered in more locations, which could alter this present belief.

In fact, two Argentinean paleontologists, Carlos and Florentino Ameghino, were convinced that Argentina, especially Patagonia, was the source area not only for the earliest mammals and their beginning radiation, but also for earliest man. It is possible that some South American mammals evolved from the endemic Triassic mammal-like reptiles (chiniquodantid cynodonts) in which case there is interest in South America as a geographic center of some early mammalian evolution.

The evidence does not appear to support a number of the Ameghino claims. Fossil evidence is narrow in the South American Paleocene, with only seven orders of mammals represented and three of these were ungulates.

According to George Gaylord Simpson (1980) a North American expert on South American mammals, there was no basic stock of insectivores found in the South American Cretaceous or Paleocene that might have bridged the gap from mammal-like reptiles. Most of the existing South American mammal orders did not spread from there to other continents, and some experts believe that several of them did not originate in South America.

In addition, there appears to be world-wide consensus that man arose in the Cenozoic of Africa, not of South America. Nevertheless, I believe it is highly probable that the lack of understanding of foreign languages hampers some Americans' knowledge of paleontological finds in other countries, and thus understanding of the world-wide fossil record.

George Gaylord Simpson (1980: 13) thought that Florentino Ameghino had misinterpreted the evolutionary significance of some of Argentina's fossils. He thought Ameghino did not understand evolutionary convergence: the similarity of structure in many continentally separated animals because of similar environmental factors, rather than from similar hereditary factors.

I believe this evaluation is an example of the inadequacy of Darwinian theory and monistic either/or logic. DNA may be recombined and-or moved by viral action in unknown patterns. Many of the structural patterns which have been thought to represent only random variations and natural selection, may in reality have resulted from new recombinations of DNA. These

DNA patterns represent various means of solving similar structural, functional and environmental problems.

Although I accept American scientists' consensus opinions as having great weight, I insert certain significant counter claims suggested by complementarity dynamics and supported by credible evidence. And I continue to question which DNA patterns may be responsible for the fossil variations we see.

The Tangled Heritage of North American Animals

Fifteen orders of mammals appeared in the North American Paleocene: Multituberculata, Marsupialia, Insectivora, Dermoptera, Tillodontia, Taeiodonta, Pholidota, Rodentia, Dinocerata, Perissodacatyla, Primates, Condylarthra, Creodonta, Carnivora and Pantodonta (Simpson, 1978: 71). The boundaries between the body characteristics (morphology) of these orders are difficult to determine and are set differently by different writers.

During the Paleocene, the earliest pantodonts evolved into Bary *(Barylambda,* Figure 23.1). No known predecessors to pantodonts have been found. Bary's heavy construction and large size, nine feet in length, seems strangely incongruous as a mammal of the Paleocene. But it is interesting to speculate that the configuration of its tail could be from a DNA recombination similar to the DNA for the tail configuration of a large mammal-like reptile such as Kanne (*Kannemeyeria,* Chapter 19, Figure 19.5), or even DNA which patterned the tail of one of the dinosaurs such as Campto (*Camptosaurus,* Chapter 21, Figure 21.4).

The marsupial opossum (Figure 23.2) was one of the best known of the animals that carried over in similar form from the Cretaceous into the Paleocene of North America and lives on into the present. It was known in the later Paleocene of South America and Europe. The ancestors of opossums are unknown but are thought to be of early Cretaceous age.

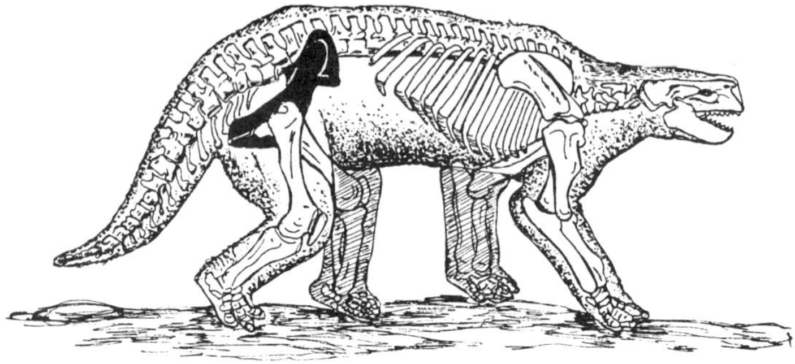

Figure 23.1 Bary (*Barylambda*, Pantodontia). (After Patterson).

Figure 23.2 Opossum (*Didelphis*, Marsupialia).

The fossil record supports the concept that salmutations occurred many times in evolution from mammal-like reptiles to the variety of mammals known in the Paleocene.

William K. Gregory (1963: 48-49) had the imagination to consider the morphological relationship of an extinct early opossum and a recent opossum. He superimposed the skull outline of a recent opossum on an extinct opossum jaw bone, and showed a close fit of the contours. In addition he showed a close relationship between the skulls of an advanced mammal-like reptile and a modern opossum, raising the possibility of evolutionary ascent (Figure 23.3).

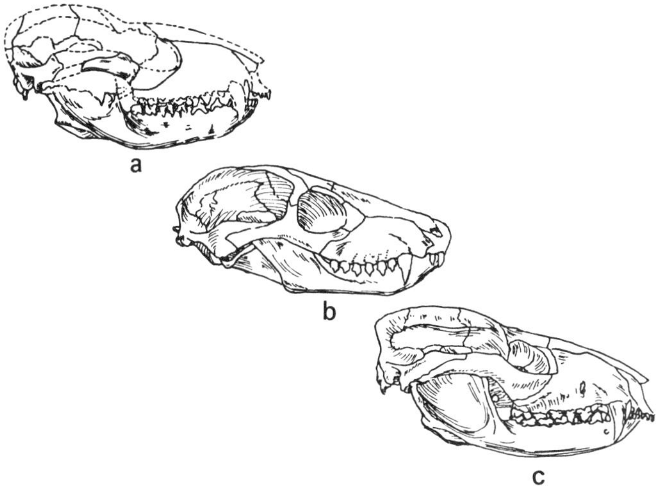

Figure 23.3 (a) Skull outline of recent opossum, superposed on jaw of extinct opossum. (b) Skull of advanced mammal-like reptile. (c) Skull of modern opossum. (After Gregory, 1963).

One who uses monistic logic may infer from these similarities that opossums have changed very little in an evolutionary ascent from mammal-like reptiles. However, one who thinks in complementarity dynamics might reason that there were many evolutionary connections which were not direct sequential routes from animals of the Paleozoic and Mesozoic to animals of the Cenozoic. Some of the paleontologists of the early 1900s were, perhaps, more flexible in their judgments of relationships than later scientists and may have identified relationships which can now more realistically be called DNA similarities rather than evolutionary sequences.

Dominant mammals in North America in the Paleocene were insectivores such as shrews, and archaic ungulates such as condylarths, length 5.5 feet (*Phenacodus*, Figure 23.4). There were fossil indications of condylarths earlier, in the Cretaceous, but the first relatively complete and numerous skeletons were unearthed from Paleocene deposits. Their fossils have been found in several areas of the Northern Hemisphere. Condylarths are thought to be descendants of insectivores (like shrews) but there is no identified fossil connection. It is thought that condylarths were the precursors in North America of amblypods (*Uintatherium*, Figure 23.5) and later appearing perrisodactyls: odd-toed, hoofed, herbivorous, ungulates (*Eohippus*, Chapter 24, Figure 24.3) and artiodactyls (even-toed, hoofed, herbivorous, ungulates).

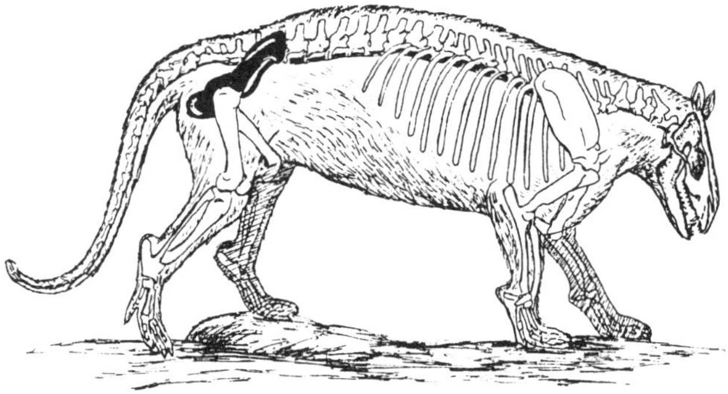

Figure 23.4 Condylarth (*Phenacodus*, Condylarthra). (After Osborn, 1910).

Condylarths exhibited dentition which was considered carnivorous, but tending toward herbivorous. Having five-toed clawed specimens before them, some scientists posited that condylarths might have been predecessors of members of the cat family as well as ungulates. The structure of condylarths gives rise to my opinion that they represent salmutations from a number of different DNA sources and provided DNA to a variety of different lines by further salmutations.

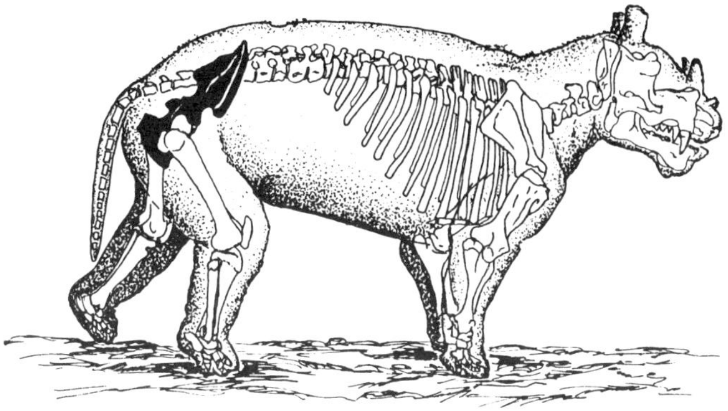

Figure 23.5 *Uintatherium*, (Amblypoda). (After Osborn, 1910).

Scientists are beginning to make more biomolecular camparisons of animals across order classification lines, especially comparisons of blood (serum) proteins. However, I have found little information from this field pertaining to perissodactyls.

Some paleontologists have posited that five different condaylarth ancestors gave rise to perissodactyls, artiodactyls, proboscideans, cetaceans (whales and their relatives) and also South American ungulates in the Paleocene (Carroll, 1988: 506).

Previously, scientists had posited that condylarths gave rise only to perissodactyls and artiodactyls in the Paleocene, and to the South American ungulates later, in the Eocene. Probosicedeans were thought to have a different origin in Africa in the Oligocene and cetacea were thought to have arisen from artiodactyls.

Until the 1980s, scientists thought almost unanimously that condylarths were ancestral to early North American and European Eocene horses and other perissodactyls. A recent discovery in China of a new genus of Embrithopoda, *Radinskya,* has initiated a reevaluation of this ancestry. It has been posited (Prothero and Schoch, 1989: 504) that perissodactyls may not have originated in North America from condylarths but rather diverged in Asia and-or Africa from tethystherians (a new classification that includes early probosideans, sirenians, desmostylians and arsinoitheres).

It is premature to revise classifications so extensively, and posit new evolutionary pathways based upon a few fossil teeth and bones of one animal, pending further knowledge of DNA relationships among the living genera of hyracoids, proboscideans and perissodactyls. However, the find does point out the importance of determining initial incidence of genera in specific locations and considering its relationship to migration routes that were then available.

The location in which a particluar species or genera of animal arose will always remain speculative. It cannot be pinpointed in the fossil record because an earlier fossil may always be found which could change the entire picture. It happened with *Radinskya* and it happens frequently. As knowledge about place of origin changes, the ideas of migration routes must also change. When the unknowns of origin and migration are compounded by the unknowns of different environments, by incomplete fossil skeletons, undiscovered characteristics of biomolecular structure and uncertainty of viral action, it is a miracle that we know all that we do.

We know much because many scientists continue to produce succeeding revelations. The existence of unknowns and the search for the intelligence inherent in the life of the world are stimuli that will forever give rise to increases in human knowledge.

Creodonts to Carnivores? Insectivores to Mammals?

Some authorities believe that early creodonts were precursors of carnivorous cats and dogs. Creodonts were known from the Cretaceous and Paleocene but these fossils were not as complete as some from later Oligocene deposits (*Hyaenodon*, Chapter 25, Figure 25.2).

William D. Matthew (1909: 504-506) recognized close relationships between condylarths and creodonts and posited a common ancestor. He noted similarities in dentition of creodonts, condylarths, insectivores, perissodactyls and of artiodactyls, but believed that teeth alone were an insufficient guide to determine ordinal relationships. He wrote that primates approached condylarths in tooth structure but were very different in foot structure. He saw similarities between multituberculates and marsupials. He thought that placental mammals evolved from insectivores, not from marsupials, but Matthew noted:

> The modern Insectivora approximate the less specialized Marsupials in certain important characters of the skull, and in the general construction of the teeth, more nearly than do any other Placental orders... But for the most part the characters regarded as definitive of the Insectivora are shared by one or more of the other order of mammals, especially by rodents, lemurs, creodonts and condylarths, which approach nearest in grade of organization.

The morphological similarities identified by Matthew and William K. Gregory may be of more assistance in future research to identify DNA patterned structures than some of subsequent experts' positing of proposed evolutionary cladistic relationships. Some modern writers discount earlier work as naive, on the basis that recent fossil evidence and restudy of previous fossil finds has added significantly to understanding of evolutionary relationships. However, I believe that not enough has been done to view the fossils from the perspective of possible DNA recombinations as well as from the perspective of variation due to gene recombinations in sexual reproduction.

New kinds of insectivores, or early primates, thought to be descendants of shrews, appeared in the fossil record of both Europe and North America. The arboreal primate Plesi (*Plesiadapis*, Figure 23.6) a lemur-like or squirrel-like fruit eater with well developed eyes and primate-like hands, but with claws instead of the flat nails of most primates, was present in the North American Paleocene.

Figure 23.6 Plesi (*Plesiadapis*, early Primate).

Plesi and other lemur-like placental primates, thought by some to be precursors of higher primates, expanded in range. In the American West, three families of lemur-like primates have been found as fragmentary fossils, as well as the earliest known tarsioid primate (*Tetonius*, Chapter 24, Figure 24.9 b). *Lemurs* and *tarsiers* exist in the wild today only in Madagascar and in a few scattered areas of Indonesia.

Unique South American Fossils

South America had been moving on its tectonic plate, as an island, separated from Africa for tens of millions of years, when the Paleocene began. It was shifting northward from Antarctica, which was covered with pine and beech (*Nothofagus*) forests. South America was much more moist and lush in vegetation than it is today.

Central America was mostly below sea level, but there was a chain of islands between North and South America. Some scientists believe that this allowed early North American opossums, known to cross water barriers, to migrate to South America. Scientists who believed that these animals originated in South America posited a migration from south to north. Opossums may have arisen in North America but then evolved in Gondwana as revealed by thirteen genera of South American fossils. However, the sea soon intervened to prevent additional migration and other Paleocene and Eocene mammal fossils show little or no evidence of further intermigration.

In addition to Marsupialia, other orders of mammals that appeared in many of the South American Paleocene deposits are: Trigonostylopoidea, Notoungulata, Litopterna, Xenarthra, Xenungulata and Condylarthra.

Scientists believe that South American condylarths gave rise on that continent to litopterns, astrapotheres, pyrotheres and xenungulates. Some authors believe that they also gave rise to notoungulates, but that is disputed by others. Some evidence has been presented that condylarths were ancestral to aardvarks (tubulidentates) of the African Miocene, although these are not ungulates and have an ecological niche closer to that of anteaters.

The notoungulates (southern ungulates) are characteristic South American animals. However fossils of notoungulates have been found in North America, and one group, (palaeostylops) in Asia. Some authors believe that notoungulates originated in North America, while others think that they evolved from condylarths in South America. A few fossil bones and teeth, classified in the family Notostylopidae were found in the Paleocene of South America. By the Eocene a fuller notoungulate skeleton of Thomas was present (*Thomashuxleya,* Chapter 24, Figure 24.6). A range of many and varied notoungulate genera radiated from the Paleocene into the Pleistocene only in South America (*Toxodon,* Chapter 28, Figure 28.4).

The order Litopterna is thought to represent South American descendants of condylarths, although only one fossil condylarth has been found, in Peru. Some of the litopterns were very much like North American horses, though it is believed that they originated in South America earlier than the North American horse, *Eohippus*, first known in the Eocene (Chapter 24, Figure 24.3). There was no known land connection between the two continents, on which a hypothesis of migration of horse-like animals could be based. Diad (*Diadiaphorus*, Chapter 26, Figure 26.2) is a Miocene fossil representative of the order Litopterna, which is much more complete than Paleocene fossils.

Other litopterns of the Miocene such as Theo (*Theosodon*, Chapter 26, Figure 26.3) which evolved into the Pleistocene, were more like camels than horses, but it is believed that they had a short proboscis, suggestive of a tapir or elephant trunk. Darwinians call upon convergence and parallelism to explain such similarities as the fleshy proboscis of litopterns, tapirs and elephants. But the division of litopterns into horse-like and camel-like animals and into camel-like animals with tapir-like or elephant-like proboscis supports the speculative idea of DNA recombinations.

The earliest members of the order Xenarthra (or Edentata) found in South America in the Paleocene were armadillos (Figure 23.7 a). Although all families of Xenarthra may have also been found among North American fossils, glyptodonts and sloths were first discovered in Eocene deposits of South America and in much greater numbers there than elsewhere, so xenarthrans are thought by some experts to have originated in the south.

The fact that armadillos, glyptodonts, sloths and anteaters have been placed by scientists in one order illustrates the problem of positing sequential, morphological relationships of animals. It would be difficult to envision one common ancestor of these diverse animals. However, possible DNA elements can be seen in earlier animals which could have been recombined in the patterns which are demonstrated in these unusual beings. Armadillos, glyptodonts, sloths and anteaters have been considered to be related because they all lack incisor teeth and they have some similarities in spinal and hip structure. I believe it is more probable that such similarities as lack of incisor teeth — and absence of any teeth in anteaters — is a similar DNA pattern carried in differently evolving lines.

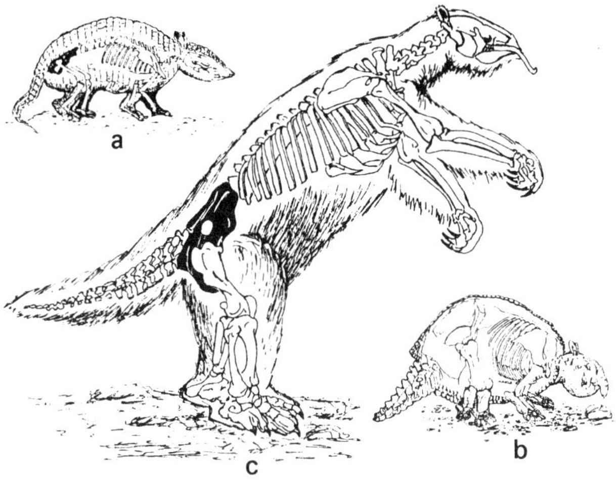

Figure 23.7 (a) Armadillo (*Dasypus*, Xenarthra). (b) Glyptodont (Xenarthra). (c) Sloth (*Nothrotherium*, Xenarthra, Pleistocene genera). (a, b, After Osborn, 1910).

Armadillos have few other characteristics in common with sloths or anteaters. Their articulated shell (carapace) is not known among other mammals, but there are invertebrates with such articulation. Could this DNA pattern have passed from invertebrates to these mammals, bypassing long vertebrate lines?

Perhaps the glyptodont (Figure 23.7 b) the most heavily armored of all mammals, and the only known mammal with both internal and external skeleton, received some of its DNA pattern from armored dinosaurs (such as *Euoplocephalus?*). Maybe turtles with a somewhat similar carapace contributed a DNA pattern for the glyptodonts, which have no known mammalian ancestors.

Sloths (Figure 23.7 c) have a long history from the Paleocene to Recent but have no known evolutionary connection from armadillo-like or glyptodont-like predecessors. They did not develop from small size to large size over time. Extinct giant ground sloths were as big as elephants — but living tree sloths are small bear-size — among the smallest of known sloth genera. There are some authors who do not believe that tree sloths evolved from ground sloths, but this line seems to be more reasonable than to posit a common ancestor for all xenarthrans.

There appear to be few significant extinctions in the Paleocene, but rather beginning radiations of modern lines of plants and animals. The fossil record of the Paleocene suggests that the small mammals of the Cretaceous did not evolve in expected Darwinian patterns. Representatives of Paleocene mammals rather suggest the necessity to search for possible DNA patterns across taxonomic lines.

24

Cenozoic Era—Tertiary Period—Eocene Epoch Rise of Horses Early Primates

The Eocene, about 60 MBP was a pivotal time in the existence of numerous archaic animals. Some, such as the multituberculates, became extinct while many others were heading for gradual extinction. Ancestral types of many modern animals appeared in the fossil record for the first time, often without evidence of direct progenitors. However, apparent transformations occurred from condylarths (Chapter 23, Figure 23.4) to perissodactyls and artiodactyls, and from insectivores, creodonts or miacids to a variety of carnivores.

The Eocene may be one of the earliest and best epochs to search for DNA clues to evolution because the bones of such animals as condylarths and early horses are numerous and available in North America. In time, perhaps some of these bones can be processed to reveal their DNA patterns.

It is believed that during the epoch, Australia continued to move northward from Antarctica with its crew of monotreme and marsupial ancestral species aboard, although little or no fossil evidence of these animals has been found from this epoch. Marsupials are thought to have migrated from South America through Antarctica to Australia before it began to split from this Gondwana connection.

New Zealand probably split off from Australia earlier because there is no evidence of mammals there before their introduction by humans in the Recent Epoch. Huge ground birds and reptiles filled the predator niche in New Zealand.

Tasmania blocked the circum-Antarctic current for a time, but deep water currents began to bring a cooling trend near the end of the Eocene when Drake's Passage, south of Cape Horn, opened. About the same time, the Greenland-Norwegian Sea expanded to the Atlantic.

These tectonic movements and cooling trends occurred while explosive volcanism changed the face of Nevada, Montana, Wyoming, Idaho and the gulf states of North America. There continued to be little or no evidence of interchange of species between North and South America. South America moved slightly west-southwest (relative to North America) probably disconnecting further the island arc which the opossums may have previously used to navigate between the continents.

In the mid Eocene, fauna of Europe and North America were evolving in their own courses. The Northern Hemisphere land masses were connected in the Bering Strait area, and there was some migration between these continents in the late Eocene.

Animals of North America

The first North American fossil whales were found in Eocene deposits but there is evidence that whales arose earlier — fossil fragments in Egyptian deposits of the Mediterranean region. Darwinians have posited that whales evolved from land mammals returning to the sea. This seems rather unlikely to me. It seems more reasonable to assume that an insertion of DNA into early sea creatures perhaps like Triassic diapsids (see Carroll, 1988: 221) by viruses carrying mammal-characteristic DNA is a better explanation of the evolution of whales and other sea mammals such as porpoises and dolphins.

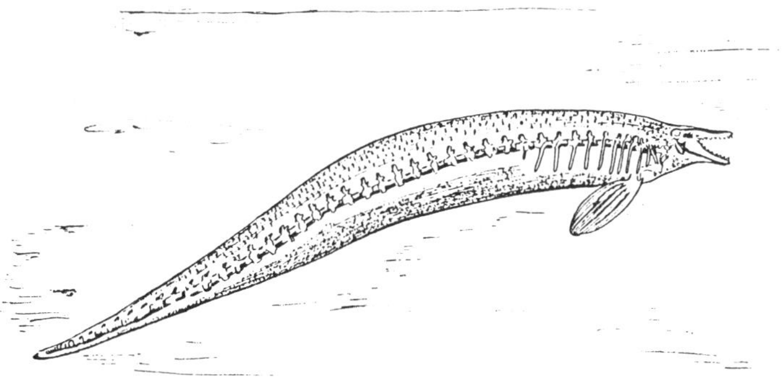

Figure 24.1 Basil (*Basilosaurus*, Cetacea, early whale). (After Gidley).

Basil (*Basilosaurus,* Figure 24.1) was a huge North American whale, 55 feet long, structured for life in the sea. I can conceive of its evolution from DNA of huge diapsid sea reptiles but not from small land-living artiodactyls (descendants of creodonts or condylarths) as posited by some paleontologists.

At the beginning of the Eocene there were lush woodlands and subtropical climates in most of North America. The tectonic movements which disconnected the northern hemispheric continents proceeded from south to north so that a northern American-European connection remained at the beginning of the Eocene. Many animals were very similar in North America and Europe in the early Eocene. Both perissodactyls and certain rather rare artiodactyls appeared almost simultaneously in the fossil records of Europe and North America and then followed different developmental pathways by mid-Eocene.

By the end of the Eocene, 20 families of artiodactyls, that represented all the major groups living today, were present in North America, Europe and Asia.

Animals such as *Uintatherium* (Chapter 23, Figure 23.5) arose in North America, then arrived in Asia in the late Eocene, probably by way of the Bering connection. They never appeared in Europe. In North America taeniodonts, tillodonts and armadillos were present, yet none of these were known in Europe.

By the mid Eocene, a cooling trend leading toward Pleistocene glaciation was occurring. However, new lifeforms continued to radiate. The Bridger Basin of Wyoming (Matthew, 1909) revealed ten orders of Eocene mammals, including 15 genera of lemur-like Primates, 16 genera of Insectivora, 20 genera of Carnivora, five families of Perissodactyla, two families of Artiodactyla, and also Amblypoda, Rodenta, Tillodontia and Edentata (now classified Xenarthra). The Perissodactyla included early horses, tapirs and titanotheres.

Horses and other Odd-toed Ungulates — Perissodactyls

The Eocene is known for the sudden rise and abundance of horses in both North America and Europe, although earlier fossils were found in Baja California, Mexico, deposits. Darwin had posited long-term trends of evolving species. Later scientists illustrated these long-term trends by tracing the fossil record of the horse.

Horse evolution is often illustrated as in Figure 24.2 showing the gradual increase in size of genera from *Eohippus* to *Equus*. *Eohippus*, the dawn horse (also known more correctly as *Hyracotherium*) about 18 inches long and 10 inches high, was a browser of leaves and fruit. Although *Eohippus* is usually illustrated as the smallest of the series, species of the *Eohippus* genus ranged to about twice that size, some of the larger ones appearing in earlier deposits.

Illustrations sometimes show changes in tooth structure from browsers to grazers. They usually show the gradual change in the feet from the four-toed forefeet and three-toed

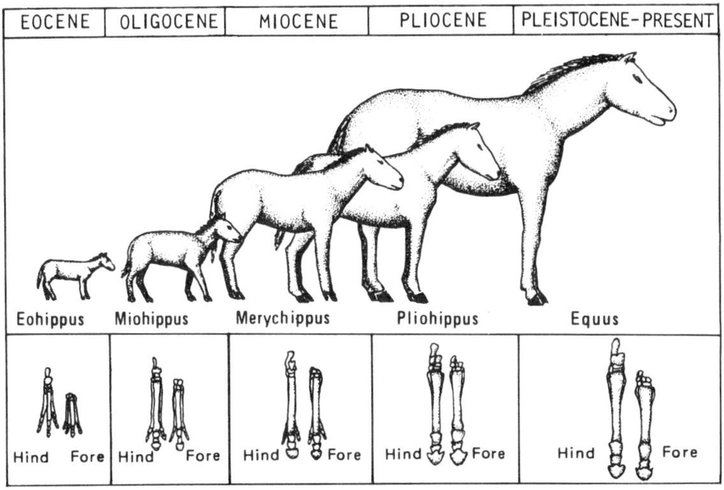

Figure 24.2 *Eohippus* (*Hyracotherium*, Perissodactyla), *Miohippus, Merychippus, Pliohippus, Equus*. (After Storer, Usinger, Mybakken and Stebbens, in Denton).

hind feet of *Eohippus* to the one-toed hooves of *Equus*. *Eohippus* had padded feet rather like those of condylarths (from which they are thought to have descended) but condylarth feet had five digits. Was *Eohippus* an experimental recombination of DNA genes for two different digit number combinations on front and rear feet prior to the division into the odd-toed perissodactyls and the even-toed artiodactyls? (Figure 24.3.)

Conventional paleontology has recorded that *Miohippus* was found in the Oligocene, 40 million years ago, and radiated into the Miocene, 20 million years ago. These horses were about two feet high with three toes on each foot. They remained browsers with low-crowned teeth. One offshoot, *Merychippus* (pony size) with three toes, became a grazing form in the Miocene. It has been theorized that during the Pliocene, ten million years ago, *Pliohippus* emerged, a one-toed hoofed animal with teeth very efficient for grazing. *Equus*, the modern

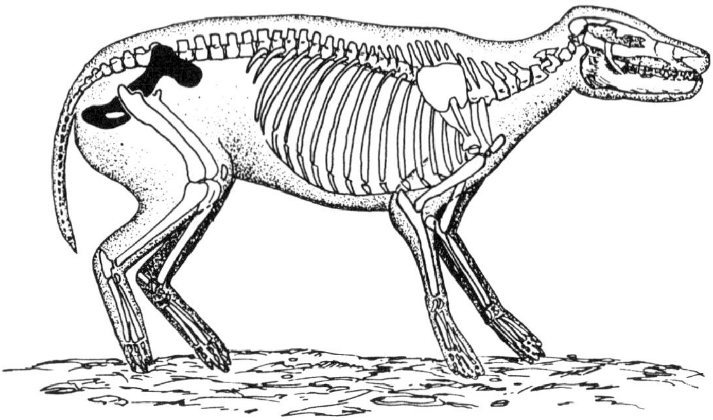

Figure 24.3 *Eohippus* (*Hyracotherium*). (After Osborn, 1910).

horse, was thought to have evolved from *Pliohippus* at the close of the Pliocene, two million years ago.

The changes in the feet of horses from the Eocene to the Pleistocene were both a decrease in digit number and a fundamental structural change from padded feet to sprint-tension, ligament-controlled tiptoe hooves. As horses grew larger, this fundamental change in structure occurred to accommodate the increased weight of *Merychippus* while still retaining the speed of *Eohippus*. This change, along with the evolution to grazing dentition in *Merychippus* was not a matter of random mutations. It appears to be intelligent engineering of each new functioning animal species!

The five horse illustration (Figure 24.2) gives the impression of a smooth evolutionary transition during at least 40 million years. Evolution of the horse is often used to illustrate sequential development: radiation from browsing species into grazing species, and growth in size from rabbit size to modern horse size. Examples of the horse line which do not fit this sequence are often ignored, not a proper scientific practice. Those horse taxa which do not fit the sequential pattern, that rather represent unusual configurations and dead-end lineages, should also be considered and explained as did Romer (1966) and Simpson (1951).

There were many other horses in addition to the five illustrated here. George Gaylord Simpson (1951: 111-112) classified the family Equidae into three subfamilies: Hyracotheriinae, with two genera following *Eohippus* in the Eocene; Anchitheriinae, with seven genera; and Equinae, with eleven genera from *Merychippus* to the modern horse, *Equus*. However, Simpson (1951) did not include *Palaeotherium* (Figure 24.4) in this classification. The "horse-like mammal... prematurely large and advanced in structure" (Romer, 1966: 267) was a giant of the Paleocene and Eocene of Europe. It was 10 feet long and six feet high at the shoulder. It has been thought to have arisen from *Eohippus*, but *Palaeotherium* existed earlier than *Eohippus* and evolution from *Eohippus* to the large Pliocene *Equus* required at least 40 million years.

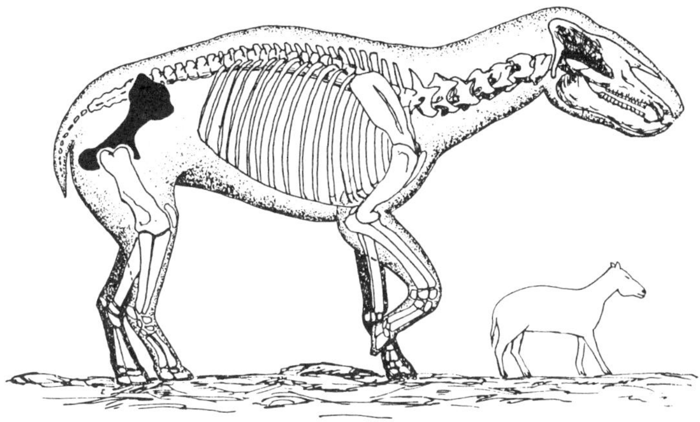

Figure 24.4 *Palaeotherium*, with *Eohippus* in approximate size relationship. (After Abel).

Note the similarities and differences in structure between *Eohippus* (Figure 24.3) and *Palaeotherium* (Figure 24.4). If *Palaeotherium* could have evolved from a progenitor of *Eohippus*, where did it acquire its foot structure and neck bones?

If *Palaeotherium* did not evolve from a progenitor of *Eohippus*, why are *Eohippus* and *Palaeotherium* generally so similar?

Understanding of horses and other perissodactyls was greatly enhanced by papers from an international conference of 1985, published as *The Evolution of Perissodactyls* (Prothero and Schoch ed., 1989). The authors presented evidence that *Hyracotherium* (*Eohippus*) was very common in the early Eocene, not only in North America but also in Europe, Mongolia and China. They showed that perissodactyls, often with similar characteristics, diverged in the Northern Hemisphere into equids, chalicotheres (Chapter 26, Figure 26.9) tapirs, titanotheres (Chapter 25, Figure 25.5) and rhinos (Chapter 25, Figures 25.6 and 25.7).

Mongolian deposits contained more tapirs, rhinos and chalicotheres than horses, but horses were also present there in the middle or late Eocene. There were migrations of horses from North America to Europe prior to the isolation of Europe from North America in the mid-Eocene. Paleotheres were a dominant animal in Europe by the late Paleocene, found there much earlier than Darwinian theory proposed.

Titanotheres dominated North America and Asia in the late Eocene. The relationship between horses and titanotheres is puzzling. Their skeletal structures have some similarities but also many differences. Horses do not have horns, but huge titanotheres had horns. Horses have teeth which are adapted for browsing or grazing, but titanotheres had teeth poorly adapted for either. Yet they lived from the Eocene into the early Oligocene. They probably could not win a competition for efficient teeth but nevertheless, they were some of the largest animals of the perissodactyl line and lived a long time.

Eotitanops (Figure 24.5) was a small Eocene titanothere, thought to have a condylarth predecessor as did *Eohippus*. The evolutionary route from *Eotitanops* is thought to lead to the giant *Brontops* by the Oligocene (Chapter 25, Figure 25.5). This huge horned titanothere, fourteen feet in length and eight feet high at the shoulder, lived not only in North America but also in Europe and Asia into the early Oligocene.

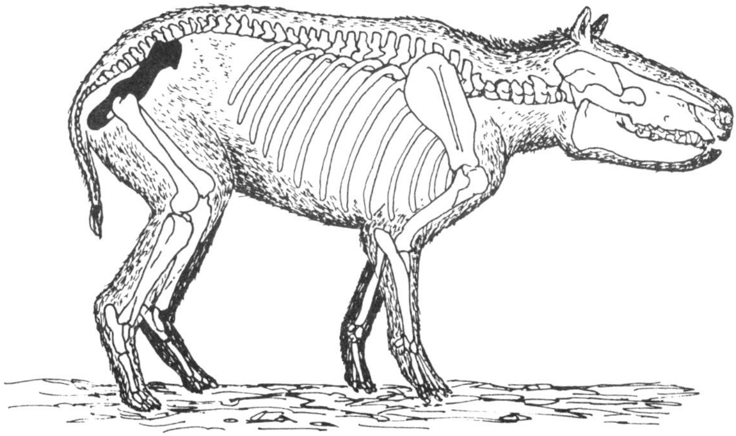

Figure 24.5 *Eotitanops* (Titanotheriidae, Perissodactyla). (After Osborn, 1896).

If *Eotitanops* and *Eohippus* had a common ancestral origin, it is difficult to imagine that *Brontops* and *Equus* could have evolved from that ancestor without salmutations. Natural selection is not adequate to explain the horse and titanothere lines. Mutations and salmutations are evident.

Tapirs appeared in North America early in the Eocene and reached their maximum diversity in Asia late in the epoch. Late in the Eocene, rhinos diversified into three groups, with the more modern type rhino becoming a dominant animal in North America and Eurasia in the late Eocene.

Donald Prothero and Robert M. Schoch (ed., 1989) summarized new evidence that hyracoids (hyraxes) were divergent in Africa by the mid-Eocene. They ranged from cat-size to rhino-size. The largest were as large as Egyptian proboscideans, giving support to the idea that hyracoids had evolved for a long time, perhaps migrating to Africa. These authors posited that hyracoids were closely related to both early proboscideans and perissodactyls and may have originated in Asia, but this remains problematical, as the perissodactyls have also been thought to have originated in North America. There are many differences of opinion about the evolutionary relationships among the perissodactyls and their connections to other orders.

The work of Prothero and Schoch is very valuable not only for raising interesting questions about evolving lineages, but also for bringing together in one volume, the names and descriptions of many animalsthat were unknown to earlier paleontologists.

South American Ungulates

The South American Eocene notoungulate Thomas (*Thomashuxleya,* Figure 24.6) is thought not to be a horse according to classical theory, but to have evolved from condylarths of South America. It has a neck structure and foot structure rather similar to *Eohippus* but it was much larger, about five feet long. Thomas' skull and dentition are not similar to those of *Eohippus,* nor to *Palaeotherium.*

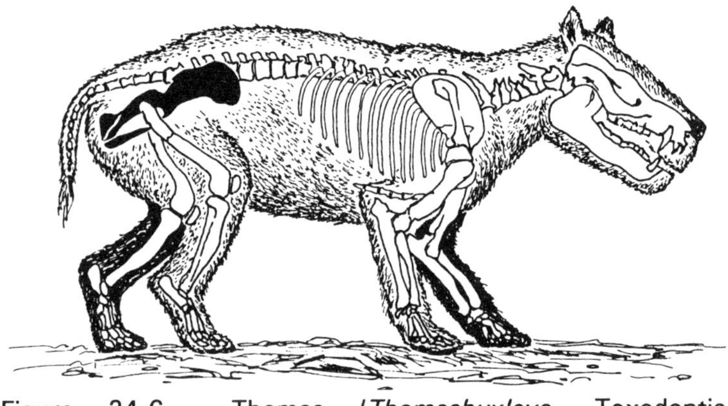

Figure 24.6 Thomas (*Thomashuxleya,* Toxodontia, Notoungulata). (After Simpson).

Darwinian theory views notoungulates as parallel or convergent with the horses of North America. That theory posits that similar environments adapted these animals in similar structural patterns. How this could occur is unexplained.

We know now that structural patterns are governed by DNA. In this case, Darwin's idea of adaptation puts the cart (the environment) in a sequence, before the horse (its genetic potential). Complementarity dynamics enables one to comprehend the manifold interrelationships of cell DNA, of viruses, horse and notoungulate morphology and of the environment.

Another South American ungulate *Thoatherium,* a horse-like litoptern of the Miocene, is also thought to have evolved from South American condylarths. *Thoatherium* had single toed hooves much like *Equus,* long before the single toe evolved in the North American horse line. The evolution of animals from three-toed to one-toed hooves is thus not only representative of an evolutionary sequence in North American horses from Eocene to Pliocene, but may also include salmutations of this characteristic in different animals of an earlier time period in South America. Other South American Miocene litopterns such as camel-like Theo (*Theosodon,* Chapter 26, Figure 26.3) have three-toed feet, rather like *Eohippus.*

It is interesting to consider that South American horse-like animals such as Thomas and *Thoatherium* are not classified in the order Perissodactyla, while many non-horse-like North American animals *are* considered perissodactyls.

I believe that no series of small mutations beginning with condylarths, could lead to such different end products as horses, titanotheres, rhinoceroses, tapirs and chalicotheres in North America and Eurasia, and to the many and varied notoungulates and litopterns in South America. Salmutations are obvious.

Even-toed Ungulates — Artiodactyls

Condylarths (Chapter 23, Figure 23.4) have been considered to be ancestral to artiodactyls as well as perissodactyls, although the concept of such a large and varied progeny by means of minor mutations is becoming an embarrassment to modern paleontologists. The classification into even-toe and odd-toe orders seems an artificial division. There is overlap from one group to the other, although there are many animals which follow this division, as a general rule.

There were massive radiations of new kinds of animals in the Eocene. Fossil evidence shows early camels, peccaries and pronghorns in North America and camel-like litopterns as well as horse-like litopterns in South America.

Pigs and hippopotami are found throughout Eurasia beginning in the Eocene. By the end of the Eocene the fossil record shows the first ruminants: chevrotains, deer, okapi and early giraffes.

Insectivores and Rodents

The insectivorous shrews and moles developed in Europe during the Eocene. Primitive rodents were present on both North American and Eurasian continents. It is thought that rodents and rabbits (lagomorphs) may have evolved from insectivores but they were all independent taxa as early as the Paleocene. Despite similar dental structures, there is no serological evidence that rodents and lagomorphs are closely related. The lack of identified DNA relationships between these two orders is a puzzle which remains to be explored.

Early Creodonts

Fossil fragments of Eocene creodonts of many kinds have been found in Europe, Asia and Africa as well as in North America. Creodonts are thought by some scientists to be ancestral to dogs and cats and other carnivores. Other writers believe carnivores arose from insectivores.

A relatively complete North American fossil creodont of the Oligocene, *Hyaenodon* (Chapter 25, Figure 25.2) was four feet high at the shoulder and had many characteristics similar to condylarths. It had a larger skull than condylarths and dog-like premolars. It's dentition was similar to that of saber tooth cats and the dentition of artiodactyls, primates and proboscideans. Its foot structure has been thought to be superior to that of later cats. Its shoulder bone (scapula) was most similar to that of wolves (Mellett, 1976). Most experts believe that *Hyaenodon* was not in the direct ancestry of fissiped carnivores, but rather a dead-end carnivore offshoot.

Most paleontologists now believe that the order Creodonta was separate from the order Carnivora in the Eocene, perhaps even in the Paleocene. The extinct miacids, rather than creodonts, are thought to be ancestral to the living carnivores.

It is not easy to visualize that condylarths evolved into the hoofed mammals, and that insectivores, creodonts or miacids evolved into carnivores. There are too many close similarities in structure of all these animals. One can visualize characteristics of each passing by DNA insertions, into various descendent animals. Many innovations would be needed for any of the insectivores, creodonts or miacids to evolve into various carnivorous dogs, civets, cats, bears, weasels, hyaenas and the aquatic pinnipeds. I believe much work is needed to distinguish the DNA patterns of carnivore evolution.

The Mysterious Source of Bats

Contemporary texts say that bats appeared fully formed in both North and South American Eocene strata. Little fossil evidence can be found concerning their ancestry. In *The Origin of Species*, Charles Darwin reviewed, at some length, objections to his theory by a distinguished zoologist of his day, St. George Mivart. Darwin (1872: 180) wrote:

> Mr. Mivart is further inclined to believe, and some naturalists agree with him, that new species manifest themselves 'with suddenness and by modifications appearing at once.' For instance, he supposes that the differences between the extinct three-toed *Hipparion* and the horse arose suddenly. He thinks it difficult to believe that the wing of a bird 'was developed in any other way than by a comparatively sudden modification of a marked and important kind;' and apparently he would extend the

> same view to the wings of bats and pterodactyls. This conclusion, which implies great breaks or discontinuity in the series, appears to me improbable in the highest degree.

The sudden appearance of pterodactyl and bat wings cannot be ignored even though these animals appeared in different time periods. Both similarities and discontinuities are possible. I believe Mivart and Darwin were both right. Paradox logics allow it. The sudden modifications, envisioned by Mivart, could have been salmutations mediated by microorganisms. Salmutated DNA may have been carried from generation to generation of animals. That Mivart and Darwin knew nothing of DNA and viruses is beside the point. They each saw an essence of the evolutionary process.

George Gaylord Simpson (1967: 235) also recognized a time problem in the development of bat wings. He wrote:

> The rate of evolution of the insectivore forelimb into the bat wing...must have been many times more rapid than any evolution of the bat wing after it had arisen. The whole record attests that the origin of a distinctly new adaptive type normally occurs at a much higher rate than subsequent progressive adaptation and diversification within that type.

In *Tempo and Mode in Evolution* (1965) Simpson called this "quantum evolution." I consider this process of change may be instantaneous salmutations. Salmutations may be understood via analysis of the quantum logic of Niels Bohr and the logic of my reciprocal moment.

Assuming an evolutionary movement from the insectivore line to the bat line, the question arises: from whence came the leathery wings? Did Mivart have in mind some possible connection between the wings of pterodactyls (Chapter 21, Figure 21.6) and bats? (Figure 24.7.) Could similar DNA patterns have produced leathery wings in both pterodactyls of reptilian ancestry and bats which arose from insectivore ancestors? Since pterodactyls were extinct for many millions of years before the appearance of bats, could viral crystals have held the pterodactyl-like leathery wing gene combination?

Figure 24.7 Bat (*Palaeochiropteryx*). (After Revilloid).

Or could an energy pattern such as that recorded by Kirlian photography or a morphogenetic field, envisioned by Rupert Sheldrake (1988) add a parameter to the process of recombination of DNA necessary to produce leathery-winged bats from genetic material of insectivores of the Eocene? Kirlian photography and Sheldrake's morophogenetic field experiments provide evidence of an energy field that exists before and after the physical entity. Kirlian photography shows that an energy field remains after physical matter is removed from an organism. Morphogenetic field experiments show that an embryological chicken wing reforms in a normal pattern when (end of wing bud) tissue is removed. These findings could evidence complementarity of patterning mechanisms: salmutations within parameters of energy fields.

Our generation is exploring these frontiers of possibilities which should not be closed off due to preconceived ideas and the straightjacket of orthodox logic.

Evolution of Birds from Dinosaurs

Some paleontologists complain that too little research has been done to determine evolutionary lines of birds from such thecodonts as Salty (*Saltoposuchus*, Chapter 20, Figure 20.2) or from theropod dinosaurs such as *Compsognathus* or Celo (*Coelophysis*, Chapter 20, Figure 20.3) to Archy (*Archaeopteryx*, Chapter 21, Figure 21.7) and thence to birds of the Cretaceous, Paleocene and Eocene. Many bird books have little discussion about evolutionary pathways. The reason may not be that research on fossil specimens is lacking but that fossil evidence of intermediates between clear bird groups have not been found.

Early gulls, auks, ducks, flamingos, albatrosses and petrels were in the fossil record before the Paleocene. By the Eocene, fossils identified as early pigeons, parrots, pelicans, penguins, swifts, loons and turkeys were present. These ancestral birds were similar enough to modern birds to be classified in the same orders. Thus evolutionary lines were distinctive very early, some at first appearance. Intermediaries have been thought to be missing due to inadequacies of the fossil record, but scientists are beginning to realize that the fossil record alone will never supply the answers to the mysteries of evolutionary lineage.

Darwinian theory has no explanation for the sudden appearance in the Eocene of giant ground birds such as *Diatryma* (Figure 24.8) known from North America, Europe and Asia, and *Phorusrhacus* known from South America. *Diatryma* was over two meters in height and is thought to be a significant predator of mammals, perhaps of *Eohippus* and other early horses.

There is speculation that Hesper (*Hesperornis*, Chapter 22, Figure 22.7) of the Cretaceous is related to penguins, but by what route? And why is the ground bird *Phorusrhacus* considered on the same branch of an evolutionary tree as Hesper, rather than on the branch with other giant ground birds such as emus, rheas, cassowaries and kiwis that appeared in the Pleistocene?

Figure 24.8 *Diatryma* (Aves, giant ground bird). (After skeleton, Natural History Museum of Los Angeles County).

The synthesis between the concepts of physical sequential continuity of birds from their origin to the present and the concept of formation of new bird lines by DNA salmutations requires complementarity dynamics thought processes. In 1980, a leader of evolutionary thought from Harvard University, Stephen J. Gould (1980: 115) wrote an article entitled "Is a New and General Theory of Evolution Emerging?" In it he stated "...if Mayr's characterization of the synthetic theory is accurate, then that theory, as a general proposition, is effectively dead, despite its persistence as textbook orthodoxy." This willingness to reevaluate Darwinian theory is needed to accelerate understanding of the fossil lifeforms that paleontologists have discovered.

It is interesting to note that Simpson (1978: 269) although a confirmed Darwinian and a founder of neo-Darwinian synthetic theory, anticipated the changing climate in paleontology. He retained his belief in the synthetic theory but stated in his autobiography:

> Still other recent contributions are indeed quite outside the previous body of theory, but do not contradict any aspect of the synthetic theory and thus can, and I am confident will, become an extension of that theory, which is not bound to a static dogma. For example, I believe that to be true of the possible role in evolution of recombination of genes from widely different species.

The best minds revise and expand existing knowledge and anticipate future developments. Instead of old monistic, either/or reasoning which traditionally results in discard of existing theories in favor of new ones, complementarity dynamics combines some of the old and the new to increase understanding of evolution.

Enigmas of Primate Evolution

Primitive primates like lemurs and tarsiers persisted in North America and Eurasia but were reduced in number near the epoch's end. Most of the known wild lemurs are found in Madagascar, where, it is believed, they have evolved in island isolation since the Eocene.

Paleontologists continue to dispute whether shrews or animals such as Plesi (*Plesiadapis*, Chapter 23, Figure 23.6) should be considered to be the first primates. They haven't agreed which animals were ancestral to the prosimian lemurs or to tarsiers, or to both. Some primatologists posit an ancestral line from shrews to lemurs, others from shrews to tarsiers. Some believe tarsiers rather than lemurs have led to monkeys (simians).

Some of these disputes hinge upon the fact that both lemurs like *Notharctus* (Figure 24.9 a) and tarsiers like *Tetonius* (Figure 24.9 b) have structural and behavioral characteristics seen in higher primates (anthropoids). Again, patterns of sequential thought have limited the opportunity to posit mosaics of DNA-carried characteristics, jumping and leaping, and spanning the centuries between lifeforms.

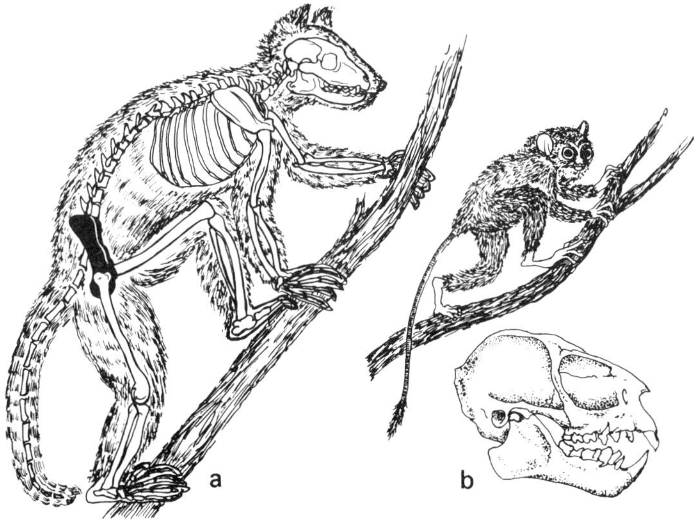

Figure 24.9 (a) *Notharctus* (lemur, Primate). (b) *Tetonius* (tarsier, Primate). (a. After Gregory, 1920; b. After Osborn, 1910).

The toothless gap (diastema) between canines and cheekteeth which occurs in Plesi is also characteristic of multituberculates, rodents, and some perissodactyls, while other structural characteristics appear in different combinations in these animal groups. Such a diastema is *not* a characteristic of lemurs, tarsiers or simians. It becomes quite impossible to believe in lineal lines of descent without positing jumping genes!

The grasping hind feet and divergent first toe of prosimians are seen not only in certain higher primates but also in some arboreal rats, chameleons and marsupials. The advanced hand to eye coordination of prosimians and simians is seen also in squirrels and cats. Primates and cats have visual systems with stereoscopic depth perception not present in squirrels and tree shrews. Characteristics such as these do not appear in animals in ordered taxonomic patterns. The concept that direct sequential lines of descent can produce an orderly primate fossil record is obsolete. Concepts of DNA patterning and jumping genes are making their indelible mark on theory.

Efforts to integrate the primate fossil record and plate tectonics have met with some successes. The earliest plesiadapiforms have been found in North America and Europe. Eocene prosimians have been found in North America, Europe and Asia, thus suggesting possible evolutionary connections between plesiadapiforms and prosimian lemurs or tarsiers. The continuous life and evolution of lemurs on Madagascar until the present is thought to have occurred because that island was isolated by plate movements during the Eocene.

The African fossil record is very poorly known for the Paleocene and Eocene. However, because forest dwellers are not easily fossilized, the lack of an African fossil record does not necessarily negate the possibility of an African pathway of ancestral shrews or plesiadapiforms moving from Europe to Madagascar in the Mesozoic when the continents of Europe and Africa were joined. Such animals could then have provided some ancestry for Malagasy lemurs.

Plate tectonics has been of little or no assistance in resolving the question of the origination sites of the higher anthropoid primates. The only known Eocene higher primate fossils are from Burma (*Amphipithecus* and *Pondaungia*) but they are not thought to be directly ancestral to other higher primates. There are generally considered to be no confirmed anthropoid primates from Africa until the Oligocene.

In 1987 Russell L. Ciochon and John G. Fleagle edited *Primate Evolution and Human Origins* in which 40 authorities contributed articles to update the known facts and theory of primate evolution. It is striking that only a few chapters, especially those of V. Sarich and M. Goodman, dealt with comparative molecular processes of selected primates. It appears that the field of paleontology is only beginning to encourage biomolecular contributions to evolution theory, although as early as the 1960s, Linus Pauling suggested the advantages of molecular *dating*.

The study of DNA affinities does not necessarily provide direct evidence for dating a separation time between sister genetic lines. Rather, the differences in DNA nucleotide sequences may give evidence of recombinations at the time of viral activity, which occurs at random intervals.

The best near-future contributions of paleontologists will be (1) continued identification of fossils, with dating based on stratigraphic and radiometric methods, (2) comparisons of animal similarities across phyletic boundaries, with awareness of possible DNA patterns, and (3) abandonment of efforts to classify and interpret fossils according to rigid Darwinian theory. In the long perspective, it may become possible to reinterpret the fossil record in conjunction with molecular knowledge and clarify many points of theory.

Expanding Perspectives

Linus Pauling has not been the only physicist to urge biologists and paleontologists to expand their perspectives. Since the time of Niels Bohr, physicists have attempted to expand the dialogue of evolution. To think of an animal as coevolution of a macrosystem of body and a microsystem of molecules or cells is an exercise in complementarity dynamics for which physicists have been prepared. They have dealt with the many complements of atomic processes and have seen the necessity to use similar reasoning to contemplate molecular processes. Thus they have dealt with ever expanding concepts.

While paleontologists generally have continued to use monistic reasoning in evaluation of the structure and function of teeth and bones, physicists and mathematicians Ilya Prigogine and Erich Jansch (1980) elucidated principles of self-organizing structures, of catalyzing processes (hypercycles and ultracycles) of dissipative structures, which function in an interplay of matter and energy. It is as though many paleontologists have neglected to walk across the hall to the physics department to inquire about the changes in the way physicists and molecular biologists think about evolution.

F. Eugene Yates (et al, 1987) edited *Self Organizing Systems, the Emergence of Order* in which 36 experts examined the causalities within and of nature's system. These experts believe that self-organizing systems produced the universe in its infinite variety with transformations of energy to matter to life. The contributors included only one paleontologist, Stephen J. Gould, who is pioneering paleontological movement from the position of modified neo-Darwinian (Gould, 1977) to expanded view of transformationism (Gould, 1987).

Evolution theory is no longer the realm only of biologists and paleontologists, nor only of physicists and mathematicians. It involves the combined intelligence and insights of all those who think deeply. The Eocene Epoch is a transitional time frame to explore the beginnings of DNA recombinations leading to modern animals, as we synthesize the new scientific information and methods of modern thought.

End of Eocene Epoch

The Eocene closed with evidence of more modern types gradually replacing many primitive species. The Eocene fossil record revealed 25 new families. Thirteen families became extinct, including multituberculates, many of the more primitive insectivores, condylarths, plesiadapids, oxyaenid creodonts, adapid lemurs and the uintatheres. The series of extinctions in the Eocene were probably related most closely with explosive volcanism and lowering temperature, with increasing aridity and corresponding vegetational changes from woodlands to grasslands.

25

Cenozoic Era — Tertiary Period — Oligocene Epoch Age of Cold Seas, Volcanic Eruptions, and Rising Mountain Ranges

Forty million years before present (40 MBP) the Oligocene Epoch began, with continued separation of North and South America but continued interrelationships of animals of North America and Eurasia across the northern Bering reaches. The climate continued to become more seasonal as seas retreated from land areas. Algeria, Tunisia and Morocco emerged as the sea withdrew from North Africa. Flowering plants and grasslands extended their range.

The lowering of sea level is believed to be associated with expansion of ocean spreading ridges as volcanism and mountain building increased. The coastal mountains of California and the Cascades of Oregon and Washington, USA, upthrust in the Oligocene. An intensive period of volcanism began. From 38 to 27 MBP, explosive volcanic episodes rocked wide areas of Colorado, New Mexico, Nevada, Utah, Oregon, and Mexico in North America; Peru in South America; the Philippine Islands; and Ryukyu Island of Japan.

Extensive volcanic eruptions occurred in the Sierra Madre Occidental field of western Mexico between 35 and 27 MBP. The area, 250 kilometers wide by 1,200 kilometers long, evidences 200 to 400 volcanic centers (calderas). There was a precipitous drop in temperature as thousands of cubic kilometers of ash were again added to the atmosphere. The resultant temperature drop was not necessarily a completely negative phenomenon. Decreased ocean temperatures stimulated upwelling of resources from the bottom and encouraged new aquatic species radiations.

Sea ice occurred 38 MBP, as bottom water temperature dropped five degrees celsius (C) causing a major crisis for some of the deep sea creatures. As cold ocean currents circled the Antarctic, weather patterns changed. Marine mammals were rare in the Oligocene, but there are fossils of early whalebone whales.

The frost frequency increased where the evergreen dicot broad-leaved rain forests gave way to deciduous hardwoods in areas such as Oregon, USA. Shifts from dominant conifer forests to dominant juniper woodlands resulted from increasing aridity in other areas such as Colorado (Axelrod, 1981: 11).

Development of "Modern" Animals in North America and Eurasia

Oligocene animals of Europe and North America became much more modern in appearance as many of the more primitive orders and families of typical Paleocene mammals became extinct. Opossums continued to proliferate in North America, as did insectivores.

Rodents — Puzzling Similarities

Small rodents, rats and mice radiated quickly and profusely. They are thought by some specialists to have evolved from the tiny multituberculates, but others believe that rodents arose from Cretaceous insectivores. Beavers were first found in the fossil record of this period.

In the Oligocene of North America 12 families, 41 genera, and about 80 species of rodents are identified. One family (Cylindrodonta) is considered to be possible ancestors of South American caviomorph rodents, first known there in the Oligocene (Wood, 1980: 3-4).

There are very puzzling dental similarities among multituberculates, rodents, insectivores and early primates. Perhaps DNA analysis will identify various morphological patterns that have been recombined in different ways in these early mammals.

Carnivore Superfamilies

Paleontologists who consider miacids and fissipeds to be progenitors of living carnivores have identified fossils of two superfamilies from the early Oligocene of North America (Carroll, 1989: 481; Stern and Carroll, 1989: 296). Superfamily Arctoidea includes dogs (canids) weasels and otters (mustelids) and raccoons (procyonids). Superfamily Aeluroidea includes civets (viverrids) hyaenids, and cats (felids). Placental cats, both saber-tooths such as *Smilodon* (Chapter 27, Figure 27.1 b) and more modern types such as Dini (*Dinictes*, Figure 25.1) about 3.25 feet long, arose in the Oligocene in North America.

The cat family is thought by some specialists, to have evolved from creodonts known from fossil fragments as dominant carnivores of the Paleocene and Eocene. Relatively complete skeletons of creodonts (*Hyaenodon*, Figure 25.2) about 5.5 feet in length, are known from Oligocene beds of North America. *Hyaenodon* was a large animal, four feet high at the shoulder with a very large head. Hyaenodontids were a dominant carnivore in Africa throughout the Oligocene, but their number was reduced in North America and Europe.

Most specialists believe *Hyaenodon* was not in a direct line to the cat family, but rather, was an offshoot of earlier creodonts which may have been ancestral to some of the fissipeds. However, *Hyaenodon* and saber-tooth cats were very similar in their dental condyle (Mellett 1977: 98).

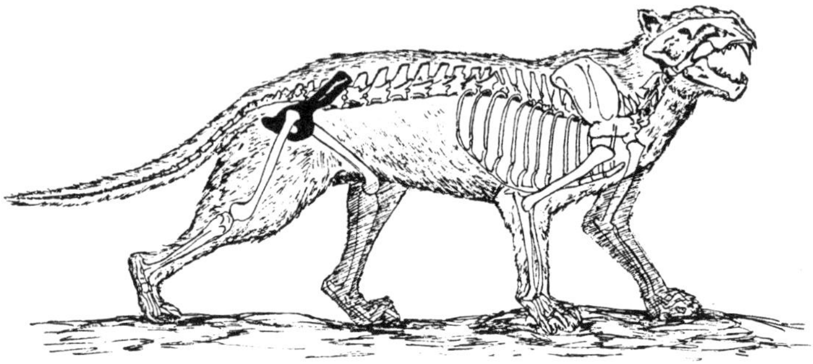

Figure 25.1 Dini (*Dinictes*, Fissipedia, Carnivora, early cat). (After Matthew, 1909).

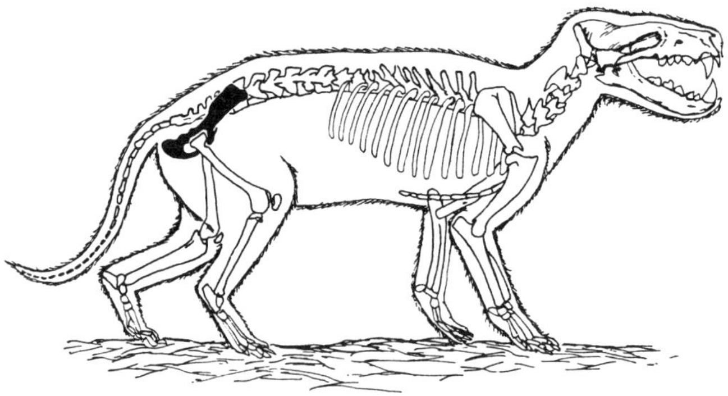

Figure 25.2 Creodont (*Hyaenodon*, Creodonta). (After Scott, 1895).

Bears (*Ursula,* Arctoidea) appeared in Europe in the middle Oligocene. An extinct lineage, bear-dogs, about five feet in length, lived from the late Eocene into the Pliocene. Although some experts believe they were related to canids, others posit that they were related to neither bears nor dogs. The geneology of bear-dogs has not been adequately investigated, but I believe these strange animals may exemplify salmutations of DNA (Figure 25.3).

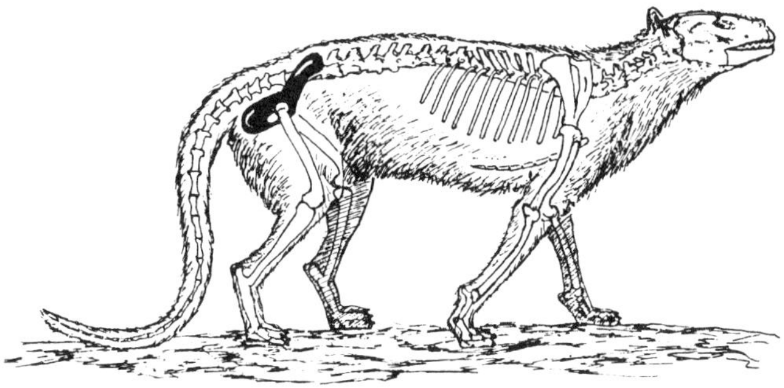

Figure 25.3 Bear-dog (*Daphoenodon*, Fissipedia, Carnivora). (After Peterson).

The extinct cat-like nimravids (paleobelids) may have evolved from miacids other than Arctoidea or Aeluroidea according to some experts. Their saber-tooth specialization, however, is known in many felids and in South American *Thylacosmilus* (Chapter 27, Figure 27.1 a) as well as in certain amblyopods and artiodactyls. The relationships of creodonts, miacids and other early carnivores, and their relationships with early condylarths, perissodactyls and artiodactyls remain open questions. The extinction of some of the archaic carnivores during the early Tertiary, possibly due to climatic deterioration, allowed radiation of more modern kinds of carnivores.

Perissodactyls — Horse Relatives Contradict Darwin

Small three-toed horses such as *Miohippus* (Chapter 24, Figure 24.2) and *Mesohippus* (Figure 25.4) were common in North America. Note the skeletal features of hip, neck, vertebrae and ribs of *Mesohippus* which have many similarities with *Hyaenodon* and condylarths as well as with other horses. By the Oligocene, horses were very diverse in North America.

Donald Prothero and Robert M. Schoch (1989) recorded not only a full range of horses beginning with *Eohippus*

(*Hyracotherium*) but also corrected the myth of a single lineage of horses graduating from *Miohippus* through *Mesohippus* in the Oligocene. They found that both of these genera included many species living in the same areas for almost five million years. These genera have not intergraded from one to the other. They are quite separate. *Miohippus* is distinctly larger than *Mesohippus*. In the late Oligocene, the *Miohippus* genera split rather suddenly into browsers (anchitheres) and grazers (equines). This rather radically challenges the Darwinian picture of minor random mutations in horse evolution.

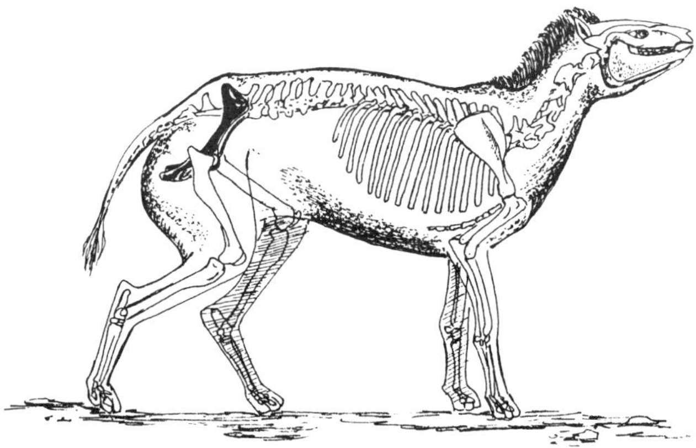

Figure 25.4 *Mesohippus* (Equoidea, Perissodactyla). (After Scott, 1937).

Horses were not as numerous in Asian deposits, but many other perissodactyls: tapirs, rhinos, chalicotheres and titanotheres were found. Numerous, huge, North American titanotheres (*Brontops*, Figure 25.5) perished in the mid-Oligocene.

Two very different kinds of rhinoceroses are found in North American Oligocene deposits. The running rhino (*Hyracodon*, Figure 25.6) was five feet in length and somewhat similar in structure to *Mesohippus*. The amphibious rhino

(*Metamyndodon*, Figure 25.7) 14 feet in length, was heavily built, more like *Uintatherium* (Chapter 23, Figure 23.5) and titanotheres (Figure 25.5).

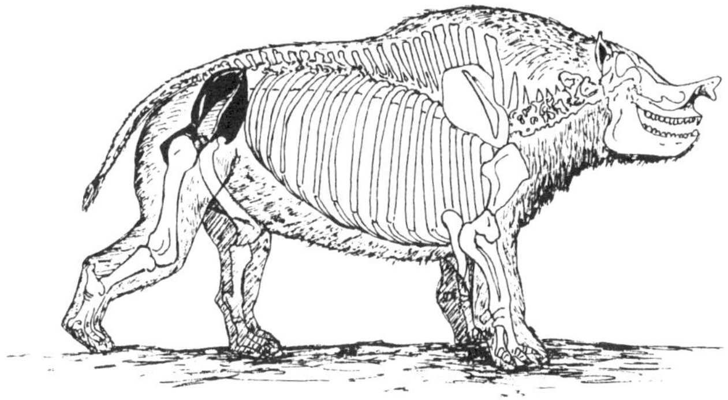

Figure 25.5 *Brontops* (Titanothere, Brontotherioidea, Perissodactyla). (After Osborn, 1910).

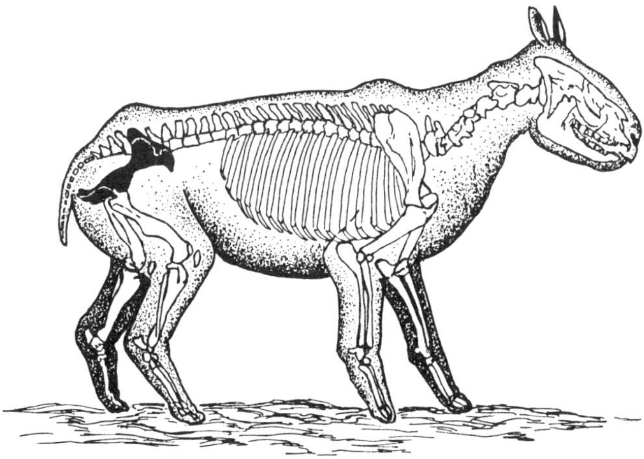

Figure 25.6 Running rhino (*Hyracodon*, Rhinocerotoidea, Perissodactyla). (After Osborn, 1910).

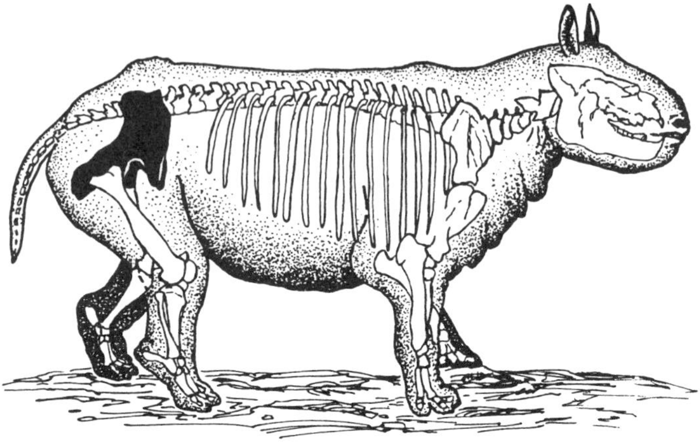

Figure 25.7 Amphibious rhino (*Metamyndodon*, Rhinocerotoidea, Perissodactyla). (After Osborn, 1910).

These animals illustrate the fact that accepted classification of animals, based on agreed upon features, may dissolve when other important features are considered. If hip structure identifies different lines of dinosaurs, why is hip structure not crucial to identity of rhinoceroses and titanotheres? If foot structure is crucial to identity of perissodactyls and artiodactyls, aren't amphibious rhinos and titanotheres closer in relationship than the two rhino lines? Why are the long-spined vertebrae above the clavicle of titanotheres, camels and bison very similar and not at all like some other animals in their own ordinal classification? These questions cry out for analysis of DNA patterns across ordinal lines.

Artiodactyls —

Mystery of Hoofed, Clawed, Thumbed Oreodonts

In addition to perissodactyls, artiodactyls were found in North American deposits. Some unusual primitive animals such as the first oreodonts arose. They were called ruminating swine as they were pig-like in appearance, but some of them had clawed, not hoofed feet and inexplicable *thumbs* on their front feet (*Agriochoerus*, Figure 25.8). What is the source of such thumbs? Some paleontologists have considered the oreodonts to be closely related to early camels. Oreodonts arose and quickly became the most common of the American ungulates, outnumbering all others in the fossil beds of South Dakota — more than 12 genera.

In the Oligocene, artiodactyls proliferated faster than perissodactyls in both the Old World and in North America. The evolutionary lines of artiodactyls are uncertain at best but seem to diverge into the non-ruminants and ruminants. The non-ruminants: pigs (such as *Propalaechaerus*) were found in Europe; peccaries (such as *Perchaeras*) and sheep-size camels were prevalent in North America. Hippopotomi, thought to be related to pigs, arose inexplicably in the Northern Hemisphere and in Africa.

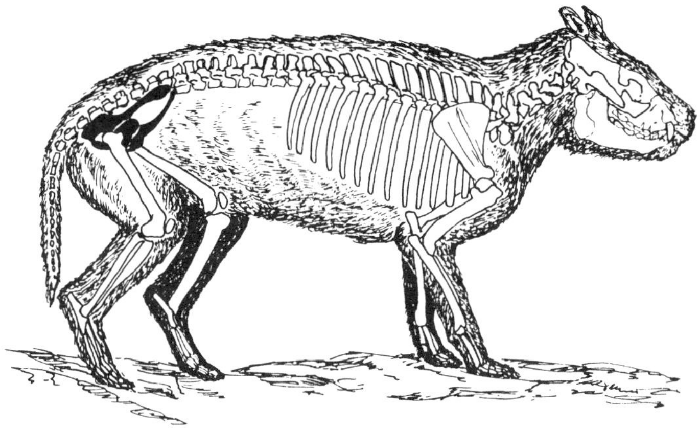

Figure 25.8 Ruminating swine (*Agriochoerus*, oreodont, Artiodactyla). (After Osborn, 1910).

Ruminants first found in Oligocene deposits radiated in the Miocene: deer, giraffes, chevrotains and okapi in Europe; prongbucks and the antelopes of North America. Although artiodactyl fossils showed some similarities in morphological development, North American and Eurasian genera are thought to exhibit little evidence of interchange during this epoch.

Isolated Curiosities of South America

Opossums reached their highest diversification in South America and radiated into other marsupials, such as primitive kangaroo-like seed and insect eaters, mouse-like marsupials and marsupial carnivores, ranging from opossum size to cave bear size (*Borhyaena,* Chapter 26, Figure 26.1). One of the strong-clawed carnivores was a saber-tooth hyena.

South American strata deposits continued to preserve armadillos with heavy body armor, glyptodonts and sloths. Anteaters developed powerful claws to tear termite nests apart. The fossil record shows no indigenous placental carnivores in South America, and a limited number of herbivorous placental mammals.

Notoungulates ranged from rabbit-size to rhino-size, radiating into the larger sizes. Litopterns of horse-like and llama-like aspect expanded onto grassy plains (Chapter 26, Figures 26.2 and 26.3). Additional orders of ungulates appeared in some of the South American island continent deposits. *Astrapotherium* (Figure 25.9) about nine feet in length, had some characteristics similar to North American amphibious rhinos, but is thought to have had a proboscis (thus is said to be convergent with proboscideans, although not related to them nor to rhinos). To explain the origin of the morphological characteristics (the phenotype) of these animals, we must consider similar DNA configurations — in addition to competition and convergence.

Pyrotheres also appeared in South American deposits. Some experts believe the fossil fragments are too few to draw a complete animal. However, others have proposed that

Pyrotherium looked rather like Figure 25.10. It had a proboscis as did animals in several other South American orders.

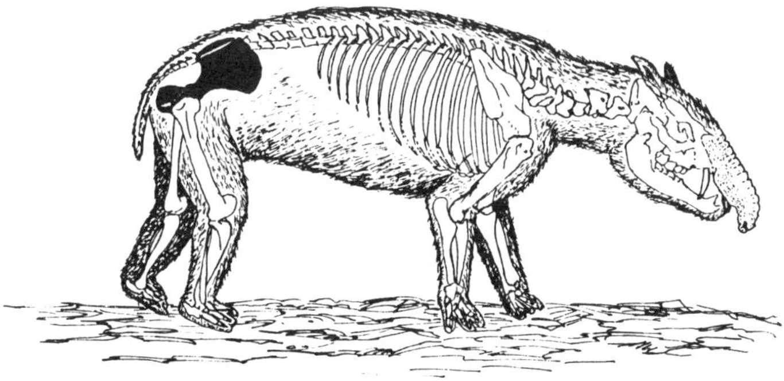

Figure 25.9 *Astrapotherium* (South American ungulate). (After Scott, 1928).

Figure 25.10 Posited *Pyrotherium* (South American ungulate).

Australia — Barren Fossil Record

South America apparently had no indigenous placental carnivores, and Australia had no placentals at all. The fossil record of Australia is very limited in Tertiary strata. There are no early fossils and few in the later Tertiary. Recently, however, one skull was found, the only Oligocene specimen, a marsupial squirrel.

Abundant Fossil Treasures of Africa:

DNA Recombination Evidence?

Oligocene river bed deposits in Egypt reveal crocodiles, turtles, fish and ancestral mammals. The main mammalian fossil finds of these lower Oligocene deposits in Fayum Quarry I represent the orders Artiodactyla, Primates and Hyracoidea. Other mammal orders represented by descending number of fossil elements were Creodonta (including Condylarthra) Rodenta, Proboscidea, Chiroptera (bats) Sirenia (sea cows) and Embrithopoda. The artiodactyls were mainly pig-like and hippo-like fossils.

The hyracoids were herbivorous animals. They had hoofed and padded feet: four toes on front feet, three toes on hind feet. They had many similarities to rodents. Their teeth had some similarities to those of rhinos. They varied in size. The largest (*Titanohyrax*) was the size of the early proboscidean Moe (*Moritherium,* Figure 25.12 a). Living members of the order are conies, hyraxes and the pig-size capybara. Such a strange constellation of characteristics in one order cries out for a DNA recombination explanation.

Hystricidae rodents are known from the Oligocene of Africa. They are thought to have evolved in the Eocene from Phiomyidae rodents. Some specialists believe Hystricidae rodents rafted from Egypt across the Atlantic Ocean to give rise to the caviomorph rodents which appeared suddenly in South American deposits in the Oligocene. Other scientists believe caviomorph rodents arose from Paleocene rodents of North America, and rafted from there to South America.

The only member of the order Embrithopoda, found in Africa, was *Arsinotherium* (Figure 25.11). That such a large animal could be the single known representative of its kind, suggests a possible salmutation from another large animal. Some authorities have recently suggested that these animals are related to both the proboscideans and perissodactyls.

The Oligocene fossil sites of Egypt contained fascinating evidence of the earliest proboscideans (known also as Order

Subungulata). Moe (*Moritherium,* Figure 25.12 a) is thought by some paleontologists to be ancestral to the mastodonts, mammoths and elephants.

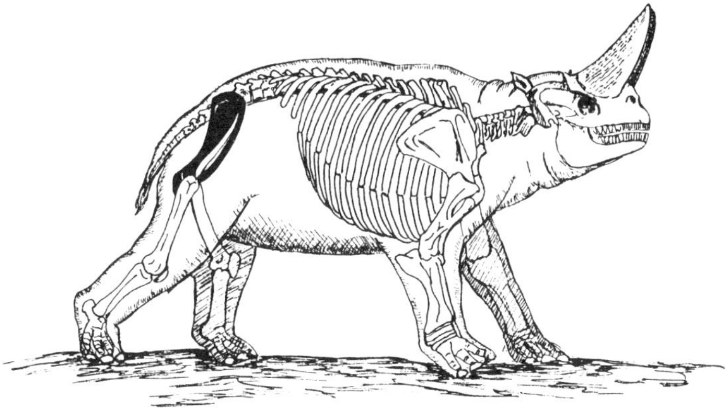

Figure 25.11 *Arsinotherium* (Egyptian Embrithopoda). (After Andrews).

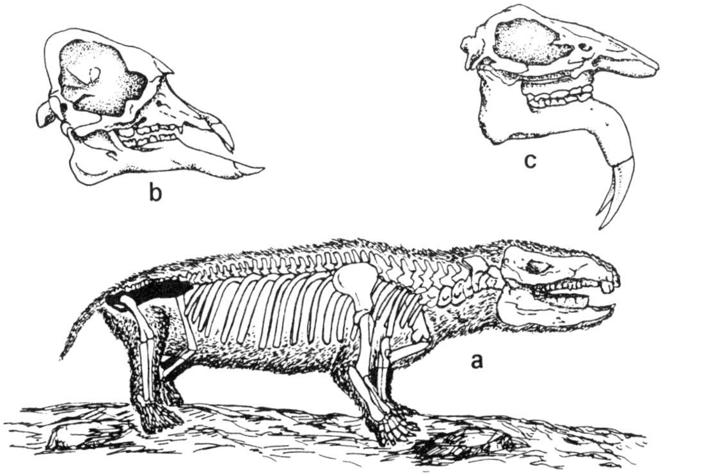

Figure 25.12 Proboscideans (a) Moe (*Moritherium*). (b) *Phioma.* (c) *Deinotherium.* (a, b, After skeleton and exhibits, University of Nebraska State Museum; c. After Gaudry).

Some experts think that Moe had a tapir-like snout. Others believe that it had no proboscis at all. It would seem strange, viewed from a Darwinian perspective that an order of animals named for a characteristic proboscis — originated from a genus without that characteristic, yet earlier genera of litopterns in South America and tapirs in North America *did* have proboscises.

Perhaps the inferred lack of a proboscis is one reason some scientists would not place Moe in the classification Proboscidea, but give it a separate classification (Moeritherea). Other animals classified as proboscideans, from the same Egyptian quarries, include *Phioma,* the earliest one with a trunk, and *Deinotherium* (Figure 25.12 b, c).

Deinotherium is a fascinating genus because it had some characteristics in common with *Barytherium, Palaeomastodon* and *Phioma,* but it had a most distinctive difference. *Deinotherium* had a lower set of tusks that turned downward in a 45 degree angle, instead of having two sets of tusks projecting forward, as in other early proboscideans.

This seems to be such an abnormal condition that it is a wonder how they survived. But their tusks must have been useful, for these animals lived for millions of years, into the Pliocene. This characteristic is an unusual salmutation. Nevertheless, the tusks of deinotheres evolved from incisors, not canines, the same as other proboscideans. Again we see evidence that DNA moves in patterns which do not illustrate a sequentially evolving lineage. Many paleontologists believe that the Fayum of Africa was the place of origin and radiation of the proboscideans that spread, in subsequent epochs, to almost all parts of the world. Others have recently posited that ancestral proboscideans may have arisen in Asia, although the evidence is slight.

The first sirenians, sea cows, classified as ungulates, are known in the Oligocene of northern Africa, the West Indies and eastern Europe. Some scientists think sirenians may be related to proboscideans, because sirenians have skulls with many resemblances to the skull of Moe. However, they resemble whales in their body form, their forelimb paddles and their flattened horizontal tail fluke. Their thick ossified ribs are similar to those of aquatic reptiles.

One living form of sirenians, the manatees have eight pairs of cheek teeth which are continually replaced, a condition found in duck-billed dinosaurs, but not in other *mammals*. This chimera, the interesting combination of characteristics of these animals, makes the sirenians a good choice for DNA analysis.

It seems to me that resemblances in Oligocene sirenians and proboscideans do not necessarily indicate a land animal return to the sea, but rather suggest that DNA patterns may move between animals in the two mediums.

Africa has never been a habitat for Dermoptera, Edentata, Monotremata nor Marsupialia. By the end of the Oligocene, Embrithopoda was absent from East African deposits.

Fossil evidence of vertebrates, invertebrates and plants shows closer relationships between South America and Australia than between Africa and either of these continents. But even the marsupials of South America and Australia evolved in different patterns.

The Rise of Higher Primates

The first evidence of higher primates in Africa was from a tropical rain forest area of Egypt. These primates are known only from dental fragments and a few skulls. They are considered to be catarrhines, the classification which includes Old World monkeys, apes and humankind. The fossil record shows that they evolved separately from the smaller South American prehensile-tailed monkeys of the Oligocene, the platyrrhines, which never evolved into apes.

In his search for catarrhine and platyrrhine primate origins, R. Hoffstetter (1987: 105-106) hit upon the idea of identifying and comparing their parasites. He discussed the work of several researchers and thought the evidence should be interpreted cautiously, but he concluded that there was a common origin for catarrhines and platyrrhines because different worms (nematodes) infest lemuroids and tarsioids. And he believed that only one of these lines could have been ancestral to the platyrrhines, catarrhines and later anthropoids.

Whether or not there was a common origin or whether the facts may lead to a different conclusion, the concept of comparing parasites of animals which have evolved in similar patterns in different locations is innovative and intriguing.

The technique is an initial approach to synthesis of information from two diverse fields, paleontology and parasitology. A reasonable next step will be identification of the viruses of the parasites, and virus DNA relationships of both parasites and hosts. I believe this will lead to confirmation of my principle of responsive salmuation. The DNA of viruses of host and parasite of two separately evolving lines may identify the DNA relationships of the two lines, and provide understanding of DNA recombination, and production of similar phenotypes in totally separate environments. This would explain the phenomena previously called convergent or parallel evolution.

Found in Oligocene Fayum deposits of Egypt were the primates: *Oligopithecus, Moeripithecus, Parapithecus, Apidium, Aegyptopithecus, Aeolopithecus* and *Propliopithecus* (Conroy, 1976; Simons, 1987: 146-147).

Oligopithecus is thought by some experts to be a cercopithecid monkey, the most primitive of these Egyptian fossils. Other scientists believe all the Fayum primates were apes, and that monkeys were first found in East Africa in Miocene deposits. If apes preceded monkeys in the fossil record, the accepted concept of evolution from prosimians to monkeys, to apes, may be invalid.

Parapithecus is considered by primatologists to be ancestral to most other catarrhines. It is thought to be closely related to *Apidium,* so they are included together in the family Parapithecidae and considered to be Anthropoidea.

Aegyptopithecus (Figure 25.13 a) about the size of a gibbon, is considered to be an early ape (pongid). It had a small brain, a long snout like a lemur, and it ate leaves and fruit. It had some similarities to the Miocene gibbon, *Pliopithecus* (Figure 25.13 b) and some similarities to the Miocene Dryopithecene ape, Proconsul (Chapter 26, Figure 26.12). Some scientists believe *Pliopithecus* may be ancestral to Proconsul.

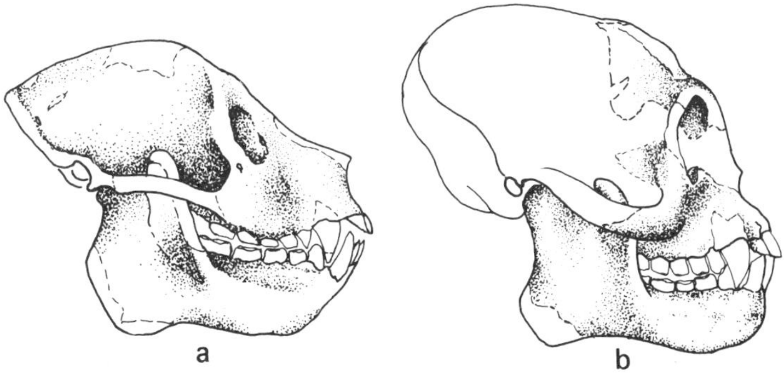

Figure 25.13 (a) *Aegyptopithecus* skull, Catarrhini, Primate). (b) *Pliopithecus* skull, Catarrhini, Primate. (After Andrews).

Aeolopithecus is thought to be a primitive gibbon and ancestral to later gibbons.

Propliopithecus is accepted by some anthropologists to be the earliest hominid, and others consider it to be ancestral to the Miocene gibbon, *Pliopithecus,* thus a hominoid, but not a hominid. The fine line between ape-like and human-like fossils has not been definitively drawn. Some primatologists believe that discovery of a fossil in Africa or Eurasia showing characteristics intermediate between *Propliopithecus haeckeli* of the Oligocene and *Ramapithecus punjabicus* of the late Miocene may verify a sequential evolutionary pathway toward humans.

Glenn C. Conroy (1976: 120) made the point that the primates of these Egyptian deposits "make good intermediaries between the Eocene prosimians and the later Miocene Anthropoidea." Although Conroy sees similarities between catarrhines and South American platyrrhines, he does not believe one evolved from the other. Rather, he judges it more likely that both African and South American Anthropoidea are descendents of some Northern Hemisphere Eocene primate group, perhaps adapids (lemurs).

I believe that the overlapping characteristics of the primate fossils require linkage of morphology patterns and DNA patterns before we can become comfortable with the evidence of higher primate ancestry.

Unknown Source of South American Primates

The primate fauna of South America was arboreal and spread at least as far as 40 degrees south latitude. Some scientists believe that platyrrhine monkeys evolved from an ancestral catarrhine, which rafted on a floating island from Egypt, along with ancestral rodents. Others posit evolution from North American prosimians. There is no known fossil evidence of either catarrhines or platyrrhines in North America, but it has been proposed that Eocene tarsiods or lemuroids may have evolved into South American ceboid monkeys, like marmosets, and then radiated into other platyrrhines. However, no intermediate stock has been found to explain the appearance of either ceboid monkeys or caviomorph rodents, which appear, fully developed, in the fossil Oligocene deposits of South America.

Although it may be possible for a few primates and a few rodents to raft across an extensive water barrier from Africa or North America to South America, I believe we should look more carefully for an indigenous origin of these fossil finds.

The first appearance of a ceboid was of a jaw fragment and teeth of *Branisella* in Bolivia. Skulls of *Tremacebus* and *Dolichocebus* were found in the late Oligocene of Argentina (Rose and Fleagle, 1987: 125). These platyrrhines shared features of both tarsiers and lemurs, and had certain features in common with the catarrhines of Africa. Again, these facts call for a search for common DNA patterns.

Platyrrhine monkeys in South America and catarrhine primates in Africa were evolving separately in the Oligocene. There is no fossil evidence that New World prehensile-tailed monkeys were ever associated with Old World Monkeys or apes, and there has been no evidence of either one in North America. Could platyrrhine monkeys and caviomorph rodents have evolved from earlier South American ancestors by means of salmutations? In the late Oligocene, prosimians disappeared from both North America and Eurasia, remaining only in Madagascar, Indonesia and the Philippines.

The predominant belief of the biological community has been that there is no feed-back loop (causal role) from the environment to the gene structure of lifeforms. This is required by a concept of linear time: that it is only possible to consider time as the framework for events from past to present to future. Scientists, therefore, have believed that the main process by which animals are adapted to their environment is by destruction of less fit individuals and species.

My conception of the reciprocal moment in complementarity dynamics, and the principle of responsive salmutation, allows that the environment can influence the gene pool directly by natural processes of DNA recombination — in anticipation of *and* in response to changing environments. The concept of time itself is part of the equation of adaptation.

Thus the beginning of the radiation of catarrhine primates in Africa and platyrrhines in South America was not necessarily two directly connected events. Rather, I believe that ancestral forms in each location gave rise to the new primate forms by means of responsive salmutations.

End of the Oligocene Epoch

The epoch closed as the ancestors of modern kinds of animals radiated in the trees and on the plains of the world. Forty-five families of animals have been identified as originating in the Oligocene. Extinct were various archaic types, particularly primitive ungulates, the titanotheres, ancestral mastodons, running rhinoceros, tapir-like perissodactyls, running swine, primitive artiodactyls including European relatives of camels, primitive Old World ruminants and several rodent families. Twenty families of animals made their last appearance in this epoch.

26

Cenozoic Era — Tertiary Period
Miocene Epoch
Miocene Apes — Ancestors of Humans?

The Miocene is a transitional epoch in which changes occurred in the environment which scientists believe influenced the rise of apes, the ancestors of humans. The epoch began 25 million years before present (25 MBP) and lasted 20 million years. The full facts of primate transitions are still a mystery, but scientists believe that the fossil record of the Miocene will eventually clarify this issue. I believe many surprises are in store before the mystery will be considered "solved."

It has been postulated that Africa was moving toward Eurasia in the Miocene. By the mid Miocene the connection of these two continents was made across the Arabian Peninsula and the Mediterranean Sea was formed from the Tethys Sea. Although a long-term cooling trend was in progress, the beginning of the Miocene remained warmer than today, with great tropical rain forests encircling the Earth, both north and south of the equator. The Bolivar Trough separated tropical Central America (an adjunct of North America) from South America, as South America moved close to its present position.

Douglas L. Smith (1985) has interpreted research on the complex movements of the tectonic plates in the Caribbean area, including interactions of the North American, South American, Caribbean, Cocos and Nazca Plates. There are many uncertainties in the related movements, but Smith determined that sporadic convergence and divergence of the North and South American plates have deformed and controlled the Caribbean Plate. These movements, coordinated with changing sea levels, brought about the rising island arc that began to link South America with North America in the Miocene. The final joining of these continents occurred about three million years ago in the Pliocene Epoch.

In the mid-Miocene, from 14 to 12 MBP, another period of volcanic eruptions occurred in the American West. Oxygen-isotope studies of foriminifera tests from sea bottom cores indicate a marked decrease in sea surface temperature. This was followed by loss of plants requiring regular summer moisture and subtropical warmth.

As world temperatures dropped there was a major buildup of glaciers in Antarctica and, by the late Miocene there were large ice masses in the Arctic and Alaska. Changes in sea level, tectonic movements, sedimentation of ocean basins and glaciation are all interrelated with climatic changes. Samuel M. Savin and Robert G. Douglas (1985: 307-308) estimated that sea level variation due to glaciation has been the most significant of these changes in the last 15 or 20 million years. They have estimated (using a 4,000 meter ocean depth average) that at the present time "97.3% of water is in oceans, 2.1% in polar ice and .6% in groundwater systems." In the Miocene, the polar ice was much less extensive than it is today, with sea bottom temperatures of 10-15 degrees C., compared with present one to two degrees C.

Miocene fossil deposits evidence temporary marine invasions of Patagonia and Columbia in South America, which brought rich and varied sea life. As seaways changed and became more restricted and mountains rose, both glaciers and semi-deserts began to spread.

There are differences in interpretation of data for climatic conditions in the Miocene; however aridity and cooling probably increased in areas of both North and South America as lush rain forests of the middle Miocene gave way to savannas. The Miocene is noted for a larger interchange of species than previous Tertiary epochs, especially between North America and Eurasia, and there were also minor interchanges between North and South America. As the Earth cooled, habitats diversified, and new forms of life arose. Twenty six new families first appeared in the Miocene. The fauna of the Miocene was greatly modernized compared to previous epochs.

During the Miocene, 17 families made their last appearance on earth. At the end of the Miocene, about half of the fauna then living represented about half of today's families of terrestrial mammals.

Australia and India

India's First Mammals and Birth of Himalayas

Fossil evidence on the Australian continent has been very sparse, yielding only a few specimens of marsupials related to living species, first located in Miocene deposits.

India exhibited the beginning of mammalian deposits in the Miocene. The tectonic movements of India are in dispute. Did it break off from Antarctica or Africa? When did it begin its northern migration? Did it arrive at its present location in the Eocene, Oligocene or Miocene? It is believed by some authorities that it was in the Miocene Epoch that India crunched into the underbelly of the Eurasian continent giving rise to the Himalayan Mountains. There is evidence of primates in India in the Miocene; but surely there is a rich paleontological history remaining to be discovered, as there has been relatively little exploration there.

Persistent Migrations

of Sea Mammals, Amphibians and Reptiles

The marine environment of the Miocene showed lingering primitive lifeforms. There was a rapid radiation of higher cetaceans. Whales had never been more varied and abundant. During the Miocene, almost every toothed and whalebone family arose, as well as sperm whales and dolphins and the first pinnipeds: eared sea lions, earless seals and walruses. A filter bridge of islands across the Isthmus of Panama is evidenced by an interchange of aquatic lifeforms between North and South America.

Many were amphibians moving to North America, some by way of the West Indies. But except for one South American frog genus, there was little or no similarity between the Miocene frogs of the two continents. New frog species arising in the north radiated into recent species. New species of lizards arrived in Florida from the West Indies. Non-poisonous snakes, with advanced jaw structure, first found in North America in the Oligocene, were found in South America by 15 MBP.

Soft-shelled river-dwelling turtles arrived in South America but failed to become established. Turtle fossils are very different in North and South. A primitive form of marsh turtles from North America are thought to have returned to an aquatic environment. They spread widely into the world seas in the Miocene. It has been postulated that some types of marsh turtles may have evolved into tortoises.

South America

Convergent Evolution Due to Similar DNA Recombinations?

Living amphibians in South America are related to Miocene genera. Crocodiles increased in diversity in Columbia, Argentina, Brazil and Peru. Primitive South American lizards crossed the aquatic barriers to the North.

In the Miocene, caviomorph rodents were numerous and varied. It has been determined from their bones that their muscle formations resembled those of Old World porcupines.

Their large, high crowned, continuously growing teeth resembled those of notoungulates. The place of origin of these rodents, whether in South or North America or Africa, remains to be discovered. Capybaras appeared in the mid Miocene. They were the largest rodents ever known — four and a half feet long and 20 inches high. The evolution of South American rodents remains a challenge for further work.

About nine MBP two terrestrial genera of ground sloths made the watery passage from South America through Central America and on to Texas and Florida. One raccoon genus (procyonid) arrived in South America from the North. Few if any other land animals made the trip across the isthmus until the Pliocene.

The South American mammalian fauna remained essentially stable, in the same orders as in previous epochs of isolation (for approximately 70 million years). Large numbers of armadillo, glyptodont, anteater and ground sloth (xenarthran) fossils have been found.

Marsupials are said to have evolved from omnivores like opossums to carnivorous forms such as numerous *Borhyaenas* (Figure 26.1). Smaller borhyaenids were known from the Paleocene in Brazil, so it is assumed that they evolved for millions of years prior to the Miocene.

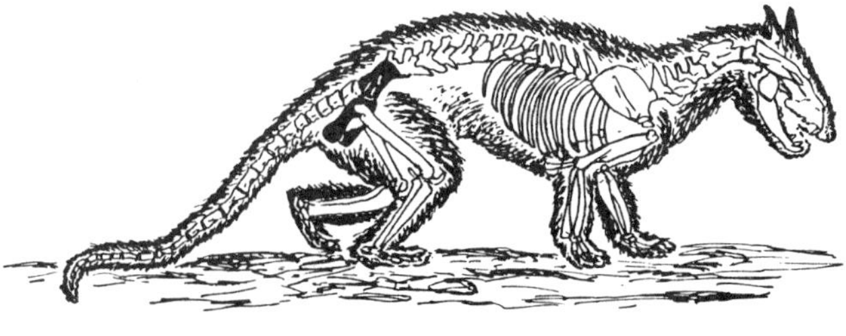

Figure 26.1 *Borhyaena*, (Marsupialia, wolf). (After Sinclair).

The so-called convergent or parallel evolution of these animals with placental wolves of North America is not an adequate explanation for their similarities. I do not believe that random variations and natural selection could structure such similar animal lineages. Scientists forged the concepts of parallel and convergent evolution because they did not then have knowledge of DNA patterning.

Reason requires similar DNA patterns for the similar bodily development of *Borhyaenas* and wolves, and different DNA patterns for their placental and marsupial characteristics. Such recombinations of morphological characteristics requires the concept of recombinations of DNA patterns in salmutations.

Theodosius Dobzhansky (1962: 203) explaining the process he called "quantum evolution," wrote:

> ...the emergence of novel adaptive designs, may involve breaks in the evolutionary continuity when the differences between the ancestor and the descendants increase so rapidly that they are perceived as differences in kind.

That such new adaptive designs could be due to recombinations of DNA mediated by viruses, and that similar DNA recombinations could occur in different lineages on different continents is an idea whose time has come.

South American notoungulates have been identified in two branches: the rodent-rabbit-like Typotheria and Hegetotheria, generally small in size but some as large as black bears, and 30 genera of toxodonts. The toxodonts (Chapter 28, Figure 28.4) looked like short legged rhinos. Notoungulates decreased in number of families during the epoch but still represented a large part of the faunal scene.

Litopterns expanded to a peak of development. Diad (*Diadiaphorus*, Figure 26.2) had browsing dentition, and together with *Thoatherium* with well developed one-toed hooves, was representative of the horse-like litopterns which predated the one-toe development of North American grazing horses of the

Pliocene. The camel-like branch of litopterns was represented by Theo (*Theosodon*, Figure 26.3) which seemed to parallel the development of North American camels. Theo was more heavily built than Diad, and had a short proboscis, giving evidence of possible DNA relationship with African proboscideans or North American tapirs.

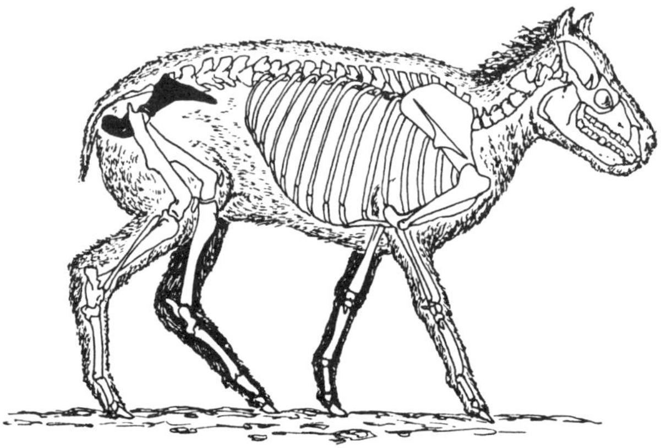

Figure 26.2 Diad (*Diadiaphorus*, Litopterna). (After Scott, 1937).

Unknown Points of Origin of North American and Eurasian Animals

Most of the Miocene fossils of North America have been found in the Great Plains, Oregon and Florida, USA; fossils of Europe in France and Germany; fossils of Asia in Mongolian deposits.

As grasslands expanded and grew in rich diversity in North America, waves of immigrants arrived over the Bering Corridor. The first waves were modest but increased in volume near the end of the Miocene. There is much difference of opinion about points of origin of the animals of the Northern

Hemisphere, but some experts believe that large carnivorous cats and primitive bears arrived in North America from Eurasia, as did primitive beavers and the first grazing hares and rabbits. Late in the Miocene, proboscideans such as *Tetrabeledon* (but not *Dinotherium*) arrived and radiated in North America.

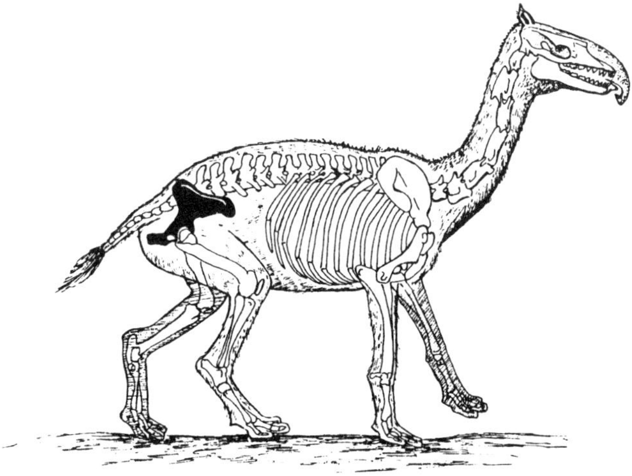

Figure 26.3 Theo (*Theosodon*, Litopterna). (After Scott, 1910).

Common in North America were peccaries, camels, saber-tooth cats, pseudo saber-tooth cats, wolves, dogs and large bear-dogs (Chapter 25, Figure 25.3).

Startling Evolution of Artiodactyls

Artiodactyls radiated into many new genera in the Miocene. Blasto (*Blastomeryx*, Figure 26.4) is thought to be the first North American ruminant, although it is also known in Europe. It was a chewer of cud and had two separate stomachs. Ruminants lack genes for the enzymes required to digest food so are dependent upon microorganisms, which evolved separately according to Darwinian theory.

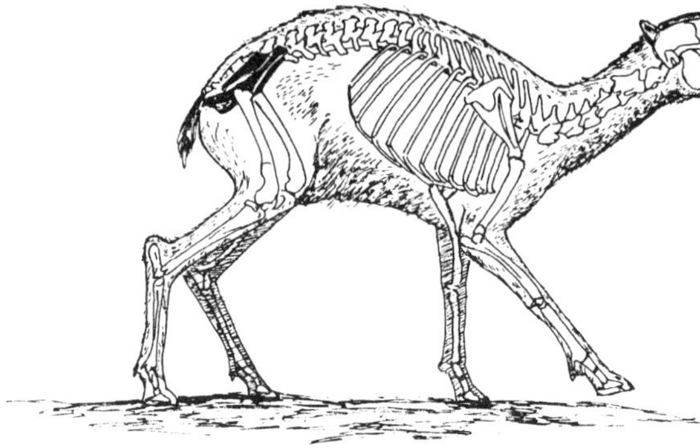

Figure 26.4 Blasto (*Blastomeryx*, Palaeomerycid, Cervid, Artiodactyla). (After Matthew, 1929).

It is totally unexpected that a *deer* should have saber-tooth-like canines. Once again we see that a characteristic which seems to be rather definitive for classification (saber-tooth = carnivore) is found in an animal in an inappropriate order. Artiodactyls are even-toed herbivores. Why does this one have saber-like canines? Why do other herbivores like uintatheres (Chapter 23, Figure 23.5) have saber-like teeth?

L. B. Halstead (1968: 155) wrote that shearing molar dentition evolved independently in three orders: Insectivora, Condylarthra and Carnivora. This molar development is often associated in the Carnivora with enlarged canines. Does this characteristic reveal DNA movements among animals?

Orthodox paleontologists might believe that the saber-tooth development evolved in sequential fashion from condylarths to uintatheres and perhaps from condylarths to this cervid, but then it should be expected in other deer as well. No, saber-teeth are not just found in a sequential continuum of morphologies. Development of this characteristic in widely diverse animals represents a DNA pattern which has appeared in discontinuous configurations due to salmutations.

The earliest fossil evidence of deer and antelope (Bovidae, Ruminantia) was found in Europe. It is believed that oxen (Bos primagenius) evolved from some member of the antelope family in Eurasia. Goats may have arisen in the Middle East although mountain goats were found in the Miocene of North America. The first pronghorns are thought to have originated in North America.

Camels (Tylopoda) of the Miocene provide another surprise. The accepted idea that animals *gradually* evolve to increased size (Deperet's law) is challenged by fossil evidence of camels. Small camels were known earlier but in the Miocene there were also Oxy (*Oxydactylus*, Figure 26.5) about seven and a half feet long and six feet in height, in North America, and Alti (*Alticamelus*, Figure 26.6) about 18 feet in height, in both North America and Asia. Alti was at least 11 feet taller than modern camels and, in fact, is reminiscent of a giraffe. Oxy is similar to the South American litoptern Theo (*Theosodon*, Figure 26.3) except, of course, Theo had a short proboscis.

Figure 26.5 Oxy (*Oxydactylus*, Artiodactyla). (After Peterson).

Figure 26.6 Alti (*Alticamelus*, Artiodactyla). (After Osborn, 1910).

Oreodonts (Chapter 25, Figure 25.8) as well as entelodonts were common in the North American Oligocene and Miocene. They are thought to have evolved from Dichobunidae but it has been difficult to establish specific relationships.

The Suiodea (pigs, peccaries and hippos) may have originated in Asia, as early representatives (anthracocheruids) were diverse in Burma during the Eocene. They probably migrated into Europe, then into North America and Africa during the Oligocene. By the Miocene, they were common in North America. During these epochs, they evolved from terrier-size (*Diacodexis*) to hippo-size. North American Dino (*Dinohyus*,

Figure 26.7) was 10.5 feet long with short, stout limbs and is thought to be the last of the entelodonts. Could its vertebral structure define a close DNA relationship to both bison (Chapter 28, Figure 28.2) and camels?

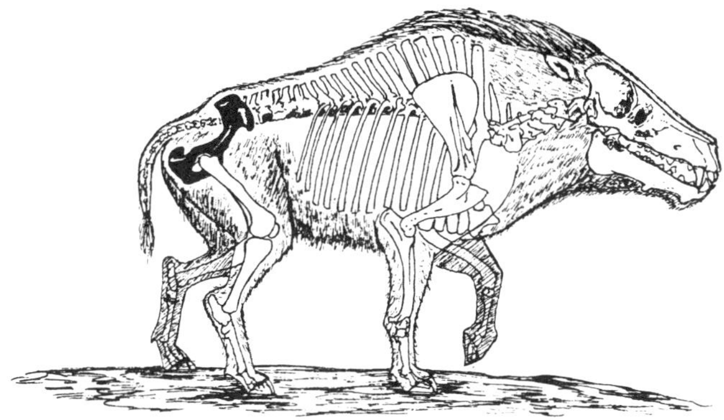

Figure 26.7 Dino (*Dinohyus*, Entelodontidae, Artiodactyla). (After Peterson).

Perissodactyls — Odd Toed Ungulate Giants

The largest land animal of recorded time is classified as a rhino. Indri (*Indricotherium,* Figure 26.8) was found in Central Asia. Reconstruction suggests the configuration of a giant horse. Indri was 37 feet long and 18 feet in height. Its antecedents are unknown.

By the Miocene, perissodactyls were widespread in the Northern Hemisphere. From Eocene to Miocene formations, fossils have been identified in at least 83 genera of rhinos, 43 genera of brontotheres, 16 genera of tapirs, 13 genera of chalicotheres, and 43 genera of horses (Simpson 1945, Prothero and Schoch, 1989).

In North America, *Hypohippus* (Anchitherinine) a browser, was about twice the size of many other Miocene horses. It was the first genus of horse since the Eocene which apparently migrated from North America to Europe and Asia where it became quite numerous. Fossils of very small horses were also found. *Archeaohippus* was a pygmy horse of the Miocene.

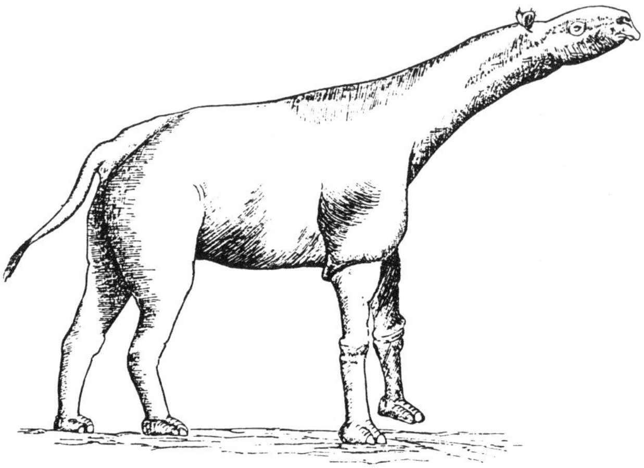

Figure 26.8 Indri (*Indricotherium,* also *Baluchatherium,* Perissodactyla). (After Osborn, 1910).

Apparently browsing did not represent an exclusive early stage of horse evolution, as previously thought. Both browsers and grazers were common in the Miocene. Nine grazing and three browsing horse genera coexisted. Many of these primitive horses were doubling in number every five million years. As dryer summers developed about 14 MBP, and the mixed evergreen-deciduous forests changed to conifer-hardwood forests of the Rocky Mountains and grasslands of the Great Plains, USA, *Neohipparion, Nannippus* and *Pliohippus* replaced *Merychippus.*

The classification *Merychippus* is an inadequate designation for animals now thought to be of many phyletic lines. Scientists struggle to classify animals which do not fit the standard taxonomy. I object to the designation "wastebasket" which some paleontologists have begun to use to describe *Merychippus* and other animal taxa of unknown relationships.

It is generally agreed that the numerous fossils of horses from Eocene to Miocene should be reevaluated. Now that it has become possible to study DNA of some of the 20 million-year-old fossils, perhaps bimolecular knowledge may assist in this reevaluation.

A horse-like animal with some similarities to the giant *Palaeotherium* (Chapter 24, Figure 24.4) first appeared in the Oligocene of Eurasia and was known in the early Miocene of North America and North Africa. *Moropus,* classified as a chalicothere (Figure 26.9) was a nine-feet-long giant, with longer front legs than rear, similar in this configuration to that of giraffes which also appeared in the fossil record of the Miocene.

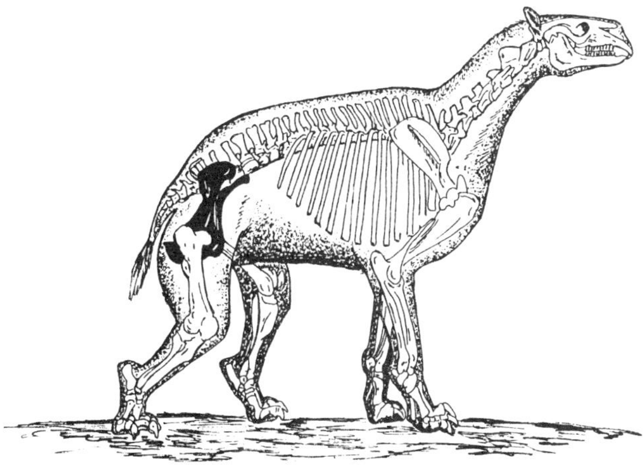

Figure 26.9 *Moropus* (Ancylopoda, Perissodactyla). (After Holland and Peterson, 1913).

Moropus had a hip bone somewhat similar to *Palaeotherium,* but heavier. Its neck bones might be considered to be closer to *Palaeotherium* than to *Eohippus* (Chapter 24, Figure 24.3). But the most interesting feature of *Moropus* was its clawed feet! How could a horse-like animal with herbivorous dentition have clawed feet? It must have received this DNA pattern from a salmutation. This feature bred true as exemplified in many specimens.

Chalicotheres lived for millions of years into the Pleistocene. They were not a fluke nor an aberration. There is something to be learned about DNA here. Horses did not just evolve in neat lines from *Eohippus* to *Equus*. Perissodactyls salmutated into a variety of horse-like forms during their long history.

One of the most interesting subfamilies of the chalicotheres, represented by *Calicotherium* (Figure 26.10) became more diverse in the Miocene when it spread to North Africa from Asia. It was unknown in North America. This strange animal had gorilla-like proportions and probably could knuckle-walk, although this idea is alien to the concept of perissodactyl.

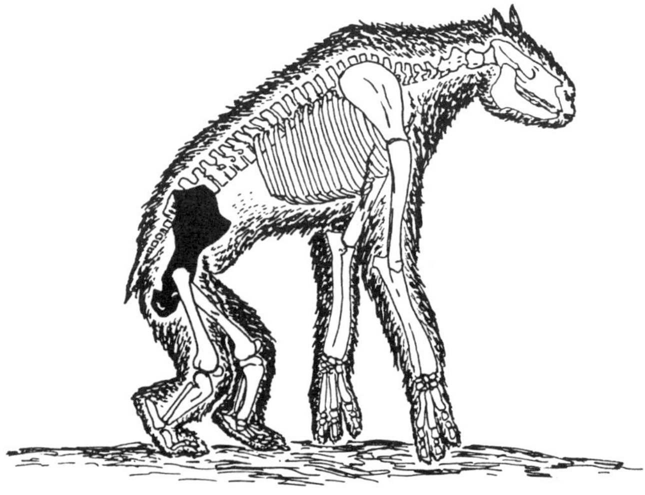

Figure 26.10 *Chalicotherium* (Ancylopoda, Perissodactyla). (After Zapfe).

There are similarities in foot configuration among chalicotheres, sloths, pangolins, oreodonts and taeniodonts, and dental similarities among chalicotheres, horses, and brontotheres. This strange combination of characteristics in chalicotheres illustrates the proposition that linear descent is inadequate as an exclusive evolutionary principle.

Proboscideans of Africa

Proboscideans radiated in Africa in the Miocene. The relationship of proboscideans to other orders of mammals is being reevaluated (see Prothero and Schoch, 1989). There is new evidence that animals classified as proboscideans: mammoths, elephants, sirenians and desmostylians (Figure 26.11) are related to hyraxes, arsinotherians (Chapter 25, Figure 25.11) and perissodactyls. Conventional classification of proboscideans did not include hyraxes nor arsinotherians, nor indicate any relationship with perissodactyls.

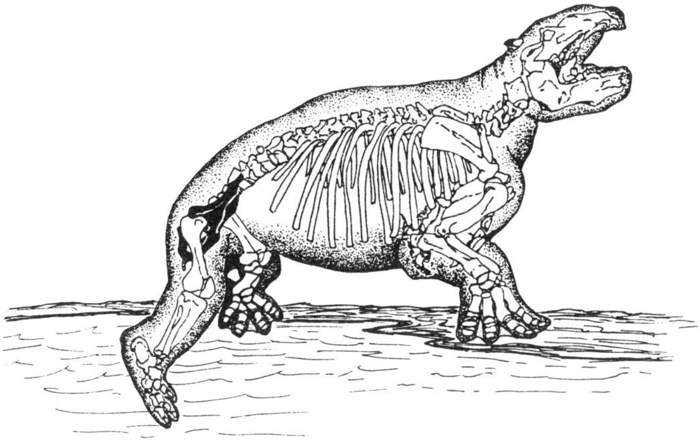

Figure 26.11 Desmo (*Palaeoparadoxia*, Desmostylia).

Since complementarity dynamics reasoning leads to the idea that similarities of structure require consideration of similarities of DNA patterns, one must agree that the taxonomy of proboscideans probably should be reconsidered. However, the whole taxonomy of vertebrates (and also invertebrates) could be thrown into chaos if all new recognition of similarities suddenly resulted in changes of classification. I believe there is need for a taxonomic moratorium until relationships of DNA and morphology become more clear.

Vincent J. Maglio (1973) presented an updated compendium of the evolution of mammoths (mastodons), gomphotheres, stegodonts and elephants. He believed that the aberrant moeritheres (Chapter 25, Figure 25.12) barytheres and deinotheres should probably not be included in the order Proboscidea, although he did not deny possible relationships. Henry Fairfield Osborn (1936) included moeritheres and deinotheres in the proboscidean order, and Simpson (1945) included all three genera as proboscideans.

There was a great expansion of gomphotheres in East Africa during the Miocene. The most primitive elephant (*Primelephas*) is thought to have evolved from *Stegotetrabelodon* in the late Miocene.

Africa was cut off from Europe by the Tethys-Mediterranean Sea except for access across Arabia through the Middle East. After the middle Miocene, a land connection was made between Morocco and Spain which continued until the early Pliocene when tectonic faulting opened the Straits of Gibraltar.

In the Miocene, mammoths first crossed from Africa into Europe, and elephants entered both Europe and Asia. The genus *Stegotetrabelodon* remained in Africa. Proboscideans may be a key for learning more about the relationships of speciation and migration but relatively few workers have apparently studied this fascinating history.

It is now thought that one of the most unusual of aquatic animals, Desmo (*Palaeoparadoxia,* Figure 26.11) is related to early proboscideans and sea cows (sirenians). Its fossils appeared fully formed off Eurasia and Africa, and also in the shore area strata of California, USA and Japan. Desmo is considered to have been rather like a marine hippopotamus, but its history of salmutations and evolutionary development can only be imagined.

Miocene Primates and Evolution Theory

When Charles Darwin published *The Origin of Species* (1859) only five primate fossils had been described. The first was an Eocene lemuroid (*Adapis*) from France, considered a primitive ungulate or pachyderm by Curvier in 1822, and only later recognized as a prosimian.

Another prosimian (*Microchoerus*) was recovered from Eocene deposits of England. The first known fossil monkey (*Mesopithecus*) from Pliocene strata near Athens, was described by A. Wagner in 1839. An ape-like form (*Pliopithecus,* Chapter 25, Figure 25.13 b) from the Miocene of France, was described in 1837 by attorney and naturalist Edouard Lartet. Lartet later named *Dryopithecus* (Figure 26.12) a large ape from Miocene sands of France, which he believed was related to man.

Figure 26.12 Proconsul (*Dryopithecus*, Hominoidea, Primate). (After Clark).

During the time frame 1822-1859, the age of fossils was only estimated by naturalists in relation to identified geologic strata. There was then no chemical nor radioactive dating system, and estimated age of strata varied quite widely. Many fossils found in these early years were thought by the general public to be the work of the devil or evidence of animals killed in the flood of Noah.

Large bones such as those of dinosaurs and proboscideans were often thought to be from a long lost race of human giants. The significance of stone age (Pleistocene) human remains, recovered from European caves during this time period, was not generally understood, and often the stratigraphic locations of these fossil finds were not recorded. Thus, much valuable information about these early *Homo sapiens* has been lost.

In the 1890s, Karl A. Von Zittle of the University of Munich, Germany, prepared one of the earliest text books which enumerated most of the then known identified fossils of the world. Max Schlosser, Charles Eastman, Lucy Bush, Marguerite Engler and Arthur Smith Woodward worked on revisions and-or translations of this work during many years to produce the 1925 edition of *Text-Book of Palaeontology*, reprinted in English in 1964. This book provides a sound beginning point for the study of the history of life. However many of our concepts have changed with recent developments in plate tectonics, genetics and microbiology.

Before about 1925, most of the work in paleontology had been discovery, recovery, preparation and description of fossils. Fossils were often only teeth and-or jaws, sometimes skulls and occasionally relatively complete skeletons, but these were often scattered and disassociated.

The descriptions often alluded to similarities and differences between the study (type) fossil and other known fossil animals. Often these descriptions were enumeration of facts — little speculation about evolutionary affinities. Occasionally an evolutionary lineage was posited, but most such speculation consisted of estimates of a possible ancestral genus, family or order. In some of the early work, *primitive* as opposed to *derived* characteristics were estimated, and potential affinities posited. Many of these workers seemed more unsure of linear relationships than paleontologists of the middle 1900s.

In the Neo-Darwinian period of the 1930s to 1960s, there seemed to be a degree of certainty in positing phyletic lines of descent. However, by the 1970s, scientists began to express some doubts about naming direct ancestral relationships. This is exemplified in the cladistic movement and in the work of recent primatologists such as Elwyn Simons and David Pilbeam of Yale University.

Pilbeam has studied the fossils and fossil record of Miocene apes as his specialty for more than 20 years. He has discovered some very revealing information about these animals, although the teeth and bones of Miocene apes are fragmentary and limited to a relatively small number of fossil finds.

Pilbeam and Simons worked to reclassify the known primates from a plethora of overlapping and redundant scientific names. Simons (1972: 283) tabulated 148 accepted primate genera: 96 extinct, and 52 living. He identified 360 known species: 206 extinct, and 154 living. He also listed the synonyms of these genera (Simons, 1972: 290-293) which are very helpful when reading earlier paleontology literature.

A single tarsioid fossil in Miocene deposits was the last of the native North American primates discovered. Ceboid monkeys as well as primitive owl monkeys, believed to be close to the base of South American monkey evolution, perhaps from a tarsioid line, were found in South America strata.

The African Miocene is noted for a variety of ape-like primates, some of which give firm indications of hominoid characteristics. Although the first Dryo (*Dryopithecus*) ape fossil was discovered in France, it is now believed that the earliest members of this subfamily arose in East Africa in the early Miocene, and in the later Miocene migrated into the Northern Hemisphere, to Turkey, Greece, India, Pakistan and France. They lived from 20 to 10 million years ago. Lack of fossilization in the acid soil of tropical forests limited the number of discoveries of these apes, and they seemed to die out as the forests shrank in size.

The best known and earliest member of Dryo (subfamily Dryopithecinae) was found near Lake Victoria by Mary Leakey and named *Proconsul* by Louis Leakey (Figure 26.12). Three species have been identified from early Miocene deposits of Kenya and Uganda, Africa.

Dryo fossils indicate that these animals varied in size and appearance. They had some similarities to chimpanzees; they had ape-like limb bones. They definitely could not be mistaken for today's chimps, although some species have been identified as having an ancestry similar to that of chimpanzees.

Some have been thought to have ancestral characteristics of orangutans or gorillas. Their beginnings of hominoid-like dentition led them to be called "dental apes."

David Pilbeam (Johanson and Edey, 1981) has classified Miocene apes in two subfamilies: the dryopithecines which were identified toward the ape end of a continuum and the ramapithecines, identified toward the more human end of a continuum. The two subfamilies lived in an overlapping time frame. The dryopithecines arose about 20 MBP and died out about 10 MBP.

The Ramas (ramapithecines) originated about 14 MBP without known connection to the Dryos. These apes were not considered hominids, nor defined as full upright walkers, but their hominid-like teeth hinted to paleontologists that this line could lead to humans.

The subfamily Ramapithecinae is classified in about five genera, ranging in size from about 30 pounds to large gorilla size. The type fossil, *Ramapithecus*, known from East Africa, India, Europe and China is considered the most human-like of the Ramas. Its ancestors are thought to have lived in the jungle, but it roamed and fed on the plains. It ate tough vegetable foods. It was believed to be quadrupedal though fossil evidence of this is lacking. Some experts consider the Ramas to be primitive Hominidae. They died out about 10 million years ago, long before the rise of the first anthropoids now classified as hominids, the Australs (*Australopithecus*) which arose about five to three and a half million years before present. David Pilbeam (1987) speculates that neither known dryopithecines nor known ramapithecines were direct ancestors of humans.

Mainstream paleontologists found a virtual blank of primate fossil evidence from approximately nine million years ago to three and a half million years ago. There are almost no fossils known which could give evidence of the transition from quadupedal apes (Dryos or Ramas) to the first upright walkers of three and a half million years ago (Australs).

However, there are theorists and scientists who have proposed some extraordinary concepts to fill the multi-million year gap. Elaine Morgan, an Oxford scholar, in *The Descent of Woman* (1972) and *The Aquatic Ape* (1982) reviewed the work of Frederic Wood Jones and Oxford Professor Alister Hardy (1960, 1977) Leon P. La Lumiere Jr. and others. She postulated that humankind evolved during this time period between about eight to three MBP, on an island in a marine environment. She believes that drastic changes in the environment were required to account for the change from ape to human within a few million years. Morgan wrote (1982: 115), "When conditions change very rapidly, then evolution goes into overdrive."

Leon La Lumiere of the U. S. Naval Research Laboratory (Morgan, 1982: 118-135) reasoned that inundation of parts of northern Africa in the late Miocene was the initiating event for the evolution of *Homo aquaticus* (Hardy's term). La Lumiere postulated that a population of apes became isolated on an island, separated from other apes of their kind for a prolonged period. This isolation could have come to an end in the Pliocene with tectonic movement. *Homo aquaticus* may have evolved from apes into the Austral species found as fossils of the African Pliocene.

La Lumiere discovered that the area of Danakil Island in the north central Afar triangle of Ethiopia had been cut off from the rest of Africa during the late Miocene, and was rejoined in the Pliocene in the northern area of the African Rift Valley. He postulated that the newly evolved species could have travelled south up the water course of this valley to the Afar where they evolved into the Australs that have been found as fossils.

He proposed that a population of apes was isolated on Danakil or adjacent islands about 6.7 MBP. It seems reasonable that these apes may originally have been marooned on high ground or in treetops as the sea rushed in from the proto Red Sea and the Gulf of Aden into the Afar of Africa, about 6.7 MBP. Scientists calculate that the release of the sea was initiated by the detachment and uplift of the Danakil Crustal Block (horst) as the Nubian tectonic plate to the west and the Arabian plate to the east separated — releasing massive volcanic eruptions along the African Rift from southern Turkey through Israel and east Africa to South Africa.

In the late Miocene, Danakil Island may have been thus encircled by the sea to the east, north and west. When northern and southern portions were covered by lava flows an area about 90 miles long and 45 miles wide was cut off from the rest of Africa.

These theorists say the Danakil Island could have been the place where apes became swimmers and upright walkers through their conquest of the sea and the sea-shore life. Fossils are lacking, but La Lumiere has drawn a map to indicate where they should be found. The surface of the Afar Triangle is awash with paleontological resources, but fossils of *Homo aquaticus*, if they existed, may be under 600 feet of sediment. Perhaps future fossil discoveries will be found to validate or negate these theories.

In the meantime, let us examine the evidence and assumptions that Morgan and the others have presented to support the idea of *Homo aquaticus*. If the apes of Danakil were left in an island environment which had been inundated by salt water, leaving death and destruction in its wake — the life of the forest would then have been no longer tenable, with trees dying, animals decimated. Many apes probably died. What could be done?

Fortunately, the ocean is not only destructive; it is also bountiful. The apes (Dryos or Ramas?) of Danakil Island could have evolved into *Homo aquaticus* in such an environment. Scientists agree proboscis monkeys of Borneo became expert swimmers and divers, and then found their sustenance in the life of the ocean.

Morgan postulated that *Homo aquaticus*, similarly, began to live on the shoreline and eat the sea life found. Gradually becoming less fearful of water, they began to spend more and more time turning over rocks, searching for worms and other sea creatures. The aquatic apes began to hunt for clams, crabs and lobsters, and found they could capture turtles and sea birds.

Morgan postulated that these apes began to develop characteristics which were different from those of predecessor apes. In substantiation of the feasibility of her new theory, Morgan discussed myriad possible adaptations of these animals to a water environment, too many to review here but some of them are:

1. Many marine mammals originated as terrestrial animals and adapted to the sea. Apes could have done so too. These marine creatures returning so completely to a terrestrial environment is unique, and evidence of their aquatic period is carried by their descendents.

2. Marine mammals of tropical areas are hairless. Aquatic apes became hairless (actually shorter and finer haired compared to forest apes) because this was an adaptation to a water environment. Morgan anticipates that they may have spent up to six hours a day in the sea.

3. Marine mammals have subcutaneous fat which keeps them warm in an ocean environment; apes do not. Humans have both the subcutaneous fat characteristic of marine mammals and sweat (sebaceous) glands which cool them in hot environments. (Note: Forest apes have these glands but the glands do not serve this purpose. Apes are quite dry in the hottest weather.)

4. Marine animals such as crocodiles and turtles and mammals such as seals and sea cows excrete salty tears from their eyes. Apes do not. Man is the only primate that weeps salty tears.

5. Marine mammals and humans can copulate face to face and the females have ventrally directed sexual canals. Forest apes copulate from a rear position.

6. The ventrally directed sexual canal is only feasible if the flexible spine and hind limbs are aligned in a straight line, necessitating a change from quadupedal to upright posture when moving on land.

7. Humans became upright while living in the sea, treading water and adapting to swimming in a straight position, eliminating any potential for the shuffling knuckle-walk of evolving apes.

8. Marine animals have webbed feet. Seven percent of humans in many cultures are born with webbed feet. Webbing between fingers is rare but there remains a vestigial triangle between the thumb and first finger of humans.

9. Human infants are born with fatty tissue which makes them more buoyant than any of the apes. Human babies can be born under water and can swim instinctively as infants before about age two.

10. Humans can connect or disconnect the nasal cavity from the other supralaryngeal air passages by control of the velum, thus hold their breath when diving and swimming under water. This is also a speech-producing feature. Apes do not have it. Their larynx and vocal cords are constructed differently. Conscious control of breathing and production of sound are interdependent.

11. In all aquatic mammals as in humans, olfactory lobes are diminished with respect to body size.

12. Aquatic mammals grow larger, live longer and nurture their young longer, than terrestrial mammals. Humans fit this pattern.

13. An aquatic environment demands a new locomotor repertoire which is associated with increasing brain size and development.

Hardy thought that a significant test to determine whether humans evolved in a water environment would be to check the diving reflex of apes. Do they have the same diving reflex as humans in which the blood supply is substantially cut to most of the body, except to the brain and muscles of the heart, when the face is in the water. This is typical of whales, seals, penguins and diving ducks. Hardy thinks it could not have evolved in humans unless they had been diving under water for an extensive period of their past.

I think this fantastic theory should be considered seriously. Paleontologists may then search deeply in the Afar Triangle for fossil verification. But I also believe fossils will not tell the whole story. How do the DNA configurations of aquatic mammals compare to terrestrial mammals? How does DNA change animals to adapt them to different environments? What salmutations could be involved in the possible evolution of aquatic apes?

Were the apes of Danakil our ancestors? If there were no Danakil apes in our past, could the many DNA characteristics of marine mammals which we carry have entered out genetic heritage more directly, perhaps carried by viruses of the Miocene? Could Lucy (*Australopithecus afarensis*, Chapter 27, Figure 27.2) discovered by Donald Johanson in the Afar region of Ethiopia in 1972, have been a successor to the aquatic apes of Danakil? We will discuss Lucy in the next chapter.

There is no evidence that Donald Johanson or David Pilbeam have considered the possibility of an aquatic ape ancestry of humans. Thus, mainstream paleontologists are continuing to look for the evolution of Dryos or Ramas (or other apes) prior to Australs.

For those who take no cognizance of the possibility of aquatic apes, the questions which are asked and the environments explored are different from those considered by Morgan.

In the middle Miocene of Africa the tropical rain forests began to die out as woodlands and open savannas expanded in a cooler climate. Why did some ramapithecines of Africa begin to show development of teeth more characteristic for processing foods of woodlands and plains (or of the ocean?) rather than the foods of the forests? Classical theory calls upon the concept of dental "adaptations" to new conditions, but what is the process by which these adaptations occur?

Ramapithecines are said to have developed molars capable of grinding seeds and husky grains. They are said to have developed smaller canines as these apes, considered early hominoids, may have substituted elementary tool use for biting and slashing.

Pilbeam was fascinated with one very exceptional genus of ramapithecines called Giganto (*Gigantopithecus*) that originated about nine MBP and may have lived until about 500 thousand years ago, well into the time range of early humans. But they definitely were not considered to be ancestral to humans; they were too large and too ape-like.

Gigantopithecus weighed perhaps 150 pounds, according to Pilbeam. Some scientists have estimated a weight up to 600 pounds and an upright standing height of eight feet. Perhaps the existence of *Gigantopithecus* has found its way into the racial memory of humankind, being reawakened in the myths of the Abominable Snowman and King Kong.

Pilbeam is conservative in his opinion of *Gigantopithecus*, probably because the fossil evidence is very limited. Pilbeam has been intrigued by the complexity of the Miocene ape record and suspects that more fossil finds will only make the picture more complex before it becomes clearer. Only tooth and jaw fossils of ramapithecines are known, but the known teeth tell a story. Pilbeam (1970: 216) wrote, "The ramapithecid thick-enamelled cheek teeth probably evolved for the powerful crushing and grinding of tougher kinds of vegetable foods typical of more open, nonforest settings."

This example of a dental adaptation in conjunction with an environmental change is similar to statements about adaptations of animals to environments made throughout the literature of paleontology. I believe these adaptations cannot be explained using monistic logic requiring a strictly sequential time frame. The adaptations of ramapithecines require a concept of interaction between initiation of new patterns of DNA for new kinds of teeth, and changes to a grassland and woodland environment. DNA characteristics do not always follow sequential patterns. A major change in tooth structure needed to cope with a new kind of food supply is a purposeful change from an existing pattern.

I believe this change from teeth adapted to forest foods of fruits and leaves to teeth adapted to tough savanna vegetation (or ocean resources) is caused by a purposeful change in DNA patterning, perhaps catalyzed by viral participation. The evolution of ramapithecine tooth structure requires consideration in a complementarity dynamics time frame of reciprocal interactions between DNA and the food supply.

The Miocene appears to be an experimental period for primate adaptations. Knowledge of DNA and viruses will help us develop a broader understanding of human heritage from the apes of the Miocene.

Some paleontologists believe that the radiation of living hominoids began in the Miocene, and that dryopithecines exemplified the ancestral characteristics for all future anthropoids. Some scientists believe that dryopithecines were ancestral to ramapithecines which led to man. Others believe that dryopithecines may not be directly ancestral to ramapithecines, but rather a branch leading only to the other higher primates.

The controversies over human ancestry hinge upon the definition of humanness. To determine humanness scientists assess:

1. Tooth development (This may be the only fossil evidence available).

2. Relationship of teeth to food sources and thence to habitat.

3. Skull, its development and especially its relationship to upright posture, identified by position of attachment site of skull (foremen magnum) and vertebral column.

4. Upper leg bone (humerus) which is diagnostic of full upright walking ability.

5. Hip bones determinant of upright walking and in females diagnostic of head size of newborns.

6. Structure of hands and feet, as related to methods of movement: quadrupedal walking, quadrumonal climbing, vertical clinging and leaping (in most prosimians) slow climbing (in potos) brachiation (in gibbons) bipedal tottering (in apes) bipedal striding (characteristic of man).

7. Brain development; brain size increase relatively greater than body size increase is generally considered to be indicative of advancement in intelligence, though not considered to be directly comparable.

8. Tool use and tool making and relationship of structure of hands to their uses. The evolution from tool use, to tool modification to tool making has been considered to be evidence of development of intelligence. Use of tools is the first diagnostic criterion which is considered both a cultural and a biological adaptation.

The uses of hands for food gathering and preparation and child care have been somewhat neglected in favor of consideration of tool use and tool making for hunting, because the evidence has not been divided into probable male-made and female-made tools. Perhaps this is the case because more males than females have studied the stone tools and written the science.

9. DNA identification. Since 1952, the fields of genetics and microbiology have added this new criterion to the previous eight classical diagnostic categories for determining evolutionary relationships.

DNA and Evolution Theory

Classical paleontology continues to embrace the idea that variations in germ plasm (now known as changes in DNA) led to new animals, different from their parents in minor ways.

After many generations, with a change in environment, a new species evolved and could be identified. This belief, based on the concept of a linear continuous time frame, still requires an explanation for the very different end products we see today.

The concept Darwin developed to explain the existence of very different lifeforms was natural selection, the differential birth rate between less fit and the fittest of the lifeforms produced. This classical belief never directly confronted the question: "What is the cause of the variation which made some animals better adapted than others to a particular new environment?" Paleontologists have begged the question by discussing the evidence rather than the reasons for the evidence. The reasons involve interaction of DNA and the environment.

Understanding of DNA is in its infancy. We are only beginning to understand its significance. It is essential that the public acquire knowledge of DNA as it develops, and that democratic controls be applied to use of this knowledge. It must not become the province of any elite, whether scientific, political, military or economic.

In studies of evolutionary biology, DNA from living primates has been processed to try to determine recency of common descent of orangutans, gibbons, gorillas, chimpanzees and humans. I believe DNA has not answered this question because DNA (of humans) is three billion nucleotides long and one minor section (say a gene for blood hemoglobin) can give information about that gene, but not about the whole animal.

While it is helpful to determine any pattern which can be identified in different animals, one pattern, or even several are not definitive for determining evolutionary relationships. More is needed. Real progress may be made when the DNA patterns from different animals for such morphology as tooth structure and hip structure can be compared. Not only do we have more paleontological knowledge of different tooth structures than any other structures, but these features can be coordinated with other structural and functional features, and environmental influences.

DNA analysis may be misleading when separated from fuller understanding of viruses and their complementary relationships with other lifeforms. Therefore, in the search for evolutionary knowledge, it is essential to coordinate work on cellular and viral DNA interactions.

27

Cenozoic Era — Tertiary Period Pliocene Epoch Upright Walkers and Vast Migrations

The Pliocene was short, but it was extremely important because "Lucy" arose, one of the earliest known Australs (*Australopithecus*). Lucy and other fossils of her genus confirmed for some scientists that these were the first upright walkers, the near-human, possible ancestors of humans.

The Pliocene began five million years before present (five MBP) according to recent texts. Estimates of the boundaries of Miocene-Pliocene and Pliocene-Pleistocene Epochs are being redrawn as new knowledge accumulates. Authors contributing to the Stehli and Webb (1985) overview, *The Great American Biotic Interchange* have proposed a beginning range for the Pliocene from 10 to five MBP and an ending range from three to two MBP. I will here use the five MBP to two MBP range for the Pliocene evidence which places the "Great American Interchange" of animals, three MBP, firmly in the

Pliocene. This designation of time frame, it must be realized, causes some problems in interpreting data from different authors who use different dating ranges. The day must soon arrive when evidence is reported more conclusively in numbers of years before present (from the established "present" standard of 1950) as dating techniques are perfected.

Glaciation and the Seas

By the beginning of the Pliocene, mountain building was relatively complete but glaciation accelerated. Both warming and cooling trends were reported by Samuel M. Savin and Robert G. Douglas (1985: 303-324). The Antarctic ice cap was one half its present size during the early Pliocene and increased to nearly its current volume in the late Pliocene. About three to 2.5 MBP continental ice in the Northern Hemisphere began periodic expansion and recession. Relatively rapid, periodic alternations of glaciation and warming trends occurred throughout the Pliocene and Pleistocene.

The first known glaciation south of Alaska occurred about 3.1 MBP in the Sierra Nevadas of California, USA. Glaciation was known in the Andes Mountains of South America and in Northern Europe. Glaciation history of this epoch is becoming more precise due to development of oxygen isotope evaluation methods.

Sea level was much higher in the early Pliocene than it is today. It receded by the mid Pliocene, as the Arctic and Antarctic ice caps grew. By the late Pliocene, sea level had dropped as much as 54 meters. This drop — together with tectonic uplift — provided a connecting land corridor between North and South America.

As the Panama Isthmus closed, ocean currents were altered. The Gulf Stream was initiated. Atlantic and Pacific waters no longer intermingled. The closing of the Isthmus correlated with cooling weather patterns and was followed by great migrations of Pliocene animals from colder to warmer climates.

Widespread in the seas of the Pliocene were porpoise-like whales. Dolphins were fewer. Sperm whales continued, but most primitive whales (Archaeocetes) were extinct. Few genera of primitive whalebone whales survived the previous epoch when they were numerous. They were replaced by more modern whalebones, including the blue whale, which later reached 150 tons, far larger than the largest dinosaurs, though it fed upon the smallest of prey, krill. As glaciation receded, major vegetative zones began to take on today's character. Desert adaptations began in the Sonoran area of the southwestern United States and Mexico.

Eurasia and North America

Separate Worlds

As the Pliocene opened, there appeared to be little interchange of species between Eurasia and North America, indicating a discontinuity, probably an ocean gap between the continents in the Bering Strait at that time. The genera in these two regions had become distinct.

Some scientists have considered that unknown ancestors of many animals lived in unexplored high latitudes and moved south into Europe and North America as cooling increased — that these animals then evolved into similar but separate lineages. Others have thought that migrations in the Eocene, Oligocene and Miocene between North America and Eurasia explained the existence of lineages that were similar, though also somewhat different. Further investigation of DNA relationships will surely enrich this dialogue.

About 80% of Pliocene animals of the Northern Hemisphere were similar enough to living genera to be considered ancestral to them. Carnivorous canids such as wolves and foxes and mustelids such as weasels and skunks were numerous. Most cats were similar to present kinds, although saber-tooth cats did not survive into the present.

The early evolutionary history is unclear about migrations of placental dogs and cats of North America and Eurasia, and their relationships to other carnivores. Some scientists have thought that different insectivores evolved into the separate lines of creodonts and miacids. Other scientists have theorized an evolutionary pathway from insectivores to creodonts to miacids (of the order Fissipeda) and then from miacids to cats (felines) miacids to wolves, then dogs (canines) miacids to skunks and weasels (mustelids) and miacids to pinnipeds and to other carnivores.

Stanley Olsen, in *Origins of the Domestic Dog* (1985: 2) noted:

> At this early period of carnivore radiation, it is contradictory, but understandable to refer to these various ancestral animals as cat-like dogs, mustelid-like dogs, hyaena-like dogs, or bear-like dogs. Such descriptions have been used to describe animals that have some features suggesting carnivore characteristics.

It is sometimes a surprise to realize that much fossil evidence remains open to very different interpretations. The concept of salmutation, and future DNA analysis when it is possible, may unsnarl some of the creodont-miacid-carnivore relationships which led to the carnivores of the Pliocene.

Hyaenas, tapirs and pigs were common in Eurasia in the Pliocene. Varied kinds of Asian giraffes included giant, horned forms. There was a variety of large horses, although *Anchitherium* and *Archaeohippus* had become extinct. *Pliohippus* was considered to be a transitional form. *Hipparion* is believed to have migrated from North America, and became common in Europe. Rhinos persisted but became rare.

The fauna of Eurasia in the Pliocene was somewhat similar to that of Africa today. There was an abundance of early bovids: antelopes, primitive cattle, buffalo, bison, musk oxen and goats. These plains dwellers, as well as true cervids — deer of temperate forests — were mostly absent from North America. In North America, camels and peccaries remained common. Many ungulates differed from the European types.

Brown bears, deer, elk, true hyaenas and mammoths are thought to have migrated from Eurasia to North America. There was an extensive radiation of pronghorn, considered a bovid; but its horns differ from true bovid antelopes, as they have characteristics which are more like deer. Deer drop their antlers each year and more complex antlers develop each following year. By contrast, bovids have true horns in many species which are not dropped annually but grow continuously. The character of prongbuck antlers (horns) seems to be a combination of these two distinct systems, suggesting possible salmutations: bovid and cervid DNA recombination.

Pliocene strata contained the last of the primitive oreodonts and other primitive artiodactyls, and the last protoceratid, a strange deer with a horn on its snout.

South America

Relationship of Placental and Marsupial Cats

In South America in the beginning of the Pliocene, one third of all fauna were xenarthrans, tending toward large size: sloths, anteaters, glyptodonts and armadillos. There were rodents of unverified ancestry, and marsupial carnivores including the marsupial saber-tooth cat Thylaco (*Thylacosmilus*, Figure 27.1 a). Thylaco of Argentina is thought to have evolved in the Upper Miocene from the borhyaenids that developed from early opossum-like forms (didelphids) of the Paleocene (Carroll, 1988: 433). However, fossil evidence of direct connections is lacking.

George Galylord Simpson (1980: 70) thought the family of Thylaco provided a striking example of convergence with *Smilodon*, the placental saber-tooth cat of North America (Figure 27.1 b). The skulls, jaws and dentition are amazingly alike. Thylaco became extinct in South America before the northern saber-tooth cats arrived there, so it is believed that *Smilodon* did not give rise to Thylaco.

I believe it will be found that the similarities in these two genera are not just an odd coincidence of habitat similarity effecting structural similarities in convergent animals. Rather,

I believe that similar DNA patterns, in conjunction with environmental conditions, effected these similar animals through salmutation.

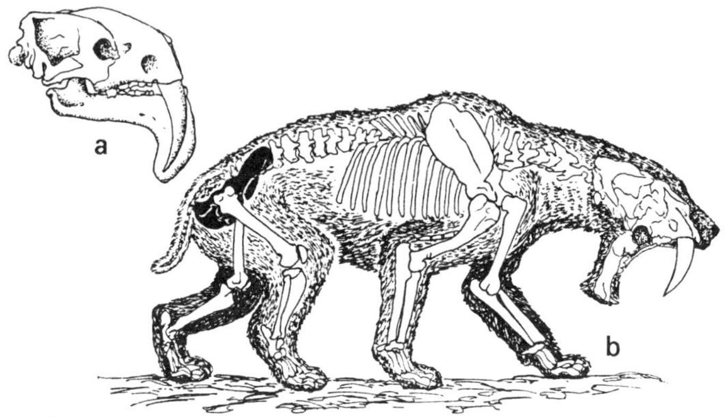

Figure 27.1 (a) Thylaco (*Thylacosmilus*, saber-tooth marsupial). (b) *Smilodon*, saber-tooth placental. (a. After Osborn, 1910; b. After skeleton, George C. Page Museum).

Monkeys apparently were swinging through the tropical forests of South America in ever increasing variety, but there is little fossil evidence, and there have been no ape forms on this southern continent.

New Pathway for Migrations

The "Great American Biotic Interchange" which occurred about three MBP involved: migrations between Eurasia and North America; vast interchanges of frogs, lizards, turtles and crocodiles between North and South America; extensive migrations of land mammals into South America; and a limited migration of land animals from South America to North America (Stehli and Webb, 1985).

The great interchange started when the Panama land bridge rose as sea level lowered. This allowed development of near-shore, grassy woodland transit corridors skirting the tropical rain forests of Central America. Insectivores such as shrews, arboreal rodents such as tree squirrels, and species of land rodents moved south.

Field mice radiated in great profusion and differentiated in South America. A. L. Gardner and J. L. Patton (1976) postulated that this radiation involved chromosomal reorganization. They noted that higher numbers of chromosomes are generally characteristic of more primitive groups and that more advanced lineages have a smaller number of chromosomes. This is a reversal of what might have been expected from classical evolution theory. It appears that chromosomes are reorganized and simplified as the evolutionary ladder is ascended — a challenging concept.

A massive modification in South American faunal history occurred as 29 genera of North American mammals migrated south. Included were saber-tooth cats, pumas and jaguars, Other placental carnivores also migrated, including skunks, weasels, otters, foxes, wolves and bears. Mammoths, tapirs, horses and peccaries relocated to South America, as did camels and deer. Most of these animals expanded their habitats in both hemispheres in the late Pliocene. After their arrival in South America, many lines experienced little or no generic differentiation, but nine families diversified quickly, including mice, cats, wolves, mammoths, horses, camels and deer. There was an exponential increase in large carnivores.

During the last three million years, about half of the land mammal species now living in South America, migrated from North America. Of the 17 mammal families which were in North America in the Pliocene, two became extinct, nine were distributed on both continents, and six are now found mostly or exclusively in South America.

Moving across the land connection from south to north about 2.5 MBP, were nine identified genera of animals. There were opossums and three kinds of xenarthrans including large and small sloths, giant anteaters and armadillos. There were two kinds of rodents including porcupines and rats.

The early Pliocene diversity of North American animals had exceeded that of South American by about 60%. The weight of the interchange from north to south either caused or followed marked extinction rate increases in South America, involving massive turnover of faunal types and an increase in evolutionary rate of radiation. On the other hand, immigrants to North America produced little diversification.

A main reason for the success of the North American migrants, generally, was that South America opened up new regions of temperate grassland ecosystems, while North America opened few tropical habitats to the southern migrants. The warmer southern climate and the associated ecosystems were essential for the success of these migrating animals.

Various Animals of Vast Africa

The originations and migrations of animals into and out of Africa are not entirely clear. Whether similar animals arose in various locations *is not known*. Abundant diversity of animals *is known*. According to paleontologists who have searched for evidence of early man in the Pliocene of Africa, more than 45,000 fossils of mammals have been found, including rats, mice, shrews, baboons, leopards, elephants, giraffes, antelopes, hares, pigs, rhinos, hippos and the horse *Hipparion* (Donald Johanson and Maitland Edey, 1981). Some Pliocene populations survived millions of years relatively unchanged. Today's guinea fowl and ostriches are similar to then existing birds. Primitive proboscideans such as *Deinotherium* continued in Africa and Eurasia, and *Stegodon* migrated from Africa into Eurasia.

Pliocene Primates — the Discovery of "Lucy"

The classification of Miocene apes, Rama (*Ramapithecus*) included not only the small forms and medium-size forms (sivapithecines) that became extinct before the end of the Miocene, but also included the apes know as Giganto (gigantopithecines) that lived in Pakistan and India throughout the Pliocene. Similar fossils, huge fossil molar teeth found in China, showed that Giganto had been a successful line, becoming extinct only about 500 thousand years before present (500 TBP). Neither the small Ramas, thought to have tooth characteristics foreshadowing hominids, nor the taxa of gigantic apes, provided a clear connection to Australs (*Australopithecus*) considered by some scientists to be the first true hominids, the upright walkers.

A large gap in the primate fossil record (from about nine MBP to at least five MBP) — that separated early apes, the Dryos (dryopithecines) from current apes (pongids) — was identified by David Pilbeam (in Johanson & Edey, 1981).

There are no fossils of apes known since the demise of the Dryos about nine MBP. Living orangutans (Pongo) chimpanzees (Pan) and gorillas are thought to have evolved from Dryos, but fossil evidence is lacking (Johanson and Edey, 1981: 363).

There is also a large gap between the ape line of Rama (ramapithecines) and the hominids. Johanson and Pilbeam believe that additional fossil discoveries may bridge those gaps, although other scientists have questioned Ramas as hominid ancestors and suggested that they had orangutan rather than hominid characteristics.

Several important questions are raised by evidence of Rama fossils in India, Europe, China and Africa in the Miocene, but none in the Pliocene. Did Ramas evolve no further in the Northern Hemisphere? Did it become too cold for them there? Did they evolve into aquatic apes in Africa? What happened to them?

Discovery of the oldest, most complete skeleton of the first known upright walker in the Afar of Ethiopia was claimed by Donald Johanson in 1974 (Johanson and Edey, 1981). This Johanson "first," popularly known as "Lucy" is officially called *Australopithecus afarensis*, and is considered by her finder to be a hominid (Figure 27.2). She was estimated to be about three and a half feet tall, and probably weighed about 60 pounds. Her skull bone fragments, and those of others of her kind, led to the supposition that her skull was rather ape-like, but she had very human-like teeth. She had small canines — a human trait, but larger cheek teeth, needed to process the coarse vegetation and small animals she probably ate. Thus her lower face was protruding (prognathous) to accommodate the large teeth, giving a somewhat ape-like appearance. Her upper leg (femur) and pelvis were similar to those of modern humans.

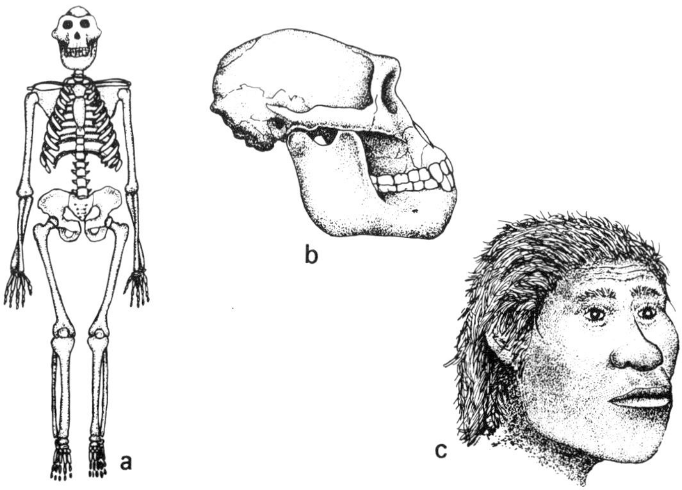

Figure 27.2 Lucy, the first known Austral (*Australopithecus afarensis*). (a, b, After Johanson and Edey).

Although Lucy's foot bones have not been found, other Austral foot bones have been found, and footprints considered by Johanson to be of the same species and in the same time frame as Lucy (3.7 MBP) were discovered by Mary Leakey at Laetoli in South Africa. The footprints show a front large toe as in humans, not a side oriented grasping toe as in all known apes. The print bed shows two series of prints, a large and smaller set, possibly a male and female, perhaps walking together across a volcanic ash deposit which can be rather accurately dated. If imagination is allowed to expand the fossil find, questions arise: Is this evidence of an early family relationship, a possible human pair-bond? The smaller prints were deeper than the larger ones. Was a female carrying a baby?

It was estimated that Lucy had a head about the size of a softball, therefore a small brain capacity, perhaps about 500 cc. Others of her kind had brains ranging from 430-550 cc. This was larger than brains of chimpanzees (300-400 cc) and overlapped later hominid fossils (*Homo habilis*, 460-775 cc).

Lucy lived about 3.5 MBP. Others of her species lived from five or four to three MBP. It is not clear from the fossil evidence whether they evolved from the Rama line of apes. There is as much as a 5.5 to 6.5 million year gap between the Rama and Austral fossils; only a few teeth, part of a jaw and an arm bone (that might be hominid fossils) represent this gap period.

Edey and Johanson (1989) believe that after Lucy's time, three branches of hominids developed, two continuing Austral lines to extinction, and one leading from Lucy-like Australs to modern man. These hominids all lived in Africa for more than a million years.

Mary Leakey and her son Richard Leakey believe that true man did not branch off from Lucy (*A. afarensis*), but was found in strata as far back as *Homo habilis,* at least 2.9 MBP. Others have dated *Homo habilis* at 1.9 MBP and 1.5 MBP. In either of these cases, *Homo habilis* might be a descendent of Lucy-like ancestors. The fossil record is becoming more complete, but interpretations of its meaning continue to vary.

Importantly, the evidence shows that competition did not cause the demise of the Australs at the hands of larger brained humans, at least not for a million years. Beings called Australs and Homos both lived during this time in eastern and southern Africa according to the fossil evidence.

Edey and Johanson (1989) received support for their evolutionary hypothesis from several colleagues. They are convinced that three species of Australs evolved from Lucy-type hominids (*A. afarensis*) which lived from four to three MBP. Gracile Austral fossils (*A. africanus*) with a brain capacity from 430 - 550 cubic centimeters were dated from 2.7 to 2.2 MBP. From them evolved *A. robustus*, a larger type which lived in Ethiopia and South Africa between 2.1 and one MBP. *A. boisai,* dated from 2.7 to one MBP may also have evolved from *A. afarensis*. All Australs became extinct in the Pleistocene.

Edey and Johanson think that the gap in the fossil record between Lucy and *Homo habilis* will, in the future, be filled by other human fossils.

It is not clear to scientists whether those of the Austral or *Homo* line were the first tool makers. The oldest stone implements found have been dated 2.5 MBP, and some paleontologists, especially the Leakeys, believe that *Homo habilis* may have arisen early enough to be the producer of these stone tools. Others believe the evidence shows that Australs were the first to make stone tools. They conclude that the evidence of close proximity of tools and Australs in time and space, in the Omo region of Africa, 2.5 MBP, means that Australs produced the primitive stone choppers that have been found there.

The long perpetuated emphasis on tool making as a main criterion of humanness has recently come under reconsideration. Monistic, cause and effect thinking previously led to either/or opinions:

1. Tool making requires large, well developed brains, therefore early tool making led to expanding brain capacity. OR: Large brain capacity led to early tool making.

2. The large, improved hominid brain led to curiosity about the developing savanna, and weapons were developed to allow movement onto the savanna. OR: Deterioration of the forest environment and reduction in size of canine teeth required a better brain in order to develop tool use to meet the challenge of savanna predators.

3. Tool making did not begin until the hands were well developed and upright walking allowed freedom for their use in hunting. Therefore hand development and upright walking led to tool making and hunting. OR: The need for more food as the forests declined led to upright walking, hand development and tool making.

Many monistically patterned ideas preceded Johanson's 1974 discovery of Lucy, who had a small brain in comparison to present-day humans, though it was proportionally larger than that of a similar size chimpanzee. Her canines were only slightly larger than a modern human. She was an upright walker on open ground, considered to have hands capable of fine movement. But she lived about a million years before evidence of the development of stone tools.

The opposability of the thumb, in humans, and the development of an opposable thumb and great toe, in apes, are not unique. It is now known that the hind foot of the American opossum also has an opposable digit similar to a human hand (Hartman, 1952: 40). There has been no suggestion of an evolutionary line from opossums to primates, yet a DNA combination which produces an opposable thumb may be sought to clarify these similarities. Stephen Jay Gould (1980) wrote a fascinating article about the panda's thumb. He noted that the thumb is not homologous to the thumb bones of apes and humans, but represents an evolutionary development of a thumb from a wrist bone. The thumb is useful to the pandas in their method of stripping and eating bamboo. What is the DNA pattern for such a thumb and how does it compare with the DNA pattern for the primate thumb? What other factors may be involved?

The existence of Lucy suggests many questions:

1. If Lucy was not exactly human and not exactly an ape, should she be considered to be a link between them?

2. If erect walking preceded a large, well developed brain, how was Lucy able to defend herself on open ground?

3. Did she use tools?

4. Did she make implements of rushes, wood, bone, or hair which did not survive?

5. Did she carry a baby in her arms? Did she carry food in some kind of container to leave her arms free?

6. Did she retire to trees at night? Did she build shelters on open ground?

7. Was she really hairy? Could she have been a naked ape?

8. Did she talk to her baby?

9. Did she have human consciousness?

10. Can we make an either/or choice of human or not human?

C. Owen Lovejoy (Johnson and Edey, 1981: 309-310) believes that bipedal walking should not be considered in cause and effect relationship with tool making or hunting or choice of environment, but rather as part of an overall survival design. His ideas are presented in systems reasoning which incorporated symbiotic living patterns. Lovejoy does not look for sequences of cause and effect but rather for relationships between and among the characteristics which are suggested by fossil evidence. He concluded that tool use and tool making were coordinated with continued brain development but not dependent on a certain stage of development.

He thought that bipedalism was associated with food sharing, pair bonding, and development of a home base that reduced necessity for mobility among females. The females could then give birth to more than one baby in five years (which is the natural infant spacing strategy among chimps, and probably among early apes).

He thought that continuous sexuality of females, pair bonding, home shelter development, and increased number of offspring interrelated in beginning the rise toward human social group cohesiveness and development.

Lovejoy's concepts may not all apply to Lucy, but his manner of reasoning is much more productive than the cause and effect patterns of some earlier writers, that must be continually revised as new fossils are found.

I believe fossils alone will never answer all these questions, but they can be used to discover truths about ways of life, which may be more important than just the physiology of Lucy and her kind.

The reason some have searched so diligently for evidence which might supply the answer to the question, "Was Lucy an ape or a human?" is that they were in a mind set requiring an either/or decision; and also, they desired to know the real meaning of "human." "When did an ancestor evolve into a human?" implies the question, "How is human different from non-human?" We try to find the sequence from the first life on the planet to ourselves. We find it difficult to accept the synthesis of the ideas of continuous lines of ascent, and of discontinuities represented by different kinds of lifeforms. When we find a line of continuities, we look for discontinuities. When we find discontinuities, we look for continuities. It is the essence of human response to paradox.

If we see the paradoxes in our thinking about Lucy, perhaps we may be excused for doing some broad speculating. Answer to the question, "Which fossil was the first hominid or the first human?" may forever elude us, but if we learn lessons from the fossils, and speculate from these — using different reasoning patterns — we may expand our knowledge and (in complementarity dynamics) perhaps our minds also.

In *Evolution and the Human Mind*, Norman Munn (1971: 152-160) recognized the continuities and discontinuities of evolution. He thought that the invention of human language came "as close as anything to supporting the argument for discontinuity." In reviewing the observations of D'Arcy Thompson, Theodosius Dobzhansky and Henry Nissen, he identified discontinuities, especially noting marked continuities and discontinuities in psychological processes.

Color vision, found in the different evolutionary lines of insects, birds and primates, but not in other mammals, is an example of continuity in a DNA concept and discontinuity in its expression in different animal lines. Munn found such continuities and discontinuities in the development of depth perception, instinct and habit, learning ability and reasoning power.

I believe that the logic of complementarity dynamics is required to enable perception of the relationships of these continuities and discontinuities and to accept the paradoxes which continue to confound those who labor in the either/or mind set.

From Paleontology to Anthropology and Beyond

New Sciences Evolve

With the study of early fossil hominids and their suggestions of human ancestry, the scientific specialty of paleontology, the study of ancient lifeforms; merges into the specialty of anthropology, the study of humankind.

Anthropology, a division of the social sciences, was separated from the biological sciences at its inception. It now includes not only paleoanthropology, the study of the earliest hominid forms; but also archeology, the study of their physical works, their tools and dwellings — which suggest their ways of life, their culture. Anthropology now overlaps biology including specialties in palynology, the study of spores and pollen of plants (often found with human habitations) and paleoecology, the study of the relationships of prehistoric lifeforms to their environments. Anthropology is merging into cultural anthropology, the study of human civilizations.

Despite the advantages of communication among experts in these overlapping fields, gaps between these social sciences and the biological sciences are creating problems. Recent anthropology-classified books (1960-1985) on the whole, show lack of coordination with microbiology and genetics (classified as biological sciences). It is perhaps indicative of this lack of coordination that biological science books may, on some university campuses, even be found in library buildings separate from anthropology books.

The almost total separation of the study of physical-cultural aspects of early humans from the sciences of microbiology and genetics has led to the writing of numerous books about early humans that have no mention, or very minimal mention, of DNA. The possibility that Lucy and her kind evolved from the *DNA* of an earlier genus (possibly DNA of Dryo or Rama or their successors) has been virtually ignored. Lucy must not be left hanging as representative of a possible progenitor of humankind without exploration of her possible DNA ancestral connections.

Scientists have found that living chimpanzees have 48 chromosomes. Humans have 46, 23 in each egg and sperm. Biomolecular studies have suggested that these two species, chimps and humans, have very similar DNA. These studies have not measured the similarity of the whole genome, all the DNA of both species. Identification of the DNA pattern of the whole human genome may be far in the future. Estimates made in 1991 suggest that exploration of this genome structure will take hundreds of scientists as long as 20 years and cost perhaps three billion dollars, even using the best computer technology available. We are not a simple species!

In the meantime, short identified sequences of proteins, patterned by DNA, have been analyzed. Blood serum and cytocrome C, a protein enzyme of mitochondrial origin, and other gene (DNA) products have been analyzed. Researchers have postulated that the DNA of chimps and humans is very similar, though it is evident that such an opinion is a vast extrapolation, and may be found to be unwarranted when full DNA patterns have been interpreted. Other apes also are judged to have DNA products similar to humans though not having as many similarities as between humans and chimps. Other primate DNA varies more. Other mammals vary generally in DNA differences as they vary in bodily construction (morphology) differences, although there are some surprises.

But what of Lucy? There is now some hope that DNA may be recoverable from some lifeforms that lived as long as 20 million years ago. Lucy may or may not be one of them. It is sometimes necessary in the course of science and social science to break a logjam of intellectual compartmentalization and inertia by proposing hypotheses which may be considered daring speculations. Let us speculate.

Perhaps Ramas or their successor apes had 50 chromosomes, and Lucy had 46 (a human). What evidence suggests this? Lucy was the earliest known relatively complete fossil hominid (40% of bone structure recovered) who was an upright walker. Not enough evidence of Ramas have been discovered to determine whether they had any human-like bodily (post cranial) characteristics. But Lucy did have such characteristics.

I speculate that Lucy walked upright because she had a DNA complex of 46 chromosomes, which coded for an upright walker. I think the chromosomes of her kind salmutated within the fertilized eggs of Dryos or Ramas (or perhaps an aquatic group). This was not just a minor gene mutation, because no one or a few genes could create the new complex of characteristics which was Lucy.

Lucy did not just walk in an upright stance. She had the DNA for pelvic structure and femurs of an upright walker and therefore the musculature and internal visceral arrangement needed to sustain her upright position. Her genitalia must have been repositioned to a forward position (suggested from the work of Elaine Morgan, *The Descent of Woman*, 1972). Lucy's muscles were recreated and repositioned from a DNA blueprint to function with the change in bone structure and the requirements of her stance. Muscles which serve as extensors in apes (and probably also in Dryos and Ramas) must function as abductors for Lucy. The large rear-end muscle, the gluteus maximus, extended the trunk so that Lucy could run climb, swim?, or comfortably sit.

From other Austral skulls we can judge that Lucy had a head which was positioned upright, with the opening for the spinal column (foramen magnum) centrally located as in living humans, not in the back of the skull as in apes. Lucy's brain is thought to be about 20% larger than that of a similarly sized living chimpanzee.

The size of the skull top (cranium) of hominid fossils has been used as the major characteristic to classify them into various genera and species. Although it has been known for a number of years that the size of the brain case or brain is not a definitive test of intelligence, paleoanthropologists, ofter having little better evidence of intelligence, continued to make these size judgments. The larger the size of the cranium, the more intelligent the hominid was considered to be.

It was 1974 when R. L. Holloway published an article which not only challenged this concept but also provided a better method for judging intelligence. He believes that brain size in hominids is generally related to body size. He found that brains of living apes and humans are very different in amount of convolution and in neurological organization. These basic differences may be determined not only from the brains of modern animals but also can be determined from fossils which have left impressions of the brain on the inner brain case bones.

Holloway (July, 1974: 109) evaluated brain casts from the skulls of all available early hominid fossils, six of *A. africanus*, one of *A. robustus* and one unclassified skull from South Africa, three of *A. robustus* and three of *Homo habilis* from East Africa. He concluded:

> ... in each instance where an endocast is available, whether the skull is less than a million years old or more than two million years old, the brain shows the distinctive pattern of hominid neurological organization."

He distinguished the cortical brain landmarks between apes (pongids) and these hominids, and concluded that the brain casts of varying sizes showed not ape-like brains but human-like brains in *every* case.

Holloway noted that the normal range of brain size of living humans averages 1,200 cc to 1,800 cc but it is known that

Jonathan Swift's brain exceeded 2,000 cc while the Nobel Laureate, Anatole France "made do with 1,000 cc."

A modern normal-size human with the abnormality microcephaly, shows human behavior although of subnormal intelligence with a brain volume of around 600 cc, less than the size of some gorilla brains. (Gorillas' brains average 498 cc ranging up to 698 cc).

Holloway (July, 1974: 110) wrote:

> It may seem surprising that *Australopithecus*, a genus that has only recently been granted hominid status on other anatomical grounds, should have an essentially human brain... Neurologically speaking, brains whose organization was essentially human were already in existence some three million years ago.

Let's make an extrapolation that Lucy, about the same size as a chimp with a 20% larger brain capacity, had a brain neurologically similar to the later Australs that Holloway evaluated. It would then be likely that she behaved much more as a human than as a chimpanzee.

The contrast of Lucy's human-like and lingering ape-like traits has led some scientists to postulate that she was an Austral intermediate between Ramas and true humans. However, if it is found that chromosome number is definitive in the determination of "human," and suppose Lucy had 46 chromosomes, as I suspect; then Lucy and her kind may indeed be considered to be not *Australopithecus afarensis* but rather *Homo afarensis* — the first known humans.

Such speculation is, of course, very tentative, but Lucy was definitely different from fossils known before her. At some point there were humans with 46 chromosomes. I happen to think Lucy was one of them.

In such analysis it becomes necessary to think of Lucy's parent species. Unfortunately, the postulated ancestral Ramas left little fossil evidence, and that little shows that they existed more than five million years earlier that Lucy. Whether they overlapped Lucy-type species remains to be discovered. The search for their fossils (and more of Lucy's ancestors) continues.

If Ramas (or other ape-like progenitors) were ancestral to Lucy and her kind, and if they were apes with 48 or 50 chromosomes — and if their progeny genomes (fertilized eggs) underwent recombination in utero, which produced Lucy and her kind with 46 chromosomes, what would be the characterisics of those ancestors? Their teeth may have had human-like characteristics, but that is about all that is known. Ramas (or other progenitors) would have needed a birth canal large enough to accommodate Lucy's head. Thus it should have been expected that the earliest hominids would be found to have had rather small heads.

Would Ramas or their aquatic successors raise a different genus from themselves? I think they would. The bonding of birth would be the same or similar whatever the number of chromosomes, and Lucy and her kind were, no doubt, charming babies! It is not unknown for one genus of mammal to raise a different genus.

Although raising of a different genus has only recently been identified in the natural history record, such an unusual occurrence has not been widely sought, so it might not have been recognized even though seen. There is at least one and possibly more confirmed cases in modern times of an ape band raising a human infant. Would Ramas do less?

It is indeed a speculative moment. I believe Lucy and her kind salmutated from Rama-like or aquatic apes in East Africa sometime in the period between seven and three and a half MBP. I believe Lucy's kind were upright walkers with 46 chromosomes, who evolved from quadrupedal apes of 50 or so chromosomes, who cared for them and raised them, although these new hominids were very different from the former offspring of the Rama (or similar) genera.

Humans arose, both males and females with human brains, with all the potential which has developed from that time to this. The implications of this chromosome salmutation hypothesis are enormous! They suggest that salmutations may be chromosomal recombination events at the genera (or above) level in both humans and other animals.

This hypothesis also suggests that the genera *Homo* of humankind goes back as far as six to three million years. The subsequent groups now called *Homo* have undergone less extensive genetic mutations than Lucy, leading to different kinds of morphological changes in relation to different environments.

Humans have migrated but there have always been groups that stayed at home in the general area of their births. The differences we see in the races of humans are mainly differences developed between those groups which stayed home. The humans who moved and mingled created the intervening types which show combinations of racial characteristics. These tended to erase the sharper distinctions between the geographically more stable groups. But all were, like Lucy, 46 chromosome beings — humans.

Classification of humans and other animals should be based on chromosome number — genera — rather than emphasizing the search for minor mutations which no longer can be considered to be definitive of those animals which started new lines of ascent. The continuities of breeding and the discontinuities of DNA recombinations will remain maddeningly paradoxical until we can think in paradox logics while still holding fast to established facts.

I believe that the complementarity of the physical processes of the brain and the mental processes of the mind will remain the greatest paradox to be confronted. This paradox may always remain to challenge the definition of human.

28

Cenozoic Era — Quaternary Period — Pleistocene Epoch Humans and the Ice Age

The Pleistocene Epoch is noted for the rise and evolution of humankind in a time frame known as the Ice Age, beginning about two million years before present (two MBP). Glaciers had been evident preceding the Pleistocene in mountainous regions of the world, in the Sierra Nevadas of California about three MBP; in the Antarctic, and the Arctic: Alaska, Greenland and Iceland. In the Scandinavian highlands three minor advances known as the Donau glaciation had occurred. Although glaciers also covered parts of northern Asia and Siberia during the Ice Age, American scientists apparently have not clearly coordinated these glacial flows with those of North America and Europe.

Prior to about 1800, humans thought that the rocks and boulders found scattered about in certain regions of Europe and North America were deposited there by the Biblical flood. Soon thereafter it became evident to growing numbers of scientists that it was ice, not water, that played the major role in shaping the northern European and North American landscapes, creating lakes and depositing boulders (erratics) pebbles, sand and fine soil (loess). By 1860 scientists began to look for the relationships between glacial flows and the fossils of animals that lived in glacial regions.

Earliest estimates of age of fossil human remains were made in the early 1900s in relationship to the named advances of ice cover in Europe: periods known as the Gunz, Mindel, Riss and Wurm glaciations. The discovery of an ancient human jaw, the Maur mandible, found near Heidelberg, Germany in 1907 was considered to date from the Mindel glacial period, or Mindel-Riss interglacial, but little else was known about it.

A leader in paleontological theory, Sir Arthur Keith of Great Britain, postulated a primate family tree in 1931, considering that the Pleistocene period began about 200 thousand years before present (200 TBP). Therefore any fossils considered to be of mid Pleistocene age would be considered to be about 100 thousand years old. If the four glaciation periods were of roughly equal length, the Mindel-Riss interglacial period would have occurred about 100 thousand years ago and Heidelberg Man (*Homo erectus*) would have lived about 100 TBP.

During the 1950s scientists correlating glacial, geological, stratigraphic and accumulating fossil data from around the world, realized that the Pleistocene Period was much longer than Keith had thought. It was considered to begin about one million years before present (one MBP). Mid Pleistocene would then be about 500 TBP. By this calculation, Heidelberg Man would have lived about 500 TBP, instead of 100 TBP.

In the early 1960s, the potassium/argon dating method was developed, and for the first time it became possible to assign dates to volcanic rocks, and thus calculate the dates of fossils found above or below a volcanic ash layer (tuft) in the Earth. After 1961, this and other methods were used to estimate that the Pleistocene Period began about two MBP. This would indicate that Heidelberg Man lived about one MBP, if he lived in mid Pleistocene. However, comparisons with Chinese and Java hominid fossils raised doubt of such an early date for the Maur jaw, and scientists again estimated a date for it of about 500 TBP.

By this time it was evident that age estimation of fossils by means of existence of glaciation was limited in its value, and radioactive processes, such as the potassium/argon decay rate and other recently discovered methods of dating, began.

Potassium/argon (P/A) dating is based upon the fact that potassium 40 has a half-life of 1.3 billion years, and decays into argon 40, a gas which is captured in rock at the time of volcanic eruptions. The ratio of potassium 40 to argon 40 in an undisturbed volcanic rock sample indicates the age of the rock. The method is accurate for dating in the billions of years to within about 50 to 100 TBP. It therefore can be used where there has been volcanism, to determine age of ancient fossils. However it does not measure biological materials directly, nor measure closely within about the last 100 thousand years.

For fossil material, a carbon 14 dating method is now used. The half-life of carbon 14 is 5,760 years, therefore the amount of carbon 14 remaining in a sample can be measured to indicate age from about 60 thousand years ago. Skeletal materials younger than this can thus be tested to determine their age.

From this knowledge of dating methods, it becomes evident that there have been many difficulties in judging the age of fossils, and it is not an easy task. It should be expected, therefore that many estimated dates in the scientific literature, written at different times, have different meanings, and indeed, estimated age of fossils still varies widely.

While dates given by different authors are all data to be considered, one must be aware of the changing interpretation of the age of the ice flows and the limitations of the P/A and carbon 14 methods of dating. Nevertheless, estimates must be given by authors writing about fossils and evolution, subject always to new correlations of data, new methods of time measurement, and reevaluation of previously accepted information.

Glaciers, by their very nature, come and go, form and expand, retreat and disappear. They cover earlier glaciers, erasing evidence of their movements. To name a glacial advance may say nothing at all about the time of its earliest formation. Therefore, there are great differences of opinion about the timing and movements of the glaciers of the Pleistocene. The following estimated glacial dates are based upon my correlations of the work of L. A. Frakes (1979: 238) J. B. Birdsell (1981: 274) and others. The dates should be considered as very tentative.

The Pleistocene advances seemed to start in North America with the Nebraska flow from 1.6 to 1.3 MBP, during a time in which Australs (*Australopithecus*) and *Homo habilis* lived in Africa.

The second advance, the Kansas glaciation from one MBP to 700 TBP, occurred in about the same time frame as the first major European advance known as the Gunz which covered Scandinavia and northern Europe from 1.1 MBP to 700 TBP. This was about the time that the first *Homo erectus* bands were venturing from Africa into China and Java, which was then connected to the Indonesian mainland due to low sea level induced by glaciation in the north.

The second European glacial advance, the Mindel, occurred from 650 to 400 TBP (with the third North American, the Illinois glaciation from 550 to 400 TBP) the time when *Homo erectus* bands spread throughout southern Eurasia and into central and western Europe.

The third European advance, the Riss, occurred from 300 to 200 TBP, as *Homo erectus* evolved into *Homo sapiens neanderthalensis*, struggling to exist in Europe, south of the advancing ice.

The fourth European advance, the Wurm glaciation and the North American Wisconsin occurred from about 80 to 10 TBP. During this time it is thought that Neanderthals evolved into (or were replaced by) *Homo sapiens sapiens*: Cro-Magnons and modern humans.

Dates of the warmer interglacial periods are based upon European data and are known as the first interglacial, Cromer (Gunz-Mindel) from 700 to 650 TBP; the second, Holstein (Mindel-Riss) from 400 to 300 TBP; and the third, Eem or Elser (Riss-Wurm) from 200 to 78 TBP. Needless to say, these warm periods allowed humans to expand their range, and they migrated further north during the second and third interglacials.

Although it would be more understandable to postulate that humans migrated only during warm interglacial periods, rather than during glacial periods, the evidence seems to indicate that *Homo erectus* bands braved the cold. They probably followed herds of reindeer and other animals which found good grazing in both tundra and temperate grasslands south of the ice flows.

The world from about 10 TBP to the present, is considered to be within the fourth interglacial period, the Recent Epoch (Chapter 29).

Volcanism and Sea Level Changes

Volcanism occurred during the Pleistocene in the circum-Pacific region of east and southeast Asia, in Africa, Idaho, USA, Mexico and Central America. Correlations between Pleistocene volcanism and glaciation have not been adequately drawn. Some scientists believe that changes in volcanism and temperature in the Pleistocene were comparable to those at the end of the Cretaceous when the dinosaurs became extinct. This remains to be verified.

Sea levels fluctuated with the build-up and melting of the Antarctic and Arctic ice caps and the expansion and contraction of the Northern Hemisphere glacial ice. Fluctuations in sea level of tens of meters were not uncommon. Evaluation of oxygen isotopes in plankton and carbon 13 cycling between atmosphere and oceans has allowed correlations between climate and sea level.

At the Pliocene-Pleistocene boundary, the Caribbean foraminifera had consistently higher levels of oxygen 18 than the

Pacific varieties, verifying that the Central American land bridge had been closed, according to P. K. Weyl (1968).

Weyl has postulated that the closing of the Panamanian Isthmus increased the flow of warm Caribbean water into the Gulf Stream, leading to increased moisture in the air mass over North America, which in turn initiated Northern Hemisphere glaciation. On the other hand, it has been postulated that the increase in the Arctic ice cap from the lowered temperature created by increased volcanism could have led to the lowering of sea level and thus the closing of the Atlantic to Pacific Ocean flow, beginning the Ice Age.

Using the monistic logic of causation and either/or choices, some climatologists and paleontologists have searched for the single initiating cause of glacial periods. Others have postulated numerous factors which interact to influence changing weather and climatic patterns, using logic which I consider to be complementary systems. I believe the latter approach is the more productive. There can be no *single* cause which triggers glacial and warming periods. The complexities of climate are enormous because of operation of all the following complementary factors:

Complementary Factors Effecting Changes in Earth's Climate

These influences are based on an assumed 12 degree difference in average temperature between full glacial and interglacial conditions.

Complement A:

Astrophysical Factors:

1. Tilt of earth's rotational axis in relation to orbital plane — cooling effect on areas of Earth farthest from the Sun.
2. Eccentricity of ellipse — cooling effect when Earth is in orbit farthest from the Sun — aphelian.
3. Precession of equinoxes — perihelion on June 21st — summers cooler than average in Northern Hemisphere.
4. Solar shift to more short wave ultraviolet radiation.

5. Increase in intergalactic dust.

Tectonic Factors:

6. Increase in continentality.
7. Increase in polar latitude position of continents.
8. Closing of sea connections (e.g. Isthmus of Panama).
9. Increase in explosive volcanism.
10. Land mass increase at poles.
11. Sea floor spreading.
12. Marine regressions (seas retreating from land).
13. Decrease in sea level (ocean volume).
14. Increase in circum-polar cold water circulation.
15. Up-welling of cold bottom waters.

Atmospheric Factors:

16. Depletion of CO_2.
17. High latitude westerly winds moving toward equator.
18. High latitude humidity.
19. Rapid evaporative conditions.
20. Steep temperature gradients.
21. Increase in cloudiness and-or atmospheric dust.
22. Increase in low pressure.
23. Increase in precipitation.
24. Increasing ice accumulation.
25. Snowblitzes.

Terrestrial Factors:

26. Increasing reflectivity (albedo) with decrease in forest cover.
27. Increasing albedo with increase in ice areas.
28. Increase in Antarctic ice cover.
29. Increase in height of mountains (e.g. Sierra Nevadas rose up to 2,000 m during Pleistocene).
30. Increase in amount of mountain building in suitable locations.
31. Surging ice sheets.

Complement B:

Astrophysical Factors:

1. Tilt of Earth's rotational axis in relation to orbital plane — warming effect on areas of Earth closest to Sun.
2. Eccentricity of ellipse — warming effect when Earth is in orbit closest to Sun — perihelion.
3. Precession of equinoxes — perihelion on January 2nd — winters warmer than average in Northern Hemisphere.
4. Solar shift to more long wave infrared radiation.
5. Decrease in intergalactic dust.

Tectonic Factors:

6. Increase in continental break-up.
7. Increase in equatorial positioning of continents.
8. Opening of seaways.
9. Increase in out-gassing volcanism.
10. Land mass decrease at poles.
11. Relative stability in sea floor ridges.
12. Marine transgressions (seas expanding over land).
13. Increase in sea level (ocean volume).
14. Increase in warm water circulation.
15. Decrease in up-welling of cold bottom waters.

Atmospheric Factors:

16. Concentration of CO_2 (greenhouse effect).
17. Equatorial winds expanding toward temperate regions.
18. High latitude aridity.
19. Low evaporative conditions.
20. Minor temperature gradients.
21. Decrease in cloudiness and-or atmospheric dust.
22. Increase in high pressure.
23. Decrease in precipitation.
24. Melting of ice packs.
25. Increasing mild weather fronts.

Terrestrial Factors:

26. Decreasing reflectivity (albedo) with increase in forest cover.
27. Decreasing albedo with decrease in ice areas.
28. Decrease in Antarctic ice cover.
29. Wearing down of high mountains.
30. Stasis in mountain building.
31. Retreating ice sheets.

Monistic, either/or reasoning might lead one to believe that the A factors are associated with glaciation and the B factors are associated with warming. However the logic of complementary systems creates the understanding that some of the A factors may interact to create or increase a warming trend and some of the B factors may interact to create or increase glaciation. In order to discover why glaciers advance and retreat, it is necessary to envision the interactions of all of the above complementary factors.

Arctic, temperate and tropical zones became more distinct through shifting glaciation during the Pleistocene, although their boundaries and time extensions fluctuated greatly. The animals which evolved during the Pleistocene on the different continents were adapted to these zones, although many of them were versatile enough to live through major climatic shifts.

The Pleistocene is noted for a large proportion of animal lines which have continued into the present. Most of the kinds of animals living today were represented by predecessors which can be identified by close similarities in structure. However, many animal taxa of ancient lineages perished, often associated with climatic changes and the inability to migrate to escape the advancing ice, as well as by human predation.

Africa — Proboscidean Proliferation and Migration

Africa retained a tropical climate in the equatorial regions throughout the Pleistocene. There were tropical rain forests in the interior. Even the Sahara Desert was then forested. The eastern rift area and South Africa were mostly savannas and woodlands. Rainfall was heavier than today. Lakes and streams provided water for a large variety of wildlife.

Proboscideans proliferated in Africa throughout the Pleistocene and migrated to all the continents except Australia and Antarctica. Vincent Maglio (1973) in *Origin and Evolution of Elephantidae* traced Pleistocene proboscideans back to *Primelephas gomphotheroides* of the Pliocene. He examined three genera arising in the Pleistocene: *Loxodonta, Elephas* and *Mammuthus*. He postulated that *Loxodonta* was represented by two species which remained in Africa, one line becoming the living African species.

The *Elephas* line was represented by 10 species, one evolving into the living Indian elephant and at least one, *E. namadicus*, found in North America. He posited that five species of *Mammuthus* evolved in the Pleistocene, four of which migrated broadly: *M. armenicacus, M. columbi, M. imperator* (Imperial Mammoth) and *M. primigenius* (Woolly Mammoth, Figure 28.1 a).

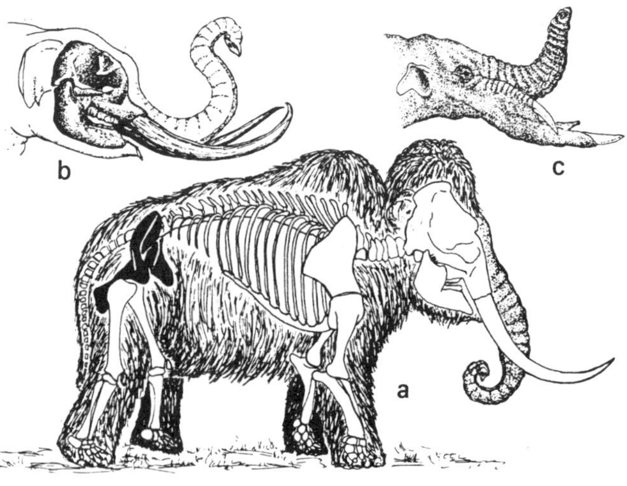

Figure 28.1 (a) Woolly Mammoth (*Mammuthus primigenius*). (b) Mastodon (c) *Platybeledon*. (After Osborn, 1910).

Figure 28.1 gives examples of Proboscideans that arose in the Pleistocene. Maglio (1973: 76) was unsure of the evolutionary lines of proboscideans. He wrote:

> When dealing with taxa above the species level it is often difficult to deduce correctly the actual ancestral-descendant sequence as opposed to successive, but discontinuous, non-ancestral offshoot taxa that form a series of grades... In most cases fossil material does not allow the degree of resolution necessary to trace actual ancestral populations.

I believe the failure to find ancestor-descendent relationships among the proboscideans does not represent a deficiency of the fossil material, but rather highlights the fact that DNA recombinations have occurred which complicate recognition of ancestor-descendent lineages.

Prior to the work of Maglio, Henry Fairfield Osborn spent a lifetime studying these animals. Unfortunately he died in 1935 before his masterwork was finished. Fortunately, Edwin H. Colbert was able to complete Osborn's *Proboscideans*, volume II, which he published in 1942. This tome, more than 1,600 pages in length, is one of the most masterful works on any line of animals and yet near the end of his life Osborn (1942: 1524) concluded that proboscidean evolutionary lines were not clear-cut.

For example, he found mastodonts and elephants: "very much alike in certain characters but...still more unlike in other characters and [they] cannot possibly be descended one from the other." He noted that several lines of proboscideans may be identified by their skulls and tusks, but their grinding teeth do not follow the same sequences. There are overlapping characteristics in the teeth of different taxa lines. He recognized the same kinds of discontinuities which were later noted by Maglio.

One can only conclude from an analysis of Osborn's great work that subsequent authors must of necessity oversimplify when writing about proboscideans. There is too much information to synthesize, yet both continuities and discontinuities are evident in the fossil record he interpreted.

Of extinct mammals, the first identified by man was a proboscidean. The Woolly Mammoth was found between 1692 and 1695 in Siberia by a Dutch traveler named Ides. It was postulated at the time that the Mammoth lived in tunnels below the Earth and broke through to the surface to die. The Woolly Mammoth was first described by a Russian explorer, Ludolf, in 1696 and named *Mammotovoi kost* (Earth ivory). Discovery of the animal was considered proof of the Biblical universal deluge.

This mammoth was the first animal to be described in Latin and the first to receive a scientific name. The scientifically accepted name, *Elephas primigenius*, was coined by Blumenbach in 1799, but a total of 30 different names were used between 1696 and 1888 (Osborn, 1942: 1123). By the year 1921, the name Mammontinae was given to the mammoths, and the Woolly Mammoth was renamed *Mammonteus primogenius*, now called *Mammuthus primigenius*.

More than 46,000 proboscidean fossils have been discovered in the last two centuries. Since 1754, they have been named and renamed and classified in more than 550 species and subspecies, and in more than 35 of the duplicating genera. Although Osborn tried to identify duplicating synonyms, the overlapping of numerous names and varying classifications have still left confusion for those subsequently attempting to hypothesize evolutionary lines.

Colbert identified 352 "valid" species and pointed out geologic successions of proboscideans, but did not go beyond Osborn's identification of migration routes to posit evolutionary progressions for the order.

Before his death, Osborn designated five superfamilies: 1. Moeritherioidea, 2. Dinotherioidea, 3. Mastondontoidea, 4. Elephantoidea and 5. Stegodontoidea. Yet he posited 26 species of North American Pleistocene proboscideans in six genera, which do not fit clearly in the designated superfamilies:

Cordillerion, three species

Mastodon, seven species

Stegomastodon, five species

Archidiskodon, six species

Parelephas, four species

Mammonteus, one species.

One must conclude that a new effort is needed to assign one agreed name to each known fossil in order to facilitate discussion of evolutionary lines in conjunction with DNA analysis.

Theoretically, the ancestor-descendant relationships of proboscideans should be more obvious than those of any other species, because these animals are so large and have left so many relatively recent fossils. Because Osborn, Colbert and Maglio found many discontinuities and few if any positive links between the different taxa lines, there must be a reason for those discontinuities.

It may be too early to expect much information about recombinant DNA patterns which could have led to the formation of the various proboscideans. Although there is early DNA evidence from work on other animal lines, which may apply broadly, only the products of the proboscidean recombinations: the fossils and two living elephant species, are currently available for study. Only when we are able to identify origination sites of each species and coordinate DNA patterns and morphological patterns will the fossil record become truly explicable.

Many African animals were very similar to Eurasian forms in the Pleistocene, including pigs, sheep, antelopes, giraffes, buffalos, cats, bears, dogs and hyenas. The horses in Africa in the Pliocene and Pleistocene were *Hipparion* and *Stylohipparion*. The modern horse, *Equus*, has been thought to have evolved from *Pliohippus* or *Parahippus* in North America and migrated to Eurasia across the Bering land bridge in the early Pleistocene, arriving later in Africa.

Asia and Europe — Land of Giants

Many mammals of the Pleistocene reached much larger sizes than those of earlier epochs. There were giant-size bison, mammoths, sheep, pigs, rhinos, hippos, reindeer and the Rama, Giganto (*Gigantopithecus,* Ramapithecine) which lived in China from 1.5 MBP to 500 TBP. Associated with Giganto in the fossil record were: orangutans, giant pandas, bears, civets, cats, dogs, hyenas, boars, deer, goats, bovids, tapirs, rhinos, mastodon, stegodon, horses and chalicotheres (Simons, 1972: 260).

Although some European hippos, hyenas and rhinos ventured northward in the warm interglacials, there were more movements southward in the Northern Hemisphere.

Camels migrated from North America onto the Eurasian continent and into Africa in the Pleistocene. Animal fossils found in France include rabbits, rodents, oxen, boar, ibex, red deer and elephants. The most numerous animals found in European caves were cave bears (Ursus spelaeus) which grew to be nine feet in length. Some animals tolerant of climatic fluctuations survived in place. Other primitive tropical genera became extinct, including archaic proboscideans such as *Deinotherium.*

North America — Huge Denizens of Icy Lands

The fauna of North America also grew to great size in the Pleistocene. It is thought that there is a correlation between large size mammals and cold climate. Larger animals have proportionately less exposed surface area and, therefore, need less energy to retain warmth. However, large herbivores need more food than their smaller relatives, and therefore require more grazing area and-or longer grazing time each day.

The DNA configurations and biomolecular processes which control the body size of animals is an area of study that is beginning to answer questions about the evolution of small and large animals of the Pleistocene.

Arctic fauna was adapted to the Alaskan climate. Woolly mammoth, woolly rhinoceros, ground sloths, musk oxen, large-horned bison, moose, short-faced bear, elk, yak, horses, antelopes, badgers, caribou, wolves, foxes and cats were all denizens of the tundra. The advancing ice sheets in the most severe periods impelled migrations of even the most fully cold-adapted forms such as the woolly mammoths, woolly rhinoceros, and musk oxen which moved as far south as New Mexico. The temperate North American latitudes were home for peccaries, horses (*Equus*) pronghorns, ground sloths, otters, saber-tooth cats, mountain lions, camels and mastodons.

Migrations to North America from Eurasia in warm interglacial periods included elephants, mountain goats and the bison, seven feet in length, which became the dominant ungulate on the continent (Figure 28.2).

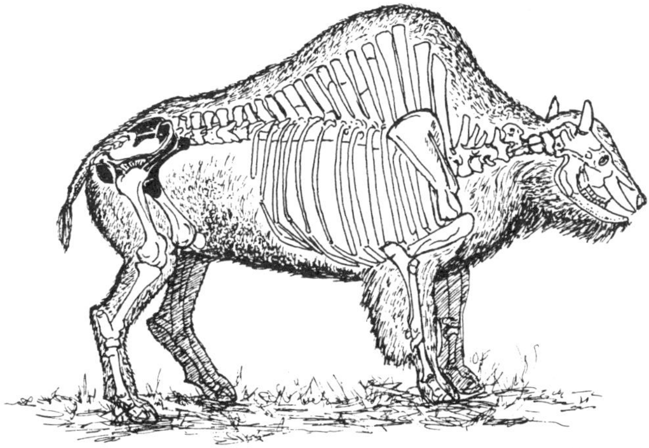

Figure 28.2 *Bison* (Artiodactyla). (After McClung).

Surprisingly, the giant beaver (*Castorides*, Figure 28.3) known from Europe and Wyoming, USA, grew from seven and a half to 10 feet in length, although later beavers were smaller. R. A. Stirton (1936) posited that Castoridae may have been derived from an Eocene squirrel (*Scuiravus*) but no intermediates have been found. Perhaps Pleistocene *Castorides* evolved from *Dipoides* of the Pliocene, but there is no verification of this.

Neither is there evidence of the evolution of these giant beavers into the smaller living beaver, *Castor*. Are salmutations involved here?

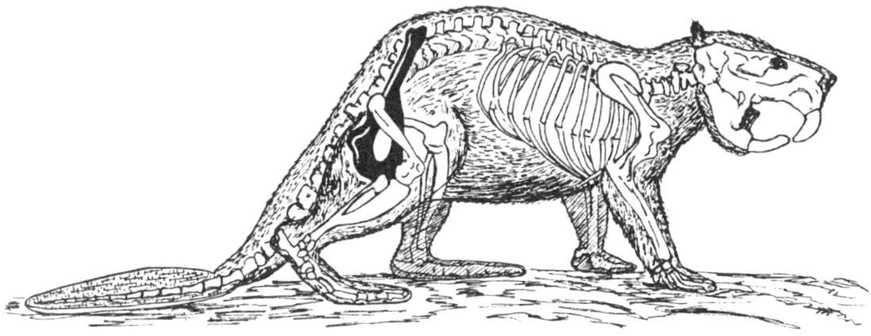

Figure 28.3 Beaver (*Castorides*, Rodentia). (After skeleton, University of Nebraska State Museum).

End of the Epoch — Death of the Giants

Although most of the Pleistocene animal lines continued into the Recent Epoch, many of the largest species perished. The giant ground sloth, *Nothrotherium*, seven and a half feet in length (Chapter 23, Figure 23.7 c) had been one of the first sloths to reach North America long before the Panamanian land bridge was formed in the Pliocene. These giant sloths became extinct shortly after the end of the Pleistocene Epoch. Fossil sloths (*Megaloyn*) dated about nine TBP, have been found with the fossil bones of humans in the LaBrea Tar Pits of Los Angeles, California, USA. Some experts believe that living tree sloths of South America did not evolve from the giant ground sloths, but it would be difficult to posit a different ancestral line leading to tree sloths (*Bradypuss*, three-toed form and *Choloepus*, two-toed form). As DNA relationships of size are unraveled, the relationships between these sloths may be determined.

Of the South American mammals which had travelled north in the Pliocene, few other than sloths, porcupines and glyptodonts survived in the new environment. Glyptodonts were an exception (Chapter 23, Figure 23.7). They reached their largest size, 12 feet in length, and were common in the southern states of the United States before becoming extinct in both North and South America prior to the end of the Pleistocene.

South America — Why Extinctions?

North American invaders of South America included dogs, saber-tooth cats and other placental carnivores. Some experts believe that herbivorous newcomers such as tapirs, peccaries and camels competed with indigenous species almost to the vanishing point, and that carnivores decimated many of the remaining South American herbivores. Others believe that many indigenous species had become extinct before the invasions. The idea of competition as the main reason for extinctions has come under fire from scientists in recent years. Michael Mares (1985: 490) wrote:

> Hundreds if not thousands of ecologists are examining living systems to see whether or not competitive effects can be unambiguously determined, yet it is not clear whether or not competition even exists.

Competitive and cooperative interrelationships cannot always be identified separately in the evaluation of living systems. How much more difficult it is to determine such relationships from fossil evidence! Surely the migrations during the Pliocene and Pleistocene initiated many new relationships: combinations of competition, predation, cooperation and environmental interaction, and may have led to some species extinctions. These processes remain open to investigation.

Litopterns such as *Macrauchea* represented the largest of their line as did the notoungulate *Toxodon*, nine feet in length (Figure 28.4) whose line dated back to *Thomashuxlia* of the Eocene (Chapter 24, Figure 24.6). Both lines became extinct by the end of the Pleistocene but some paleontologists believe that they coexisted with placental mammals for many thousands of years before the final toll of extinctions.

The similarity of characteristics of many South American mammals to those of species on other continents supports the concept that similar DNA configurations of different animals evolved separately over vast time spans.

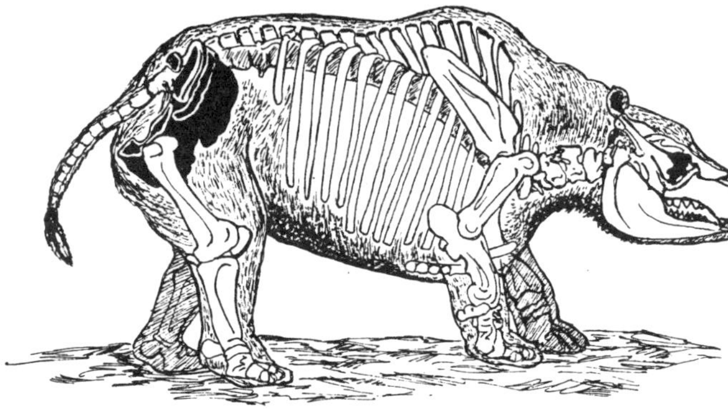

Figure 28.4 *Toxodon* (Notoungulata). (After Scott, 1912).

Indigenous rodents, opossums and monkeys survived to Recent times. Tapirs, deer, peccaries, proboscideans and camels migrated to South America. It is thought that northern camels evolved in South America into the genus *Vicugna* with one living species, and the genus *Lama* with 3 living species: *L. guanaco, L. llama and L. alpaca.* I believe it is also possible that these camels evolved, not from northern camels but from the indigenous litopterns such as *Macrauchea.*

The horses *Parahipparion* and *Onohippidium,* thought to have evolved from *Pliohippus,* migrated to South America early in the Pleistocene. The modern horse, *Equus* is thought to have migrated there from North America later in the Pleistocene. *Equus* was not found in South America after this epoch until the Spaniards reintroduced it in the exploration of the "New World" in the 16th Century. It was then reintroduced as domestic breeds into North America. *Equus* continued to live in Eurasia and Africa into the present.

Australia — Monotremes and Marsupial Giants

The Australian monotremes and marsupials continued their evolutionary patterns and, as on other continents, some of them reached their size apex. The giant kangaroo (*Macropus ferragus*) grew to 10 feet in height and, it is thought, became extinct only after human hunters arrived in Australia about 23 TBP to 12 TBP.

Possible DNA Function in Giant and Pygmy Development

DNA configurations, biomolecular and embryological development patterns, and interactions with environments that created the giants of the Pleistocene must also explain the pygmy species which existed in the Pleistocene and earlier. Pygmy elephant fossils were excavated in Sicily, Malta, Sardinia, Cyprus and Crete. Pygmy ground sloths were discovered in Cuban strata. Dwarf hippopotomi have been found in Madagascar, pygmy horses (*Archeaohippus* of the Miocene) in Europe, and ponies in the Shetland Islands. Giant and pygmy anteater fossils have been recorded. It can be no mere coincidence that pygmy species, very similar to full-size species of adjacent land masses, often occur in island environments. There is more to be learned about the interactions of DNA and environments in reciprocal moments of time.

Ohio University molecular biologists John J. Kopchick and Wen Y. Chen (1990) reported that in trying to produce large mice by modifying the gene for a growth hormone, they instead produced midget mice. These mice are about one half the size of their progenitors, and have reproduced midget mice of their own kind for three generations. Kopchick and Chen believe the process of inserting a modified gene in other mammals will also produce midgets in those cases.

Their discovery verifies my principle of responsive salmutation. This was not a gradual change in size over millions of years. Kopchick and Chen produced mice by means of a salmutation. If these midget mice had been found as fossils on an island in the past, they would surely have been identified as a different genus from their larger progenitors. Let us speculate and compare these "artificial" salmutations to the "natural" salmutations which occurred in the production of pygmy elephants.

The maximum extent of the Wurm glaciation occurred about 18 TBP. The large amount of the Earth's water captured as glacial ice would have meant that more land was exposed; there were lower sea levels than at present. Perhaps there were land connections between the Mediterranean islands and the northern mainland.

If normal-size elephants were present on the islands when the ice began to melt and were stranded there, they would be hard-pressed to find enough to eat as they continued to reproduce. However, if their fertilized eggs salmutated and pygmy elephants were produced, such small elephants would have been able to find forage on the small island land masses. It would have been a life-saving adaptation. Although normal-size elephants may not have been found on the Mediterranean Islands, I believe they were probably there, and their fossils may be found in the future. The fact that pygmy elephant fossils have been found on five of these islands supports the concept of DNA recombination in responsiveness to environmental constraints. We now know from the experiments of Kopchick and Chen that changes in one gene could have produced the pygmy or midget animals that are found in the fossil record.

Their work and other biomolecular studies may provide better understanding of how this gene has been modified and how it has produced different animal lines during evolution. However, I fear that our unwillingness and-or inability to guide these developments, through law, may ultimately lead to scientists' production and patent of a race of human Tom Thumbs, to be sent aboard a missile to explore the universe. It is a fascinating and dangerous time to be alive!

Pleistocene Primates

The island of Madagascar off the eastern coast of Africa remained a tropical rain forest during the Pleistocene. Because fossilization is very rare in such tropical climatic conditions, little is known about primate history on this island except from recent fossils (subfossils) dating from about 2,000 years ago.

Madagascar is one of few locations in which wild lemurs (prosimians) remain in their natural habitat. It may be postulated that some early prosimians arose there and migrated to Africa, or that prosimians migrated to Madagascar from Africa during the period when there was a land connection between the two, or there may be other possibilities to explain this interesting prosimian location. Madagascar is a place of special interest to evolutionists, but sheds little light so far on the evolution of the higher primates.

Numerous genera and species of monkeys are believed to have lived in the Pleistocene of South America, Africa and southern Eurasia, although there is little fossil evidence. Apes are known only from Africa and southern Eurasia. The Dryos and Ramas disappeared from the fossil record of both Eurasia and Africa long before the beginning of the Pleistocene.

The hominids called Australs (*Australopithecus*), thought to begin with *A. afarensis*, perhaps as early as six or five MBP are believed to have given rise to the Australs of the Pleistocene (Mrs. Ples, *A. africanus*, also called *Plesianthropus transvaalensis, Figure 28.5*) to *A. robustus* (also called *Paranthropus robustus*) and to *A. boisei* (also called Zinj, *Zinjanthropus boisei*). The particular lines of ascent of Australs remain uncertain.

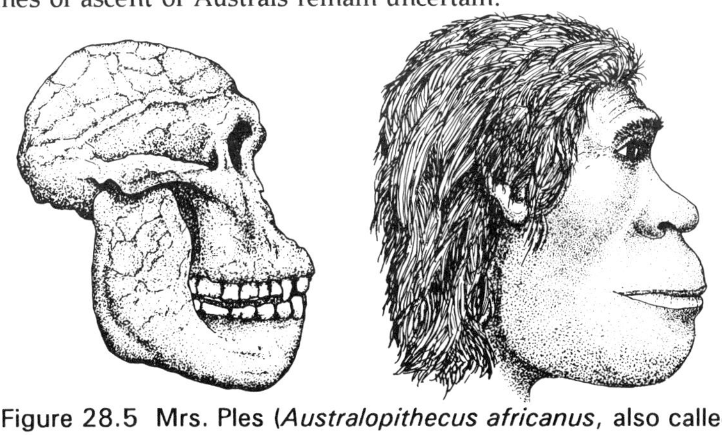

Figure 28.5 Mrs. Ples (*Australopithecus africanus*, also called *Plesianthropus transvaalensis*.

Robert Broom who discovered Mrs. Ples in a limestone cave in Sterkfontein, South Africa, found that gracile fossils of her kind came from caves of older age than the near-by caves in which he and others discovered more robust Australs. Although the nature of these cave deposits precluded fossil age estimates based on stratigraphic placement, other similar fossils of East Africa have been more precisely dated (Plate 28.1).

These time range estimates are gathered from several sources which vary according to dating techniques used and the opinions of different experts. However, fossils in each of these classifications have been dated as carefully as is currently possible. Note that the Australs lived in an overlapping time frame with the first group presently classified as human, *Homo habilis*.

I agree with Johanson that the discovered fossils confirm at least two and possibly three lines of hominids in the Pleistocene. However, I am convinced that these lines should all be considered to be human (*Homo*) because they apparently branched by means of relatively minor mutations from the first full-upright walkers, *A. afarensis,* which I believe were human and should be renamed *Homo afarensis*.

If support develops for the concept that Ramas or another intervening branch of hominids, perhaps *Homo aquaticus*, gave birth to *Australopithecus (Homo) afarensis* by means of salmutation, then the names of all Australs should be changed to *Homo*. However, I will continue to refer to *Australopithecus* (Australs) while this is the classification given to them by the scientific community.

The earliest known fossils which to date have been identified by some scientists as humans are *Homo habilis* (considered *A. africanus* by other scientists). They are known from 2.3 to 1.5 MBP in three sites in East Africa. Calculated from several individual skulls and skull fragments, the brain volume has been estimated to range from about 500-775 cubic centimeters. Skull KNM-ER 1470 (Figure 28.6) found in 1972 by Bernard Ngeneo according to Richard Leakey, in Koobi Fora deposits, East Turkana, Kenya, and painstakingly reconstructed by Meave Leakey and Alan Walker, has the largest known estimated brain volume for this classification, 775 cc, and is dated 1.9 MBP.

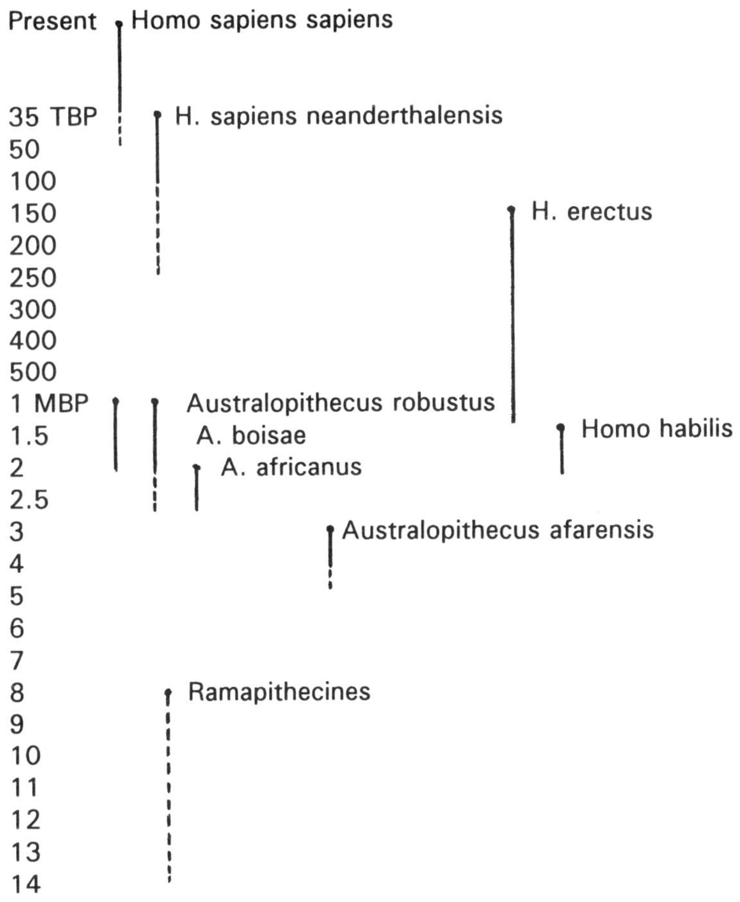

Plate 28.1 Hominid Evolution. Compilation of recent judgments of time relationships during last 14 million years. (Not in proportion).

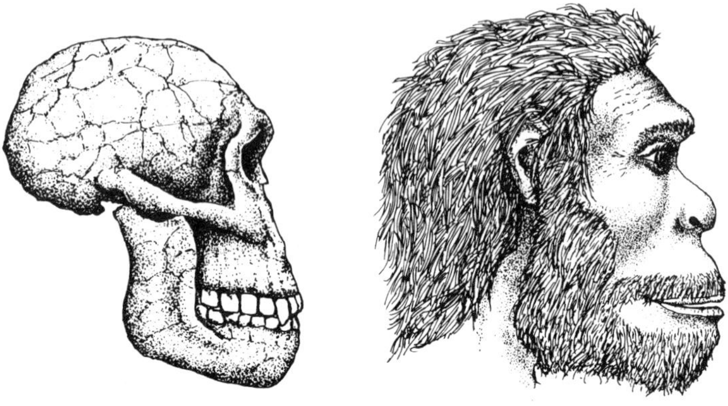

Figure 28.6 Skull KNM-ER 1470 *Homo habilis* (also called *Australopithecus africanus, Australopithecus habilis*).

Endocasts of *Homo habilis* skulls, like those of the Australs, indicate human brain development. The size of the *H. habilis* brain (range 500-775 cc) overlaps the range of Lucy and her kind (range 430-550 cc). Some scholars believe that they, therefore, should be classified together in one genus, but other experts believe the two kinds of hominids are different enough in other characteristics to require two different genus classifications.

I believe that the total morphological differences between Dryos or Ramas and Lucy's immediate ancestors (*Homo aquaticus*?) will be very profound. The skull size differences between Lucy and subsequent *Homo habilis* will then be judged to be of minimal significance. The most profound change, from ape to human, may have occurred as much as several million years earlier than the time of the first *Homo habilis*.

The configuration of the human brain is much more definitive of humanity than the size of the brain. I believe it will be found that the salmutation to human brain development and full upright walking was accompanied by the change from 50 or so chromosomes to 46 chromosomes. Such a change from apes to Lucy's progenitors, perhaps *Homo aquaticus*, also probably involved the gene mutations which created the relative hairlessness of humans and the correlative sweat (sebaceous) glands which are organized in ordered geometric patterns in conjunction with the hair follicles. Most paleoanthropologists now believe that the relative hairlessness of humans evolved late in hominid history, at the *Homo erectus*, Neanderthal, or even the Cro-Magnon stage, in the Pleistocene.

This is relatively unlikely because *Homo erectus* migrated toward colder northern climates. Why would selection favor hairlessness and development of efficient sweat glands in hominids living in colder climates, and ape-like hair and inefficient sweat glands in apes remaining in a hot climate in Africa?

Lucy must have been more like us than has been imagined. She may in the future be considered to be quite human. The differences between Lucy and the habilines may be minor in comparison with the differences between Dryos, Ramas and Lucy's immediate predecessors, when those ancestors are discovered.

In addition to Skull 1470 of Kenya, *Homo habilis* fossils have been found in the Omo area of Ethiopia. They were first discovered in the Olduvai of Tanzania and identified by Louis Leakey in 1964 as *H. habilis* (Cindy, George, Twiggy and Johnny's child).

Homo habilines were considered by Louis Leakey and others to be the first stone tool makers, and thus the first humans. Evidence is strong, however, that early Australs may also have made stone tools, the basic Olduvan hand choppers.

Tool use and tool making by Australs from more easily worked materials of tooth, bone and wood probably preceded stone tools by a million or perhaps several million years. It might be expected that these materials would leave little or no evidence.

However, Raymond Dart thought that bashed skulls of forty-two baboons found in a Makapansgat, South African cave showed that *A. africanus* used knobby leg bones as hunting weapons (Johanson and Edey, 1981: 64-65). Other scientists have postulated that the skull fractures occurred by means of compression of rocks and soil on top of these animals during the time of their burial, thus giving no evidence of human hunting weapon use.

The significance of stone tools as indicative of humanity should not be overemphasized because a thriving culture of human child care, of vegetable food gathering and preparation, and stick shelter building must have preceded the Olduwan culture which is characterized by the earliest known stone tools.

The oldest site in which stone tools (hand held Olduwan choppers) have been found is the Hadar site of East Turkana, Kenya, which has been dated to be more than 2.5 MBP. Scientists are unsure whether the makers of these tools were the Australs or *Homo* habilines found there.

A Koobi Fora, Kenya, site of 1.7 MBP reveals a base camp area about 50 feet in diameter where several hundred stone tools had been made by *H. habilis* from materials that were not found in the immediate vicinity. Animals identified from their bones at the site include gazelles, giraffes, hippos, pigs and porcupines. The evidence seems to indicate that a band of 20 to 30 persons inhabited this location and traveled some distance to acquire the materials needed to make their tools (Birdsell, 1981). Paleonanthropologists have speculated that this site provides evidence that there was food sharing and a division of labor among the men who hunted and the women who gathered vegetable foods and small game, and cared for the young, as is done in hunting societies of today.

By 1.5 MBP both *Homo erectus* and *A. robustus* lived in the Koobi Fora, East Turkana area of Kenya. Scientists have recently confirmed that *A. robustus* and *Homo habilis* inhabited Africa together for at least 600 thousand years (from about 2.1 MBP to 1.5 MBP) and *A. robustus* lived on with *Homo erectus* for an additional 500 thousand years.

This probably came as a shock to some neo-Darwinians who believed that more advanced humans would always out-compete and cause the demise of similar but less advanced humans. Surely our human ancestors would have overcome and quickly eliminated such primitive beings as *A. robustus*. They existed together, another blow to rigid Darwinian concepts.

The fossil record seems to indicate that humankind at this early hunting stage of existence was inclined to expand from region to region without evidence of much, if any, conflict between groups, probably due to the fact that food resources were essentially unlimited at the time, and contacts with adjacent groups would have been appreciated for trade, learning, social relationships and exogamous mating.

The most complete skull of *Homo erectus*, considered to be a female, KNM-ER 3733 (Figure 28.7) was discovered by Richard Leakey in East Turkana, Kenya. It was dated 1.5 MBP and had a brain volume of almost 900 cc.

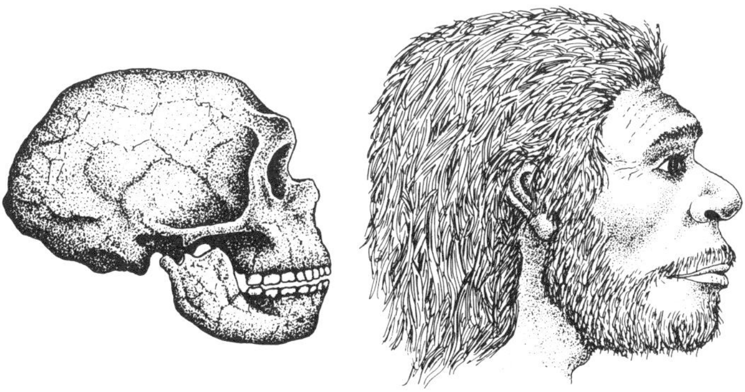

Figure 28.7 KNM-ER 3733, *Homo erectus*, East Turkana, Kenya.

Additional *Homo erectus* fossils have been found in South Africa, the Olduvai area of Tanzania, in Ethiopia, Chad, Morocco and Algeria. They have also been found at sites in Java, China, Germany, England and Hungary.

Homo erectus, Java Man (previously called *Pithecanthropus erectus*) dates in several series of fossils from one MBP to perhaps 150 TBP. The earliest evidence dates from the time of the northern European Gunz glaciation and worldwide sea regression, when Java was connected to the mainland of Asia. *Homo erectus* bands probably expanded gradually from Africa to the Mideast and then eastward to Java and China.

Home erectus, Peking Man (previously called *Sinanthropus pekinensis*) dates in several series of fossils from one MBP to about 300 TBP.

Homo erectus, Heidelberg Man from Germany (represented by the Maur jaw) dates from an estimate of 650 to 500 TBP (considered early Mindel glacial period). The northwest expansion of *Homo erectus* out of Africa may have followed a path through the Mideast, or these people may have crossed from Africa to Europe by way of Gibraltar when the Mindel glaciation of the north and sea regression created a land bridge between the two continents. Perhaps as more fossils are found in intervening locations, the routes will become more obvious.

The brain volume of *Homo erectus* ranged from 750 to 1,300 cubic centimeters, compared to Lucy's 500 cc and today's normal human range of 1,000 to 2,000 cc. It can be seen that the smallest *H. erectus* brain was only 250 cc larger than Lucy's and some were larger than brains of some modern humans.

Since size of brains is related to the size of the body, there may be more difference in the body size of various hominid groups than in their basic intellectual endowment. Considering that Lucy was about half the size of some modern humans, and her brain was about half the size also, if the relative size of brains were definitive of human intelligence, she might have been as intelligent as some modern humans.

At five feet to 5'7", some *Homo erectus* people may have been more intelligent than some modern people. Perhaps we must give up the obsessive reliance on the criterion of brain size in our evaluation of evolutionary advance and concentrate more upon learning the DNA, physical and cultural changes which have occurred.

One of the most interesting and extensive *H. erectus* sites discovered is that of Choukoutein (or Zhoukoudian) a cave 65 feet deep near Peking (Beijing) China, occupied intermittently by *H. erectus* from about 700 TBP - 300 TBP (and also about 15 TBP by modern humans). The cave site was in a high cliff overlooking a grassy plain near water where grazing animals came to drink. The evidence indicates that cave bears competed with humans for occupancy of this advantageous area, but humans assumed full possession by the time that continuous fire was maintained.

This site verifies the first known use of fire in the archeological record. Hearths have been discovered in at least 22 feet of the upper levels. There is evidence that animal skins had been hung on poles near the cave entrance to protect against the icy winds circulating from the northern glacial areas (the Mindel).

From the charred bones of more than 80 species of animals including deer, goats, boars, bison, rhinos, horses and at least one giant beaver, it is known that these people not only hunted large game but also cooked it on their hearth fires. They knew that fire also hardened antler, bone and wood for spear tips. The hardened antlers were used as hammers to split stones and create sharp cutting edges. Their stone industry, called Acheulean, included flaked blades and sharp scrappers as well as the kinds of core chopper tools known from Africa as Olduwan. Stone tools were in use among all known *H. erectus* bands.

Although these Choukoutein people knew how to keep a fire, no iron pyrite is found to give evidence that they could start a fire by striking pyrite and flint. They must have obtained naturally occurring fire, managed the firepits and carried live coals as they traveled from one camp site to the next in search of new vegetation and game.

One of the earliest of 40 skulls from the Choukoutein site, dated about 700 TBP was very similar to skull TNM-ER 3733 from Africa dated about 1.5 MBP, indicating that the *H. erectus* skull changed very little during this great intervening time and space, at least in this particular case.

These large-brained humans are believed to be cannibals, presumably due to conditions of recurring food scarcity and after death by starvation or from other causes, or by selective killings, since the cave was continuously inhabited during times of these depredations. All 40 skulls, as well as the long bones of these Choukoutein people (including the 40% who were children) had been battered and broken, presumably by their own fellows, to reach the brains and marrow. The bones of animals were similarly treated.

Homo erectus knew both feast and famine. As early as 400 TBP, European bands were using fire for hunting as well as for warmth and for cooking. At Taralba, Spain, a band of hunters set deliberate fires to stampede a herd of 25 *Elephas antiquus* into a bog where the animals were killed and butchered. In France, *Homo erectus* people made wooden bowls in which they placed hot stones and water to cook their meals.

It has been estimated that by 500 TBP there were one million humans on earth. Calculations have been made that hunter-gatherers could find enough to eat in an optimum environment, with a ratio of one or two persons per square mile. In temperate regions, vegetation was only available in summer and fall, so hunting for meat was essential in winter and spring. In tropical rainy climates, humans could continue to find food year round. Thus population flourished in Africa while Europe may have been occupied by only a few thousand humans during the time of the farthest reaches of the glaciers.

There is some disagreement among paleoanthropologists about which human fossils exhibit the next stage of advancement following *Homo erectus*. One group (including Niles Eldredge and Ian Tattersall, 1982: 148) describes a post-erectus phase of "archaic *Homo sapiens*" represented by European and Middle Eastern fossils, dating from about 400 to 200 TBP. The skull and teeth characteristics are closer to those of modern humans than to those of the European Classic Neanderthals, which date from about 160 TBP and disappear about 35 TBP.

Other scientists consider that *H. erectus* may have lived in some places until about 150 TBP, and most of the humans living from about 250 to 30 TBP should be considered to be at least neanderthaloids if not *Homo sapiens neanderthalensis*. They cite statistical analysis of the similarities and differences in structure of the Classic Neanderthals and the other humans of the same period that show no greater differences than those among modern humans today. They reason that humans evolved in different areas with minor variation in structure, but that all *Homo sapiens* evolved as one species from earlier *Homo erectus* populations.

Scientists of the British Museum (1980) and others have identified 24 Neanderthal fossil sites, including at least two in Italy, one in Gibraltar, one in Yugoslavia, seven in France, three in Germany, one in England, one in Russia north of the Black Sea, one in Siberia, three in Israel, one in Iraq, one in Morocco, one in Zambia (Rhodesia Man) and one in South Africa.

The first scientifically described specimen was found in the Neander Valley of Germany, thus the name Neanderthal, now called *Homo sapiens neanderthalensis*. The area where the most numerous sites have been found (four) is in the Dordogne Valley of southwest France. The skull of the old man of La Chapelle-aux-Saints (Figure 28.8) shows the characteristics which identify the Neanderthal group: heavy brow ridges, sloping forehead, cranium low and long but generally higher than that of *H. erectus*, more pointed occipital bone (back of the skull) than modern humans, broad nose, heavy jaw, enlarged pulp cavities in teeth, receding chin, long face and large braincase. Other fossils also show that Neanderthal bodies were shorter and more heavily built than modern humans.

Neanderthaloids were the first hominids known to bury their dead, often with artifacts which seemed to indicate belief in magic or religion. They made advanced stone tools of at least 60 kinds, called Mousterian Industry. Their bows and arrows and spear throwers were of efficient design. Using bone needles, they sewed skins into clothing, and placed skins for shelter at cave entrances and in the open. They decorated their bodies with ochre, and painted animal forms on cave walls.

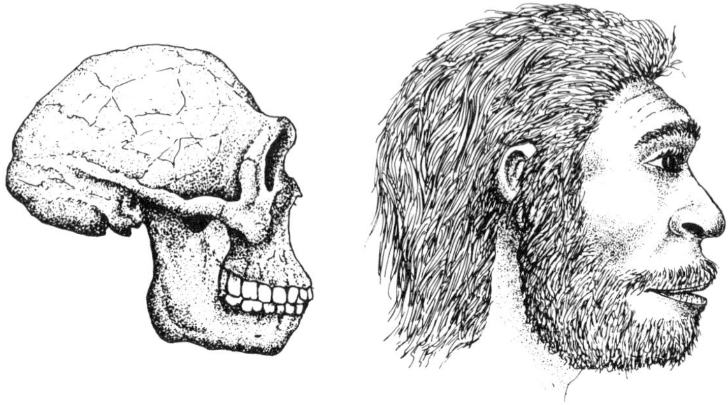

Figure 28.8 Old Man of La Chapelle-aux-Saints, *Homo sapiens neanderthalensis*.

Modern scientists now identify these beings as similar enough to us to be considered *Homo sapiens*, but *neanderthalensis* is added as a subspecies name to distinguish them from our more evolved subspecies.

Modern Humans and DNA

Paleoanthropologists collected increasing numbers of hominid fossils during the 1970s and 1980s in Africa, Asia and Europe but the question: "When, where, and from what group of predecessors did modern *Homo sapiens* arise?" remained unanswered.

The disagreements within the field continued to pit those who considered the fossils subsequent to the stage of *H. erectus* to be neanderthaloids who evolved into modern humans, against those who believed that the Neanderthals became extinct about 35 TBP and were replaced by descendants of archaic *H. sapiens* from the Middle East, Asia or possibly Africa.

Other paleoanthropologists concluded that modern humans evolved on all three continents from indigenous branches of *Homo erectus* to become archaic *H. sapiens* and then modern humans, with Neanderthals an evolving European branch.

Anthropologists *do* agree that about 35 TBP modern human beings, *Homo sapiens sapiens* arose on the Old World continents of Africa, Asia and Europe. Many deduced that humans then migrated from southeast Asia to Tasmania, New Guinea and Australia and from northern Asia to North America, populating most of the known world of the Pleistocene.

A Cro-Magnon skull from Les Eyzies, Dordogne, France (Figure 28.9) exemplifies early humans of the classification *Homo sapiens sapiens*. It has been identified with stone tool cultures of advanced Upper Paleolithic design known as Aurignacian, Solutrean and Magdalenian. Cro-Magnon art includes beautiful cave paintings of animals as well as weapons and trapping devices. Anthropologists have noted carvings and engravings in bone and ivory "of taste and sophistication" describing the Cro-Magnon culture.

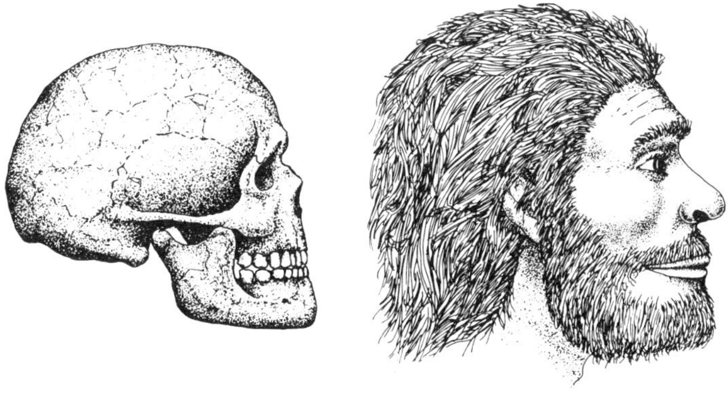

Figure 28.9 Cro-Magnon Man of France, (Modern Human, *Homo sapiens sapiens*).

These early *sapiens* carved small icons of sexually explicit female forms, thought to be associated with worship of life's essence and creation of life. These European artifacts distinguish a high point of early *Homo sapiens* culture. Other high points were reached — the first evidence of medical success — in an arm amputation in a Middle Eastern cave.

Homo sapiens occupied much of the Old World by 35 TBP and continued to increase in number, reaching an estimated 2.2 million by the end of the Pleistocene. As the Ice Age ended about 10 thousand years ago, most humans were living in small hunting and gathering bands. It is believed that the number of early humans increased with gradual expansion into unoccupied territory, moving at an average rate of about 12 miles per generation. In the late Pleistocene *Homo sapiens* began to increase their clan and tribal range by longer treks into new lands.

Thus the stage has been set for consideration of evolution in the Recent Epoch. But like a thunderclap on January 1, 1987, a group of three biomolecular geneticists broke into the monopoly of the study of human evolution held by paleoanthropologists.

Rebecca L. Cann, Mark Stoneking and Allan C. Wilson reported in the prestigious journal *Nature* (1987) that they had discovered the first female ancestor of modern humans, and that she lived in Africa about 200 TBP (actual estimate from 285 TBP to 142 TBP, later stretched to 500 TBP to 50 TBP). These scientists extrapolated from their molecular evidence that modern humans all evolved from one female ancestor and her tribe, which replaced all of the then existing hominids of the world.

The research was based upon analysis of the mitochondrial DNA (mtDNA) in the cells of 147 individuals from five areas of the world. Mitochondria, organelles of the cytoplasm of the cell, about 1,000 in number, are *not* distributed equally from a sperm and an egg to an offspring, but rather are distributed from a mother's egg cytoplasm to her offspring at the same time that the nuclear genes from both parents are joined in the initial egg fertilization event. Thus, it is theoretically possible to trace inheritance in the female line through mitochondrial DNA (mtDNA) separately from inheritance carried by the nuclear DNA contributed by both parents.

This *Nature* article and announcement that accompanied it challenged the adequacy of fossil research and offered substitution of DNA analysis to explain modern human origins.

In the continuing tradition of Aristotelian logic, either/or choices were made. Scientists argued either that fossils supplied the best evidence of human evolution or that DNA supplied the best evidence. Lines were drawn between the geneticists and the paleoanthropologists, and the battle for knowledge of the "true origin" of modern *Homo sapiens* erupted. Reputations were to be made or broken on the basis of bones or genes.

An expert on the fossil record of hominids, Milford Wolpoff fired one of the first volleys from the side of the mainstream paleoanthropologists. Wolpoff disputed the idea, suggested earlier by Linus Pauling, and now maintained by the Cann group, that mutations occur on a regular time schedule; one or two mutations per one million years. This idea was basic to the Cann group's estimate of the age of "Eve" the name publicized as the hypothetical human female ancestor from Africa.

Wolpoff also criticized the adequacy of the sample of 147 individuals and faulted the use of mitochondrial DNA from American blacks rather than from indigenous Africans. He insisted that the study did not adequately consider the effects of natural selection in eliminating certain mutations over time. He refused to believe that mtDNA evidence could show that Africans could take over Europe and Asia and become the only modern humans without any interbreeding with then existing *Homo erectus* or Neanderthals (Brown, 1990).

Wolpoff pointed out that the numerous fossils of Choukoutein cave showed continuities of skull structural development beginning at least 700 TBP. The similarities of cranial and facial characteristics proved to his satisfaction that these Asians evolved in Asia from *H. erectus* to modern *Homo sapiens*. There was no sudden appearance of African fossils replacing Asian fossils about 200 TBP.

He noted continuity of evolutionary patterns in European fossils as well as continuity of African fossil patterns. He saw "multi-regional continuity." Populations evolved in their own regions with regional characteristics, but he also saw evidence of gene flow between the regions.

Other paleoanthropologists leaped into the fray describing fossil evidence showing continuity from neanderthaloids to modern humans in the digs of Krapina and Vindija, Yugoslavia; Sala, Czechoslovakia; Spy, Belgium; Hanofersand, Germany; Shanidar, Iraq; Meadec, Maldavia; N'ame, Lebanon; Tabun, Shkul and Qafzeh caves, Israel; Morocco; the Crimea; Afghanistan; southeast Asia; Jinniu Shan and Xujiazao, China; and Omo, Ethiopia.

C. Loring Brace maintained that modern northwestern Europeans have certain skull characteristics closer to Neanderthals than to other living populations, supporting the idea of indigenous morphological continuity (Brown, 1990: 257).

This and other evidence seemed to support the concept of regional continuity rather than the Eve theory that an African population of modern humans replaced the neanderthaloids in Europe and Asia. Fossil authorities also questioned the idea that Africans replaced Asians and Europeans by means of a new culture of advanced weaponry (or by being the first humans with speech). There is no evidence in the fossil record of replacement of these populations by violence and annihilation.

The dispute among geneticists and paleoanthropologists continued to rage in the 1980s but Edey and Johanson of Lucy fame, in a popular book, *Blueprints; Solving the Mystery of Evolution* (1989) went beyond the confines of the scientific community to begin reconciliation of the genetic and fossil approaches to human ancestry.

They pointed out that the geneticists' theory of "Eve" cannot stand alone. Even assuming that an African female ancestor of 200 TBP *did* carry mtDNA characteristics of modern humans, the study of fossil bones is still required to answer some of the most basic scientific questions.

After the shock of "Eve" reached through the fields of genetics and paleoanthropology to the public, other biomolecular studies of the 1960s, 1970s and 1980s began to impinge more prominently upon the public's developing awareness of DNA and its relevance to evolution.

Biomolecular studies by Walter Fitch and Samuel Margoliash, using a variety of approaches and techniques including analyses of ABO blood groups, and analysis of the cytochrome C enzyme of a variety of animals, provided new genetic evidence of evolutionary lines. Several of these approaches depended on the concept of a "molecular clock" to estimate dates of divergence of proteins or DNA of one kind of animal from another.

This concept is based upon estimates of one or two (or three to four) mutations per one million years. If mutations occur at regular intervals of time, it would be possible to judge the length of time it would take for a DNA segment to change to its present configuration from a time in the past. However there is excellent evidence that mutations do not occur at regular intervals, even if an *average* interval may be calculated. Morris Goodman, Rodney Hunnecut, John H. Gillespie, and Carl Woese' study of RNA of bacterial ribosomes, show that mutations occur at different rates.

This is especially true when the mutations under consideration are mediated by viruses and are thus salmutations from one taxa line to another in response to environmental changes. As noted throughout *Evolution and Reason,* viruses are often directly involved in the salmutations from one taxa line to another. Therefore, I disagree with the idea of regularly timed mutations and the molecular clock concept. I believe mutations and salmutations are not regular but occur in anticipation of and also in response to both rapid and long term changes in environmental conditions. Although the research of molecular biologists is valuable as evidence of difference and amount of difference that can be statistically evaluated, the time of the changes cannot be determined by molecular evidence alone.

A University of California, Berkeley geneticist, Vincent Sarich, postulated in 1967 that the human and ape lines diverged about five MBP, rather than in the 20 MBP to 15 MBP range, previously accepted by paleoanthropologists. This new estimate was based upon analysis of the protein serum albumen in the blood of chimps, gorillas and humans. Because this protein was coded from DNA, it was used as evidence to evaluate the closeness of evolutionary lines.

Paleoanthropolgists, for more that 20 years had virtually ignored the Sarich hypothesis that the human and ape lines diverged as recently as five MBP. Edey and Johanson (1989: 367) raised one of the most significant questions to challenge the five MBP date, "So how did we get on our hind legs?"

Johanson pointed out that the knuckle walking of chimps and gorillas is a specialization requiring different adaptations of arms, muscles and hands than the adaptations required for full upright walkers. He believed that human bipedalism evolved from quadrupeds — that it was virtually impossible for humans to have detoured through a knuckle-walking stage. He concluded that the bones of the ancestors of humans and knuckle-walkers, as well as DNA studies of relationships, will be needed to shed light on his question.

Edey and Johanson (1989: 213) explored the role of DNA in making protein (heterocatalytic) and in making DNA itself (autocatalytic) — recognizing that the science of evolution is on the brink of a new synthesis of the relationships between DNA and proteins as well as the relationships between DNA and fossils.

Evolution during the Pleistocene is coming into focus. The story is told in *Evolution and Reason,* as the role of viral DNA is inserted into the discussions of fossils, proteins, mtDNA, and nuclear DNA and RNA. The paradoxes of these relationships call out for complementarity dynamics. The 1990s will usher in the most profound insights into the processes of evolution since Darwin's *Origin*. What a grand time to be alive!

29

Cenozoic Era
Quaternary Period
Recent (Holocene) Epoch
Beginning of Civilization

Human cultural evolution is the most significant development of the Recent Epoch that began ten thousand years before the present (10 TBP = 8 BC). During this time span people developed from hunter-gatherer bands to the high-rise hyper-paced civilization of the 20th Century.

The Recent Epoch began near the end of the fourth (Wurm) glacier period. Massive geological changes affected all areas of the globe (Chapter 12, Plate 12.2, # 8). At the maximum extent of Wurm, 18 TBP, so much water existed as ice-pack that the seas were 120 to 140 meters lower than today (1990s) with almost twice as much land exposed. By nine TBP, the oceans were still 60 meters lower than today (Bender, 1975: 225). As the seas continued to rise, land forms such as the Sunda Shelf, off southeast Asia, were inundated, leaving islands of Sumatra, Borneo and Java. People living in such areas were gradually forced to move inland, or became stranded on the new islands. Population densities increased in many coastal areas as tidal lands disappeared under the seas.

Today's land forms and climates were born during the changes of the Holocene. A warming trend accelerated. The Great Lakes, formed from glacial ice in North America, as well as lakes formed in the Eastern Hemisphere, were at high levels and rivers were flowing swiftly from melting snow packs. By 7,000 BC worldwide climatic conditions became similar to today's as rivers and lakes deposited rich new alluvial soils. Woodlands of oak, pistachio and olive, along with expanding grasslands began to replace the sage brush (*Artemisia*) of the Mediterranean dry belt as the climate became warmer and wetter.

Almost all orders of today's animals were present in the Pleistocene, but many animal lines became extinct in the late Pleistocene and early Holocene. However, humans continue to discover animals which appear to be hold-overs from earlier eras. Among these living fossils are coelacanth fish, from 130 million years ago, discovered living in the Pacific Ocean in 1938, and large dragon (monitor) lizards of Komodo Island, Indonesia (Chapter 22, Figure 22.6) discovered in 1910, an order probably existing for more than 230 million years.

At the beginning of the Holocene all proboscideans had disappeared from the world except two species of elephants and a few mammoths. Gone were most of the giants among beaver, deer, bison, cave bears, woolly rhinoceros, South American ungulates and the Australian marsupials. Gone were the chalicotheres, the sivatheres (a branch of the giraffe family with large branched horns) and anthracotheres (primitive water dwelling, pig-like forms).

Remaining in North America at the beginning of the Holocene were mammoths, deer, elk, antelope, saber tooth cats and other carnivores, a few ground sloths and large herds of bison.

Disappeared from North America were camels, tapirs, horses and most of the peccaries; and from most of Eurasia, rhinos and hippos. Some of these extinctions were climatically induced, but the radiation of human hunters probably was a significant factor in the extinctions of large and giant mammals (Romer 1966: 346).

The human population at the beginning of the Holocene is estimated to have been between two and 20 million, most living as hunter-gatherers in Africa but many living in nomadic bands and tribes in every continent except Antarctica. During the early Holocene the human population of the earth expanded gradually. The population is estimated to have been by one AD, 200 million; by 1600, 500 million; by 1700, 700 million; and by 1800, 900 million, despite the decimations caused by the plague of 594, Black Death in Europe in the 1300s, ravages of preliterate peoples by European diseases in the 1600s, and numerous deaths in disasters and wars. Population of the world grew exponentially to 1.6 billion by 1900, 2.5 billion by 1950, 3.6 billion by 1970, and 4.5 billion by 1980. It is estimated that world population will reach 6.4 billion by the year 2000, doubling in less than 35 years!

For most of human prehistory, population was "controlled" by natural infertility during up to four years of lactation and nursing, by infanticide, murder, head hunting, cannibalism, senilicide, war and migration; more recently by medical abortion, and only since the work of Margaret Sanger about 1921, by effective birth control methods in parts of the world. Humankind has multiplied phenomenally but has acquired wisdom and compassion very gradually.

Bernard Campbell (1976: 435) has given estimates of the average life spans of evolving peoples throughout time: an *Australopithecus* person lived only 15-20 years, Homo erectus 20-30 years, Neanderthals of Europe about 29 years, Cro-Magnon of Europe about 32 years, English people in the fourteenth century about 38 years, and United States people in the 1970s, 70 years. (The US average life span has now risen to 75.2.) Ornstein and Ehrlich (1989: 43) reported that 76 skeletons from Europe and Asia, dated about 10 TBP, showed that less than one half of these people reached the age of 21, only 12% lived to be more than 40. Not one of the females reached age 30. Most of the Holocene has been populated by very young, short-lived humans, giving evidence of the hazards of life amid dangerous animals, uncertain food supply, poor nutrition, disease, and childbirth without modern medical science.

It has been estimated that nomads, extending their range by 12 miles per generation, could cover the distance of 8,700 miles from Nairobi, Kenya, to Peking (Beijing) China in 20,000 years (Leakey, 1981: 117). At this rate, *Homo erectus* clans of the Pleistocene, expanding ever outward, could have populated much of the Eastern Hemisphere long before they apparently evolved into *Homo sapiens* in different areas, with minor different physical characteristics of one human race.

By the Holocene in certain areas, *Homo sapiens* population groups reached densities which began to create scarcities of food supply. Some of these groups by then had assumed a semi-settled way of life, especially now known from the earth mound ruins (tells) of small villages in the area of the Middle East. The earliest villages were found in Mesopotamia (Iraq) Syria, Palestine (Lebanon and Israel) and Anatolia (Turkey); and later in Edam (Iran) Egypt, eastern Europe, China, Russia, Thailand, Japan, Peru, Mexico, India and Africa. As the Holocene Epoch began, 8000 BC, humans were becoming controllers of evolution through their cultural development.

Evolution of Human Culture

Human biological evolution had reached modern identified *Homo sapiens sapiens* in the Middle East by 100 TBP, in most parts of the Eastern Hemisphere by about 35-30 TBP. There is evidence that modern humans may have evolved in South Africa as long as 60 thousand years earlier (Protsch, 1975). By 10 TBP, humans experiencing major cultural changes in several areas of the world began to accelerate the biological evolution of plants and animals. The hunting and gathering patterns that had filled human consciousness for more than 95% of hominid time on Earth began to give way to herding of red deer, gazelle, reindeer, goats, sheep, horses and cattle.

As the supply of wild game decreased, the necessity for providing new food supplies was an incentive for animal husbandry and agriculture. Humans began to gain control over their food supplies. This was not a sudden event in only one place. People developed new methods of providing food based on different environments and resources. The factors which led to cultivation and plant processing varied from region to region. In the African tropics, tree and root crops were developed first. In the Mediterranean climate of the Middle East, grains and legumes were cultivated, and small ruminants were first domesticated.

A preadaptation for agriculture is evidenced by the grinding of wild seed grains, although not all people who processed and ate wild grains later cultivated them. Grinding stones are known from 49 TBP in Florisbad, South Africa; from 44 TBP in Molodova, Ukraine; from 22 TBP in Arnhem Land Australia; from 18 TBP in Ein Aqev, Israel; and from 13 TBP in Pikimachay Cave, Peru; where people also hunted and ate horses, deer, pumas, llamas and rodents.

By 12000 BC, wild wheat and barley grains, cut by sickles made of stone and bone, were crushed on grinding stones in Egypt, but there is no evidence that the grains were planted or cultivated that early. Sheep and goats were controlled but not domesticated. In Palestine and Europe gazelles and deer were shepherded but not domesticated. Domestication of animals apparently started much earlier than the Holocene in western Europe. There is evidence from tooth wear of fossils that horses were tethered in France as early as 30 TBP, and bridled horses were shown in cave art 15-14 TBP.

Human selective influence upon sea and fresh water life was almost nil until recent decades, although the importance of aquatic animals in the diet of very early man has not been fully appreciated, especially if *Homo aquaticus* did, in fact, play a role in human evolution.

The largest human selective pressures of the Holocene were the choices made about which animals to kill, which to domesticate, and which plants to cultivate. The domesticates were developed mostly through unconscious selection of plants and animals which had undergone small mutations. In the case of grains, the mutant seeds which remained in the head — rather than the wild varieties which shattered from the head and scattered on the ground — were most likely to be harvested and planted the following year. The animals that were most calm and friendly in captivity were usually selected over their more rambunctious relatives. Human clans, as they observed significant characteristics of the plants and animals that lived in their territories, gained increasing influence over those lifeforms.

According to Daniel Zohary and Maria Hopf (1988) the first signs of cultivation occurred in areas that often still have the wild ancestral stocks of those plants. Pollen of many kinds, recovered from archaeological sites, gives evidence of changes from wild to domesticated plants. Thus it is possible to research early domestication. Barbara Bender (1975: 98, 99) presented maps showing wide distribution of wild barley in the fertile crescent, from Israel north through the upland valleys of southern Turkey, northern Iraq and western Iran. Einkorn wheat still grows wild in various areas of Turkey, mainly southeastern Turkey, and in northern Iraq and western Iran. Wild emmer wheat is concentrated in northern Israel and southern Syria. The staff of life, and the culture based upon it, evolved in this cradle of civilization.

Milestones of Cultural Development

Before Written History About 3000 BC

The following is a timeline of important cultural events which influenced evolution from 10000 BC to the beginning of history, about 3000 BC. Little is known about the evolution of animal lines which may have been influenced by human selection (other than by predation) earlier than the Holocene. Significant information is now available about selection and domestication, but it is time to also incorporate knowledge of DNA and viruses into the theory of Holocene evolution. Canids, sheep and maize are promising candidates for verification of salmutation theory, because they developed from wild species of different chromosome numbers or polyphyletic gene configurations.

> **10000 BC: Dog Domestication:** The dog is considered to be the first domesticate in several parts of the world. It is known earliest in Europe and southwestern Asia, and is thought to have traveled with humans across the Bering Straight to the New World, where it lived in Idaho by 8350 BC. According to Darwinian theory, domestic dogs (*Canis familiaris*) evolved from wild wolves, which roamed widely over North America, Europe and Asia. Wolves (*Canis lupus*) coyotes (*Canis latrans*) and jackals (*Canis aureus*) all have 78 chromosomes. Wolves and coyotes can produce hybrids and can mate with domestic dogs (Olsen 1985: 12). The pattern of dog evolution is thought to be linear, beginning only with wolves.

Under salmutation theory, it can be postulated that several dog-like genera such as hyenas, African wild dogs, wolves and foxes arose polyphyletically. Each branch could have received unique DNA or experienced recombination of DNA mediated by viruses. Although all domestic dogs are categorized as the same species, some varieties may have experienced an unidentified mutation or salmutation and may already be a new species or genus. This could excape detection unless all varieties are tested for interbreeding potential or DNA compatibility. Continuing DNA research may clarify the evolutionary lines of wild and domestic dogs and similar genera, and support salmutation theory.

> **10000 BC: Agricultural Settlement Builders:** The Natufian people of Syria, Lebanon and Palestine lived in caves and camps before a group of them became, in a village called Mallaha, the world's first known settlers. There, in the Upper Jordan Valley, a population of 150-200

people used sickles to glean wheat and barley, and used harpoons and fish nets weighted with stone sinkers to fish in Lake Hula.

The Natufian people of Jericho built the first known houses of mud bricks, stone foundations and plastered walls and floors. Some rooms contained large grain storage pits. They built the first stone walls and by 7000 BC, a tower fortress to protect their village, as their newly developing abundance of grain, and their collection of salt and asphalt (bitumen) excited both trade and also raids from their neighbors.

Scientists believe these and other early settlements were prerequisite to the rapid development of domestication which followed.

9000 BC: Sheep Domestication and Weaving: Sheep were domesticated from wild sheep (*Ovis orientalis gmelini*) at Zawi Chemi Shanidar, in the foothills of the Zagros Mountains of Iraq.

There seem to be fossil goats but no fossil sheep known prior to about 12000 BC. A test of salmutation theory may be made by considering the evolution of goats and sheep. Modern DNA analysis and experimental recombinant techniques may soon verify whether sheep with 54 chromosomes have salmutated from goats with 60 chromosomes sometime in the Pleistocene, long before either was domesticated. Only then will it be possible to evaluate clearly the relationship of the process of DNA recombination to the process of selection, among these major significant domesticates.

Charles Reed (1977: 562) speculated that the first people to domesticate lambs were probably young girls. The nurturing attitude needed was probably not then characteristic of the male hunters who only gradually appreciated the advantages of having their prey close at hand. The Natufian women and girls apparently learned to twist wool into yarn and weave yarn into cloth as they probably strove to save their sheep from premature slaughter.

Some scientists believe slash and burn agriculture occurred in Thailand, and perhaps in Japan, by 9000 BC.

> **8000-7000 BC: Trade:** Boats may have been used earlier but some archaeologists speculate that Mediterranean Sea area peoples built boats in this period to carry trade goods between islands and the mainland.

Evidence of transport of obsidian, green stones, salt, bitumen and grains between Jericho, Syria, Turkey and the Aegean Islands, substantiates this assumption.

> **7000 BC: First Agriculture in Americas:** Summer squash was cultivated in Oaxaca, Mexico — the first evidence of agriculture in the New World.

> **7000 BC: First Cities and Crafts:** The people of Catal Huyuk (central Turkey) were living in a well developed city surrounded by fields of wheat and barley, and herds of goats, sheep and cattle.

The larger of two city mounds was 32 acres in size. There is evidence of timbered and plastered houses with wall paintings, and plaster reliefs of the Great Goddess giving birth to a bull's skull (bucranium). The skull reliefs were thought by some archaeologists to be symbols of male fertility and others thought they represent a male deity. (A bull god was known in Palestine as Baal and in Mesopotamia as Gugalana, Bull of Heaven.) Weaving, woodwork, baking, copper and lead metallurgy, in addition to agriculture and stock breeding, including the earliest known domestication of aurochs was reported by James Mellaart (1967) who excavated the mound. Mellaart believes that great herds of wild aurochs (*Bos primigenius*) giant pig (*Sus scrofa*) and red deer (*Cervus elaphas*) may have attracted people to these grasslands as early as 9000

BC. Domestic pigs, barley and flax were grown in Cayonu, Turkey about 7000 BC. Cattle were also domesticated in North Africa about this time, and were milked there by 6000 BC.

> **6500 BC: Ceramics Trade:** Early ceramic ware, created in Catal Huyuk and in Ras Shamra, Syria, joined the trade in minerals and grain.
>
> **6000 BC: Irrigated Crops:** As agriculture spread from the foothills of the Zagros Mountains, the first irrigation was initiated in the Tigris and Euphrates Valleys of Mesopotamia. Agriculture then spread from Mesopotamia and Anatolia to Greece.
>
> The people of Spirit Cave, Thailand grew taro, yams, water chestnuts, the first known domesticated rice, and gourds; the latter, *Lagenaria,* may have been grown as early as 8000 BC.
>
> **6000-5000 BC: Aztec Agriculture:** Beans, squash, foxtail millet and maize (corn) were major crops of Mexico's plant domestication in the Tehuacan Valley. Gourds (*Lagenaria*) were probably domesticated earlier in west Africa, possibly entering Mexico by way of the Amazon Valley and Peru.

By this time, the combination of human hunting and adverse climate in Mexico left few animals for hunting (deer, rabbits) and none for herding. Only hairless dogs, turkeys, muscovy ducks and bees were domesticated, and human population density was very low. By 2000 BC the population had not reached 1,000. It was 3,500 years ago (1500 BC) before the Aztecs developed agriculture which could support communities. Some protein was supplied by cannibalism, which flourished until the Spanish conquest.

Most experts believe that maize was domesticated from wild teosinte in Mexico, or possibly in Peru. They generally believe that maize domestication depended upon continued hybridization from teosinte, left growing wild, in or on the margins of corn fields. The development of hybrid corn is considered to be one of the highest achievements of an early agricultural people. Both maize and teosinte have 10 chromosomes. However, there is some opinion that *Tripsacum*, which has 18 chromosomes, may be involved in the evolution of maize (George W. Beadle, 1977: 620). Under Darwinian theory, *Tripsacum* has not been considered to be ancestral to maize. However H. C. Wilkes (1972: 1075-1076) reported from his field studies in Mexico that several parallel variations in maize, teosinte and *Tripsacum* suggest gene exchange and-or parallel variation. Under salmutation theory, DNA of *Tripsacum* could have been passed by viruses into the teosinte or maize genome.

Scientists are unsure whether the agriculture of Peru predated or antedated that of Mexico. Not enough excavations have been completed to adequately trace developments in the Western Hemisphere. Between 6000 BC and 3000 BC in Peru, wild vicuna and guanaco, and domestic llama and alpaca verify mutation (or salmutation) and selection occurred.

5500 BC: Walled Villages: Populations of as many as 1,000 were in walled villages in the Balkan peninsula as well as in Egypt, Jordan, Poland, Switzerland, parts of India, and Russia.

5000 BC: Western Hemisphere — Cultivation and Domestication: By 5000 BC, stone hoes were in use in the Ayacucho Basin of Peru (Pikimachy Cave) and cobs verify that corn was cultivated, as well as gourds, cotton, sorghum, beans, peanuts, bitter manioc and pineapple. Peru was marginally richer in animals than Mexico. Guinea pigs were domesticated as early as 5000 BC.

5000 BC: China: Millet was grown in northern China, and a little later rice was grown also, but irrigated fields were not in evidence there until about 4000 BC. Dogs, pigs and chickens were early domesticates. Soy beans were first recorded about 1000 BC.

3700 BC: Domestication in India: In Rajasthan, India, there were domestic sheep, goats and cattle.

The domestication of pigs, water buffalo (*Bubalis arnee*) camels and chickens; and the training of elephants started later in the Indus Valley where there was a large settlement by 2500 BC.

3500 BC: First City-States: Sumer (southern Iraq) was the greatest, and perhaps the first true civilization, the first proximate city-states, fourteen in number. The cities had evolved from agricultural villages; the earliest known was Eridu, soon followed by Ur, twelve miles to the north (220 miles south of present Baghdad).

Knowledge of the city of Ur has become available only in the 20th Century. C. Leonard Woolley's account of the excavation of the city and cemetery of this fabulous civilization, was published in 1930, titled *Ur of the Chaldees*. The earliest graves date from 3500 to 3200 BC and later graves from 3100 BC (including those of the time of the Akkadian Semite King Sargon I, who united northern and southern Iraq into one kingdom, conquering Persia and Syria about 2600 BC). According to Woolley (1930: 86) the early graves provide:

> actual examples...that their weapons were, both in design and in manufacture, far superior to anything that their contemporaries or any other nation was to adopt for 2,000 years... By 3500 BC this civilization was already many centuries old.

Woolley's research showed that a village of mud huts (Ubaid) four miles from the center of Ur, was covered by eight feet of soil left by a great flood on the Euphrates about 3200 BC. More advanced structures, mud-brick houses of Ur were built above the buried village. One of the oldest temples known in Mesopotamia was built after the flood, about 3100 BC by the king of Ur, A-anni-pad-da for "his Lady Nin-kharsag," Mother Goddess of Ur, also know as Ki (Kee) and Inanna.

The Written Word

The most significant discoveries from the graves of Ur are not the gold plated copper beads, the precious jewels and headdress of Queen Shub-ad, not the four upright copper bull statutes, the silver goblet, the wood-inlay lyres, not the oxen drawn wheeled chariot in the King's and Queen's outer grave chamber, not even the evidence of many humans, including grooms, ministers, ladies-in-waiting and lyre players, who apparently willingly sacrificed their lives to join royalty in death.

The most significant discoveries from Ur are the writings in cuneiform text dating from 3100 BC which begin the written history of humankind. (Sumer's cultural center, Nippur, held a library of more than 30,000 clay tablets and fragments). The scribes of the kings and queens of Sumer wrote of their dynasties, their deities and their diets while Egypt was still divided into petty kingdoms, before Menes founded the first Egyptian dynasty, more than a thousand years before the birth of Abraham of Ur, and more than 2,500 years before the first books of the Hebrew Bible were written.

The first known law code was written by King Ur-Nammu during the third Sumerian dynasty about 2100 BC, at least 150 years before the law code of King Hammurabi of Babylon (Iraq). It was written at least 300 years before the Hittite (Turkey) Code which soon after 1800 BC inaugurated secular law, the right of subjects to make their own laws, and the jury system. The code of Ur-Nammu was operative for about a thousand years before the Mosaic Code, and about 2,600 years before the Roman Code of Justinian.

Ur has the greatest significance as one of the earliest sources of the written word, and the rule making which followed, both of which are essential to understanding of cultural evolution. From the time of Ur forward to the present, we have written records to assist our understanding. The scribes of the kings and queens of Sumer first evolved cuneiform writing from earlier pictographs and symbols, but it was Henry C. Rawlinson and others who only recently translated the cuneiform texts that are known to us today (Samuel Noah Kramer 1967: 121).

The rule-making of these early law codes surely influenced the development of the concept of the 6th Century BC Greeks: that just as there are laws which govern human relationships, there are natural laws which govern natural processes. The power of the written word increased to divine proportions in the centuries that followed Ur, even as the spoken word was early accorded divine power. An ancient Egyptian papyrus states:

> There was a time when no one and nothing existed except the creator. A desire came over him to create the world and he carried it into effect by making his mouth utter his own name as a word of power; and straightway the world and all therein came into being. (Briffault 1927, 1959: 32).

Prior to the Holocene, humans understood very little about their influence on the proliferation or demise of species. It is thought that humans did not understand the connection between coitus and birth until the beginning stages of animal domestication. But with domestication, learning about animal breeding led to the concept that males "fertilize" the female seed in the womb to support growth, as humans "fertilize" plant seeds to make them grow. Although these early people realized there was relationship between human sexual activity and birth, some pre-literate human populations today still do not recognize the connection.

As written records of human activity started only about 3000 BC, knowledge of the earlier Holocene is gained mainly from interpretations of archaeological and anthropological evidence. Many interpretations have been made, mostly within the last two centuries, and especially within the last fifty years. Knowledge of the *Homo habilis* stage of humankind is limited by the findings of bones, stone implements and evidence of primitive shelters. Knowledge of *Homo erectus* is limited by findings of bones, stone implements, evidence of use of animal hides, cooking by fire, and cave shelters. There is evidence both of cannibalism and cooperative living. It is only in the time of *Homo sapiens neanderthalensis* that we find evidence of human burial with the implication of human reasoning beyond that needed to supply food, shelter and clothing.

Evolution of Human Reason

Evolution of human beings began to take on new dimensions in the Holocene. Physical changes, which are the criteria for evaluating evolution of animals, are totally inadequate to evaluate the changes in humans of the Holocene — significant changes in reasoning and understanding. Changes occurred that set humans so far apart from other life, that as they began to control plants and animals directly they also began to gain increased control over themselves. For better or worse, reasoning power advanced along with belief in magic or in an afterlife, beliefs exemplified by burial artifacts.

One of the oldest known human burials, 60 TBP, was in Shanidar Cave, Iraq (Richard Leakey, 1981: 153). From pollen grains it is known that the body of a man was buried with seven kinds of yellow, white and blue flowers. We may surmise that the flowers were placed there by a woman, for women have been known from earliest times as those who cared for the dead. It seems natural to suppose also that gatherers, rather than hunters gathered flowers for a burial.

Another Neanderthal burial, of a young child of Teshik Task, Uzbekistan, was found with six pairs of ibex horns surrounding the head. Hunters probably supplied the horns

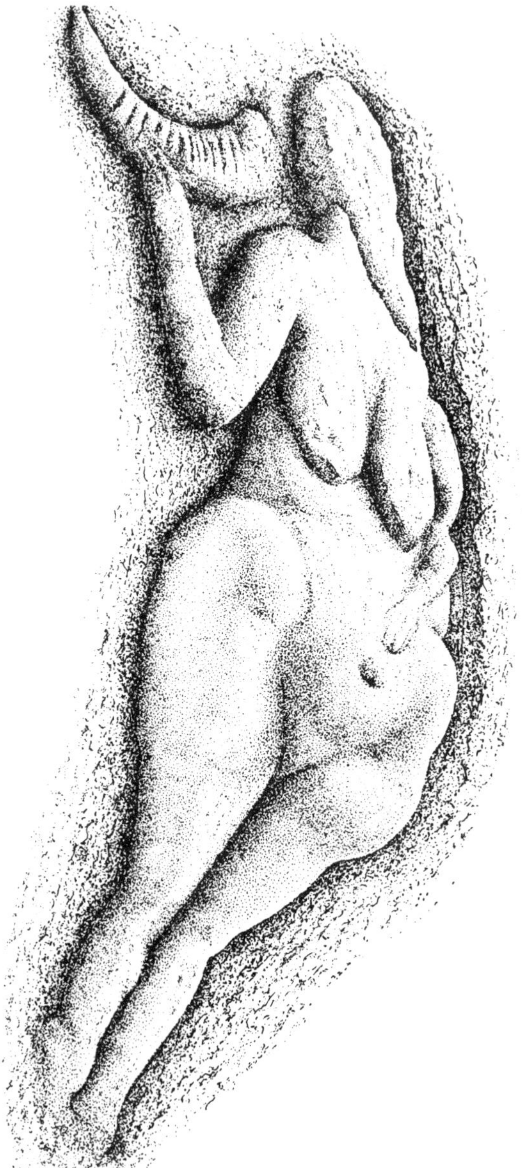

Figure 29.1 Venus of Laussel, approximately 18 inches high. (After Giedion).

which were often analogously associated with the moon, symbol of time and perhaps of eternity. This burial may also suggest female influence because, in a slightly later time period, an image of a bison horn, held in the right hand of the Great Goddess "Venus of Laussel,"was engraved in a sancuary adjacent to a mountain shelter in France. This suggests the great significance of the horn-moon symbol. In the Venus engraving, the horn had 12 carved lines, markings perhaps representing the lunar months. It may have been the first calendar, one associated with the periodicity of time as measured by a phase of the moon and possibly coordinated with occurrences of the female menses (Figure 29.1).

Earlier than 50 TBP humans often carried the skulls of their dead, or skulls such as those of cave bears, to their new habitations. Human burials became common after about 50 TBP (Denise Carmody 1979, 1989: 77). Often the red-ochre-decorated bodies or bones were accompanied by food items and jewelry or weapons, possibly evidence of belief in a life after death.

As humans evolved as *Homo sapiens neanderthalensis* and *Homo sapiens sapiens* from the Pleistocene into the Holocene, the preponderance of evidence of African, European and Middle Eastern art was cave paintings of natural appearing animals with stick figures of humans, and miniature carved statuettes of voluptuous human females. Each individual painting found in the 200 or so caves of northern Spain and southern France and 88 caves of northern and southern Africa and Tanzania is subject to the interpretation of those who study it. Expert opinion appears to be that these paintings exemplify sympathetic hunting magic and may have been used to initiate young males into the secrets of the hunting culture.

Each miniature female statuette of the thousands found from Spain to Siberia (Gimbutas 1974, 1982: 11) has a story to tell of its meaning for the development of human reason. We can never *really* know the artist's meaning. We can only speculate. Volumes of speculation and interpretation have been written. Until very recently most of the authors of these speculations have been men. In the last five or six decades, women in increasing numbers have entered the fields of archaeology, anthropology, ethnology, archaeomythology and art history, and female interpretations of the arts and artifacts of the past are adding new dimensions of understanding. It is becoming evident that human beings, male and female, have lived in complex complementary relationships and their art and artifacts need to be viewed from complementarity dynamics perspectives.

There have been a number of men, careful scholars, who cleared the pathways for contributions by women. Johann Jacob Bachofen (1860, 1967) a Swiss historian, author of *Myth, Religion and Mother Right,* was one of the first to suggest that the archaeological and ethnological evidence of many cultures shows that human beings organized their clans in matriarchal patterns for thousands of years before humans conceived of patriarchal clan organization.

Evolution of Patriarchy

Until well into the Holocene, humans generally lived in clans of female-recognized family lines and matrilocal residence. Males were clan members, and may have been clan leaders, but within most human clans they were unrecognized as contributors to the genetics of human beings until about the time of the Greek philosophers. Of the several hundred preliterate and literate cultures studied in the nineteenth century, it was found that most ethnic groups either continued to be organized in complex patterns of matriarchy or showed tell-tale signs of prior matriarchal organization, sometimes associated with female deities.

Robert Briffault (1927, 1959: 15) pointed out that Sigmund Freud's theory of projection may explain "why those who are preoccupied with the mother figure tend to envision their deity as a mother, while those preoccupied with the father tend to postulate father deities." Briffault conjectured that when perception of the moon as a goddess (Luna) changed to perception of the moon as a god (Lunus) social structures were changing correspondingly, from matriarchy to patriarchy.

With these ideas in mind, women have begun to ask: "What were women's contributions to evolution and human history? What were the activities and reasoning patterns of women that have so long been interpreted only by men?" (Lerner, 1986; Stephenson, 1988).

If the female statuettes, 30,000 or so located in Eastern Europe alone, dating from 7000 BC to 3500 BC, created in wood, bone, clay, calcite, chalk, pumice, alabaster, limestone, volcanic rock, white marble, ivory, copper and gold, were made by men and interpreted by men; then females seemed to be glorified in the fertility of pronounced bosoms, buttocks and bellies.

If the small statuettes were made by women, to be held in the palm of the hand, perhaps they promulgated female desire for safe births, for well nourished, fertile bodies, for sacred generation of human life — perhaps extrapolated as petitions to the Great Goddess for generation of abundant plant and animal life.

It is likely that both males and females participated in the production of the Mother Goddesses, beginning perhaps with the statuettes of Vogelsherd Cave, Germany, dated 32 TBP. The first known clay-fired animal and human figurines from Dolni Vestonice, Czechoslovakia, have been dated about 27 TBP. The production of goddess figures continued for those thousands of years during which only females were known to participate in the generation of life. Males probably dug the goddess-shaped tombs and raised the stone megaliths of temples from 4000 BC to 1900 BC that celebrated the Great Goddess in England, Denmark, Sweden, Iceland, Spain, Portugal, France, Greece, Crete and Malta.

The prehistory of the human race under matriarchy was essentially a balanced complementary system of male and female interactions. It could not have been as divisive as the historical transition period from matriarchy to patriarchy which occurred from about 3500 BC to 1500 BC in the Middle East and parts of Europe. This transition is continuing to occur among preliterate peoples in many areas of the world today. The great evolutionary implications of the transition from matriarchy to patriarchy have yet to fully surface. The significance of this transition is just as profound as the evolutionary significance of DNA and RNA, for the influence of the new-found male power over women and other lifeforms involved vast cultural changes leading to evolutionary dead-ends and accelerated species extinctions.

It is now generally believed that the major division of labor between males and females which occurred from at least 1.7 MBP to the beginning of the Holocene involved male weapons production, protection of the clan, and hunting of animals (estimated to provide on average about 30% of the diet) and female gathering of water, fuel (and plant foods estimated to be on average about 70% of the diet) and cooking, shelter building, clothes making and child care. Of course, there must have been variations within this general pattern, as there are variations in cultural patterns today.

There is evidence that these patterns continued until about 8000 BC, with women probably beginning the cultivation of plants, and young girls, the domestication of the young of animals abandoned by the pack or herd. Male roles evolved from hunting to herding to domestication and animal husbandry. In some cultures women raised sheep; the men, cattle.

Gradually, as development of the plow, domestication of oxen to pull it, and the rise of Aryans (plow people) overtook use of the digging stick, agriculture became productive enough to engender a surplus. From about 3000 BC, men in some areas became agriculturalists, as well as herders, husbandmen, traders and warriors.

As the male role in producing a surplus of food increased, their control over plants, animals and women increased. There is some evidence that the "races" of Aryans and Semites were first identified by occupation: plowers and herders, rather than by physiology, although some experts have spent much unproductive time trying to distinguish the long-headed Aryans from the round-headed Semites found in age old tombs.

The Egyptians identified four human classifications by use of color in their tomb paintings (Casson 1940: 23): black Africans, Europeans with blue eyes and fair beards, yellow Semites and red Egyptians. Red presumably was used to indicate everlasting life, for blood and red ochre were used as early as 200 TBP for painting bodies of the living and the bones of the dead throughout Europe, North Africa and the Middle East. Little record of different human groups developed until the time of Herodotus, the first historian and anthropologist of 5th Century BC Greece, who travelled widely and recorded his impressions of different peoples and customs.

The reasoning system in which analogies were drawn between deities and heavenly bodies, between humans and animals, between living humans and the dead, between humans and spirits, gradually began to give way to more pragmatic reasoning patterns based upon observed results of agricultural trial and error. Westerners have often thought that the most significant major breakthrough in human reasoning occurred when the Greeks philosophized about questions of good and evil and the nature of reality, and created a system of logic, supplementing and succeeding the Greek poets who wrote of the trials and tribulations of humans, heroes and gods. But the "Golden Age" of Greece, which occurred from about 600 BC to 200 BC was long preceded by the diverse cultures of Mesopotamia, Anatolia, Egypt, Europe, Thailand, India and China in which agriculture and animal husbandry were training the human mind in scientific evaluations.

It was the 6th Century BC, however, before the Greeks began to fathom that human beings did not arise solely from the coagulated menstrual blood of females, or as still believed in some quarters, originated in females due to the influence of spirits of one sort or another.

In typical Aristotelian logic of either/or choices, classical Greeks discarded the concept of female generation of life and decided that the male plants the living seed in the receptive womb of a female, thus males alone, rather than females alone, create life. Bachofen (1860, 1967: 111) noted that the Goddess

Athena (Aeschylus in *Eumenides*) "set motherless paternity in place of fatherless maternity."

The period of transition is marked by confusion regarding motherhood and fatherhod. This is exemplified in the stories of gods and heroes; in the "virgin births" of Apollo to Leto, Dionysus to Semele, Jason to Klymenes (Briffault 1927, 1959: 87, 94).

It is also exemplified by uncertainty regarding fathers of the legendary King Romulus of Rome (753-716 BC) King Ancus Martius (641-616 BC) and the Etruscan King Servius Tullius (578-534 BC). The Latins, being certain of their maternity, took the name of Latia, the wife of King Saturnus.

It was not until 1683 AD that Anton van Leeuwenhoek looked through his microscope, discovered and described spermatozoa. It was the late 18th Century before scientists realized that the male sperm joined with the female egg in generation of new life.

The earlier transition from matriarchy to patriarchy was not based upon knowledge of the male contribution to generation of life. It was based on both male physical domination of women and the new found commercial power of males. It became possible, as males gained control of food surpluses and trade, to overturn the existing Mesopotamian belief that the local Goddess owned the land, the grain and the temple herds.

The priestesses of the various goddesses fell under the control of the priest-kings of new gods and were induced first to share, then transfer the power of the Earth goddesses — to the new sky gods. This changed the concept of female spiritual power and economic ownership, to the concept of male spiritual power and economic supremacy. Patriarchy arrived with this transfer of power and control over the matriarchal agricultural and religious roles of women, to the roles of men, not only in food production and ritual — but especially in the control and redistribution of the economic surplus in trade and war.

These early societies moved from dynamic balance of the complements of male and female roles, with matriarchal philosophy of communal sharing and legitimacy of all mothers' children, to patriarchal philosophy. The new philosophy involved saving of excess production for wealth in personal property, and eventually to private ownership of real property. As fathers began to recognize their own children, the concepts of formal marriage, inheritance, adultery, illegitimacy and prostitution emerged.

The concept of prostitution changed from the male devised sacred prostitution and Ritual Marriage of the high priestess of the Goddess in the ordination and legitimization of priest-kings, to profane prostitution and slavery of foreign, poor, and unprotected women. Later, some men were also captured and held as slaves, rather than all of a group of warriors being killed by the victors in battle.

The development of patriarchal societies reduced the equal participation of women in economic activities which had characterized human prehistory prior to the Holocene. It led to class distinctions, with the few women of the upper classes leading lives of leisure, often functioning as sex objects rather than as economic partners. Poor women often could not survive, no matter how willing they were to work, without a father, mate or slaveowner.

The redevelopment of an equitable female economic role and just compensation, along with increased status of women, occurred in some agricultural societies of Europe and North America, but was repressed in the early days of the Industrial Revolution, only beginning to reappear in the 20th Century. Darwinian philosophy of natural selection has been used to maintain the unequal balance of power between male and female complements.

Evolution of Scientific Understanding

Each scientific and philosophical revolution has been won at the expense of existing orthodoxy and those initiating new ideas have often suffered persecution from those supporting the old order. The revolution against orthodoxy which marked the entrance of classical Greek science involved the successful separation of the concept of material "elements" from the grasp of the gods and goddesses. The element *earth* was wrested from the Great Goddess of a thousand names: Ki, Nin-kharsag, Inanna, Ishtar, Ashera, Gaia. The element *water* was squeezed from Nammu, goddess of the primeval sea who gave birth to heaven and Earth, and from Ea and Enki, gods of waters. The element *air* was filched from the sky god An, the air god Enlil and air goddesses Ninlil, Hera and Juno, and from the thunder god Zeus and the Pillar of Cloud. The element *fire* was snatched from Zoroastrianism, and from the Pillar of Fire, the God of the burning bush.

When it became possible to think of material "elements" as things in themselves, not as spheres of the deities, science began. Leucippus and Democrites went a step further and conceived that all things are constructed of atoms, small parts of larger wholes. This idea lay beneath the surface of history until resurrected by John Dalton in 1808, blossoming into awesome significance only in our own atomic age.

Thousands of scientists and writers have contributed parts to the puzzle of biological and cultural evolution. Since Darwin set the pattern for finding evolutionary advances according to natural selection, cultural evolution was also evaluated on the same Darwinian basis by many writers. Social Darwinians, who considered some people to be more "fit" than others, and therefore more likely and "entitled" to survive, poisoned the atmosphere of objectivity. This led to inappropriate conclusions — about the significant contributions of hunting-gathering people of early cultures, and of various racial and religious peoples of current cultures. Darwin's survival of the fittest philosophy also gave tacit approval for man's destruction of wild plants and animals as man began to influence Earth's evolution.

Futilely searching for Darwin's missing links, scientists believed the gaps in evolutionary lines were due to inadequacies of the fossil record, when in fact we know now that understanding of these saltations depends on recent discoveries of DNA mutations and salmutations. It is only now that it is becoming possible to consider how DNA and viruses have altered lifeforms in conjunction with other natural processes, often influenced by human choices. The decades-long patterns of Darwinian thought about life — that must be changed by recognition of the influence of DNA and viruses — cannot be turned around in a single action, but the process has started.

Oswald Avery first published in 1944 his tentative findings that DNA, not protein, is the product that carries the hereditary potential in both cells and viruses (Radetsky 1991: 142). This was a turning point in the history of science. It is too late to rely on the patterns of evolutionary reasoning set out by Darwin. This current Holocene time frame is too short and too imminent for us to recognize in it the long-term meanings of recent hereditary changes. It is too early to be sure of all the major implications of recent discoveries; but scientists in increasing numbers are reviewing recent developments and striving to devise new theories.

We know that living matter is made up of cells, that cells are made up of nucleus and cytoplasm, that this matter consists of DNA, RNA and proteins. DNA and RNA are the carriers of the patterns for life; proteins are the building blocks of life; and protein enzymes do the bidding of DNA and RNA. Scientists working in the medical field are gradually developing an understanding of viruses. The concept that the DNA and RNA of viruses interact with DNA and RNA of cells has only recently reached into public consciousness.

The realization of the hereditary principle of cells — that the genes do not remain in reliable order and pass in stable configurations in the germ cells from one generation to the next, was discovered by Barbara McClintock as early as 1931, more than twenty years before the discovery of the double helix structure of DNA.

McClintock's work on the genes of maize generally went unrecognized by the scientific community, and was not understood for decades, probably because she was investigating mobile elements of DNA (jumping genes) before DNA was identified as nucleotides that make up the genes.

McClintock and others found that the Mu virus functioned in the same manner as "transposable gene elements" in rearranging the hereditary material of the cell. She also discovered "controller" and "operator" elements of DNA which still are not totally understood by scientists. She belately received a well-deserved Nobel Prize in 1983. McClintock's (1984: 634) far reaching prediction is:

> In the future, attention undoubtedly will be centered on the genome, with greater appreciation of its significance as a highly sensitive organ of the cell that monitors genomic activities and corrects common errors, senses unusual and unexpected events, and responds to them often by restructuring the genome.

The mysteries of DNA functioning are still vast, and scientists are working to unravel them — to win new victories over disease, and to gain new knowledge. It is becoming possible to control disease by genetic engineering of DNA and RNA of viruses and cells. Some scientists are turning attention to the evolutionary implications of this new knowledge, but progress is slow. James Shapira (1983: xiv) wrote, "...we have not yet learned how to incorporate the basic, ubiquitous phenomena of virus-cell and parasite-host interactions into a more comprehensive picture of heredity." But contemporary writers are beginning to express a sense of the evolutionary potential of genes and viruses.

Michael Syvanen, a scientist with the University of California, Santa Cruz, said that it was 1983 when he had the first notion that "perhaps viruses do have a grand purpose after all" (Radetsky 1991: 396). Other scientists are experiencing similar revelations. Christopher Wills (1991) in *The Wisdom of the Genes; New Pathways in Evolution*, reviewed some important research and its directions. However, he continues to emphasize natural selection as the guiding force in evolution.

This thinking continues because of the scientific orthodoxy of belief in natural randomness without acceptance of natural purposefulness. It is only when the complementarity of chance and purpose is recognized that mutational and virus-induced processes will be given their balanced place in conjunction with natural selection and culturally induced selection.

Aristotelian either/or, sequential logic is impeding full understanding as scientists continue to debate the relationships of selection and mutation and the primacies of evolutionary processes: Did DNA come first or did RNA come first? Did proteins lead to nucleotides or nucleotides to proteins? Did viruses lead to "jumping genes" or "jumping genes" to viruses? Rather the question should be: "Are the processes of evolution understood better only by searching for the sequences of events and deciding between alternatives, or also by envisioning myriad complementary relationships?"

Should the concepts of mutations (or salmutations) and selection be considered as complements? The first step in examining this question is to apply the criteria for determining complementarity. Mutation and selection are (1) two concepts which (2) may appear to be mutually exclusive (3) need different criteria to judge their essence (4) are a whole (5) function together in dynamic equilibrium (6) interact in reciprocal moments of time.

The two concepts, mutation and natural selection appear to be mutually exclusive for they have been traditionally separated into the disciplines: microbiology-genetics and paleontology-evolution. The criteria for judging mutations lie in chemical, physical and microscopic analysis of organisms, while the criteria for judging selection lie in analysis of bone/fossil placement in geologic strata.

Mutation (or salmutation) and selection constitute a whole in the concept of evolution. Mutation requires selection and selection requires mutation. The two dynamic processes interact at many levels in reciprocal time frames. Thus mutation (or salmutation) and selection may be considered as complements. With the help of complementarity dynamics we can devise and evaluate new theories and appreciate the essence of DNA and RNA which patterns the evolution of lifeforms. Humankind has yet to fully grasp the complementarities of life in the universe.

I believe that it is only now, as the essence of complementarity, represented in the basic complementary functioning of DNA of cells and viruses, becomes common knowledge, that the human male and female complements may fully gain in society the integrated dynamic equilibrium which reason demands.

The discoveries of the 20th Century may yet lead to world-wide communication of knowledge and the resultant power to understand and guide evolution for the welfare of the planet. Then humanity, as a whole, may face the fact that over-population of the Earth must be controlled. The movement for voluntary population limitation will be effectively led, not by those who have only so recently recognized their generative function, and in far too many cases have still not fully accepted their responsibility for it, but mainly by women whose lives are, and have always been, involved in giving birth and nurture.

The right to choose whether or not to bear offspring is an inalienable right that women have earned through three and a half million years of "virgin" motherhood and five thousand years of suffering under the excesses of patriarchy. That suffering included the torture and murder of several million women called "witches" (from 1350 to 1750) under the aegis of patriarchal fear of women's "magical" powers, and the extensive practice of female infanticide still inflicted in some cultures.

The inalienable right to choose is accompanied by its complement: the awesome responsibility to bring human population into dynamic equilibrium with the resources available to sustain it. As women and their mates choose optimum family size, women and men will envision and plan optimum densities and organizations of neighborhoods, communities, cities, counties, states and nations.

As world population is optimized, countries can move from growth economies to dynamic equilibrium economies. In this process men's roles will change emphasis as they move the dynamic balance points of culture from exploitation toward equity, from excesses of competition toward improved cooperation — from war preparation into peace processes. The protection and nurture of humans and all other life on planet Earth must be a shared responsibility of all humankind.

30

Evolution into the 21st Century A New Theory: Complementary Systems Evolution

The foundation of this new theory of **Complementary Systems Evolution** is six assumptions:

Assumption 1. Wholeness

The universe is whole and everything is interconnected. There are no islands or unconnected units, although complements may *appear* to be mutually exclusive or paradoxical. Perceived discontinuities cannot be explained through monistic logic.

Assumption 2. Coherence

The universe, the Earth and the atom function coherently. They are open to infinite investigation of webs of complementary relationships and hierarchical relationships of subsystems, systems and superordinate systems.

Assumption 3. Intelligence

Intelligence permeates all of life. It is intrinsic in humans and all other lifeforms. Humankind, in its finite intelligence, strives to understand infinite intelligence, the essence of the universe.

Assumption 4. Mathematical Structure

The relationships in nature are frequently discovered mathematically before they are understood scientifically and-or philosophically. Nevertheless, the miracles of nature may be productively contemplated without recourse to higher mathematics.

Assumption 5. Microorganisms and Viruses

New theory must incorporate new knowledge of microorganisms and viruses. These findings are in early stages of development; only the outline of a new theory of microorganisms including virus participation in evolution is possible at this time. Except for concepts of symbiotic evolution, theory in the United States has, until very recently, mainly pertained to lifeforms above the microscopic level.

Assumption 6. Unknowns

The structure and functioning of the universe is still beyond human comprehension. It is not yet known how much humans will be able to learn. Much remains unknown about the solar system and about the Earth before the existence of its lifeforms. There is still much to learn about evolution. Humans are entitled to consider the evidence, evaluate various theories, and experience their own new insights.

General Principles

This new theory of **Complementary Systems Evolution** is founded on five general principles:

Principle 1. Complementarity Dynamics

Evolution is patterned in complementary processes. Cooperation (symbiosis) and competition function together in dynamic equilibrium. Organisms have combining tendencies just as atoms have combining tendencies. Simple organisms, with their characteristic DNA, have combined with other organisms to form symbiotic associations, both as new organisms and as symbiotic parts of cells. Many of the mechanisms by which symbiotic combinations and interactions occur, are still unclear, but the term *symbiotic processes* is an appropriate designation for them. Competitive processes, identified by Darwin and his successors, should not be neglected in the euphoria of learning about symbiotic processes. Evolutionary changes are interwoven in networks of interactions of complementary patterns and-or processes. In analyzing complements the student may focus on one or the other separately but their complementarity is in their interactions. *And-or* is the symbol for complementarity, e.g.:

positive electric charge and-or negative electric charge
protons and neutrons, and-or electrons
atoms and-or molecules
molecules and-or macromolecules
elements and-or compounds
one strand DNA and-or two strand DNA
DNA and-or RNA
DNA/RNA and-or proteins
nucleus and-or cytoplasm
bacteria and-or viruses
prokaryote and-or eukaryote
cells and-or organs
organs and-or organisms
organisms and-or ecosystems
time and-or space
matter and-or energy
body and-or mind

Analysis of isolated skeletal features such as teeth and bones was a necessary first step in paleontological research, just as the structure of DNA was a necessary first step in molecular genetics, but alone, each restricts vision. Perception of complementary structures and functions studied in biomes and laboratories, facilitates comprehension of the vast interrelated patterning of evolution.

Principle 2. Progress and Stasis

Organisms evolve progressively (toward more complex or highly specialized forms) yet some remain relatively unchanged over eons, and some regress. The fossil record proves, generally, that more complex lifeforms succeeded simple ones. *Progressive* does not necessarily mean *better* in the sense of survival and continuing evolution in this world. Most of the highly specialized or complex lifeforms have vanished from the Earth. There is a beginning and an end in the fossil record. But there are Monera and Protoctista which have continued to live from earliest times, dividing asexually as simple forms and as symbionts of more advanced lifeforms.

Principle 3. Saltations

The saltations in the fossil record represent reality. It is time to accept the evidence that at least some, and possibly many of the major gaps in the fossil record will never be filled by the discovery of so-called "missing links." I believe many of these hypothetical "links" have never existed. Gaps are gaps. Just as an electron cannot be located between two of its orbits, and there is no "missing link" insect — between the caterpillar and the butterfly — there are no organisms to be found in the saltations between some of the genera known in the fossil record. These saltations are evidence that DNA has reorganized and-or has been reorganized many times — so radically that truly new genera were formed. The DNA of those new organisms contained the same four DNA nucleotides as predecessor organisms, but their new configurations were different enough that the new organisms required a classification separate from any predecessor organisms. This process, in which recombination of DNA of genes and chromosomes occurs and produces a new taxon line, often a new genera, is a *salmutation.*

Principle 4. Coherence and Purpose

Natural processes and-or matter, pattern and-or are patterned coherently. Coherence is the function of the complements of determinism and randomness. Coherent patterning, whether of atoms, viruses, cells, prokaryotes or higher lifeforms, involves processes which are purposeful, statistically predictable and generally progressive. Because DNA moves purposefully, its processes result in purposeful functioning of other parts of the cell, of the organism and of the ecosystem of which it is an integral complement. From atoms to humans, functioning patterns are purposeful.

Principle 5. Purpose of Evolution

The grand sweep of evolution is purposeful. Although humans may not yet understand the comprehensive purpose of evolution, it is patterned in such beautifully intricate interacting wholes, that logically, it cannot be considered to be without purpose. When humans fully understand evolution, they may discover that at least one purpose is a paradoxical educational system for lifeforms that both participate in it and regulate it. Is preparation for the next evolutionary stage another possible purpose? To say evolution is without purpose is to deny the reality and significance of the infinite and exquisite patterning of life.

Specific Principles (Theory)

It is beyond the scope of this book to consider all the specific concepts of evolution which flow from the systems of logic used in constructing the above general principles. Much additional material for consideration is suggested in previous chapters. The significance of symbiotic processes should need no further amplification. Their importance is evident.

Saltations are a more complex problem. Until it is well understood how DNA reorganizes and-or is reorganized, the paradox of saltations will continue to confront those scientists who are held in the grasp of monistic logic. Perhaps a new generation of scientists with the resources of genetic understanding will face the fact of saltations. Nevertheless, I postulate four specifics which are the essence of my theory of **Complementary Systems Evolution:**

Principle 1. Complementary Origination of Bacteria and Viruses

Bacteria and viruses originated in a complementary, symbiotic system. Fossils of bacteria are the first known lifeforms. The original fossils appear to be like living bacteria. Bacteria have existed continuously, some reproducing virtually unchanged for more than 3.8 billion years. Because viruses cannot function apart from a host, and can transform a host, they must have arisen together.

Principle 2. Mutations and Salmutations

A mutation is a recombination of DNA that forms a new genotype (inheritable gene structure) and sometimes a new phenotype (characteristic body configuration) that may be identified as a new species.

Salmutation of a lifeform is macromutational and-or megamutational process of recombination of DNA in which a new phenotype, a new taxon line or genus is formed.

From the time of earliest life, mutations and salmutations, some mediated by viral insertion and-or recombination of DNA and-or RNA, have evolved bacteria and cyanobacteria into the DNA/RNA of myriad advanced organisms of Earth. DNA, RNA, viruses (plasmids, episomes, transposons, F factor and other forms of viruses) act, and-or interact through processes of nucleotide insertion, deletion, duplication, transposition and crossing over; regrouping codons, genes and chromosomes. These *natural* processes of recombination have been identified as scientists discover *artificial* methods of genetic engineering. This is analogous to Darwin's identification of *natural selection,* as 19th Century breeders perfected *artificial selection* methods.

Principle 3. Salmutation — Significance

Salmutation is a major mechanism of evolution. The results of reorganization of DNA are evident in arising new organisms, which often seem to combine characteristics of several different pre-existing lifeforms, and also show unique innovations. This fact is clearly evident in the fossil record, but its significance has not been fully appreciated.

Salmutation, mediated in reciprocal moments by viral DNA/RNA, is a process which often appears to be specified and controlled through the virus. This is in contrast to the use of cellular machinery for virus self-replication, which is a separate process not under consideration here.

However, in both cases, it appears that crucial "choices" may be made via DNA/RNA of the virus, rather than via DNA of the host. This is probably not exactly the case, because even in complementary systems reasoning, humans often do not observe or contemplate the multiple aspects or processes of nature, seeing only one at a time, and therefore may miss significant interactions. As complementary concepts are widely applied, it should become possible for individuals to switch more readily between contemplation of host DNA and viral DNA/RNA movements and their interactions. I anticipate more surprises as these interactions are revealed.

Principle 4. Responsive Salmutation

The DNA code is a self-organizing and-or organized, random and-or determined program which patterns structure and function of lifeforms in anticipation of and in response to environmental change.

The code structure of a species from generation to generation may remain essentially the same for millions of years. It may also change radically in a reciprocal moment, in a reorganization of the germ plasm which determines a new characteristic such as bacterial resistance to streptomycin. The reorganization of DNA which occurs in a salmutation appears to be in harmony with infinite intelligence, programming the new taxon to fit into a specific ecological niche, in the myriad niches of nature.

The difference between mutation within a species and a salmutation creating a new genus, may not be recognized immediately, but only become known in retrospect, after a new line of ascent becomes evident. An example of this process is the variations in the structure and function of dog-like genera such as hyenas, African wild dogs, wolves and foxes. Even modern dogs may have experienced mutations and possibly unrecognized salmutations as well as Mendelian inheritance processes. Study of the DNA codes of canids and similar genera may give clues to the processes of DNA recombination.

Humans have mediated the evolution of dogs, by selecting and breeding those individuals that had mutated or salmutated in the characteristics desired by them. So far, humans have merely encouraged the propagation of those types which represent recombined DNA. In the future, it will become possible to choose to recombine DNA in different new patterns to produce totally new animals: dogs, or other new taxa. Humans have already produced frost-resistant strawberries, pest-resistant walnuts, longer-lasting tomatoes, the enzyme interferon, and miniature mice. Scientists have inserted human genes into bacteria to produce insulin. What awesome power and responsibility: the capability to redesign lifeforms!

Evolution of Organisms and Viruses

Humans have wondered whether non-living atoms and molecules could evolve into complex living cells and organisms. They have wondered whether viruses should be considered to be a stage between non-living and living. Since viral DNA/RNA can function only within the cell of a living organism, are viruses alive (perhaps in suspended animation?) outside of a cell? Are they possibly a link between macromolecules and cells? Such questions of the relationship of non-living and living are posed in monistic logic — demanding a sequential answer.

In complementarity dynamics, it is conceivable that atoms and molecules could organize and-or be organized into macromolecules which exhibit different traits, that these macromolecules could join in symbiotic relationships to overcome hazards of the environment — forming both virus-configuration and cell-configuration DNA/RNA. The points of transition between non-life and life are reciprocal moments, not subject to definition as sequences.

Virus-like and cell-like DNA/RNA could have interacted in competitive and cooperative ways to initiate processes of evolution. DNA/RNA preserved and-or was preserved in its unique characteristics in individual organisms and viruses, and then formed and-or was formed into ecosystems of various organisms, large and small.

The transition from bacteria to more complex lifeforms involved numerous multicelled recombinations of DNA/RNA and variations of reproductive strategies. Evolution appears to be not only competitive and symbiotic but also experimental. DNA, both subject and object, evolved in the learning laboratory: Earth.

The particular ability of viruses to reorganize DNA is a major factor in the progressive salmutations of evolution, as well as a factor in incidences of cell destruction and some of the extinctions of lifeforms.

It is known that DNA moves within the nucleus of the cell, serves as a template for mRNA, which is transcribed into tRNA. Three tRNA nucleotides combine to form one amino acid. A number of amino acids provide the pattern for one protein. Combinations of proteins are the building blocks of life. Translation of the genetic material creates the individuals which are the entities of evolution.

DNA of two sex species is generally conserved in its own characteristic configuration, but in half the original number of chromosomes as it becomes the hereditary material in the negatively charged egg and the positively charged sperm. DNA changes provide variations in individuals of the same species and mutations in species pattern. It has been estimated that deleterious mutations in humans occur .01 times per generation. Pre-adaptive mutations are much less frequent. Salmutations are very few when considered in the full reach of evolutionary time.

The literature on viral reorganization of DNA provides amazing evidence of viral intervention in many taxa. Still the ability of viruses to reorganize DNA is severely limited by the cell. Cells can block viral access. Neural communication systems can instruct the body's immune system to repel the viral invaders. It appears that cells operate very well, both with and without viral intervention.

Perhaps the blow to human ego from knowledge of virus participation in evolution will be greater than the revelation that humans ascended from common ancestors of apes. But in reality, it should relieve concern about heredity because there is great comfort in knowing that all nature is united to evolve and support life. It may be no comfort to realize that viruses are capable of evolving alternatives to humans, but that fact should unite humankind to preserve a fair place for itself among all the creatures of Earth.

Gratitude and Warning

The principles of evolution presented here are speculative and evocative. Theory should never be accepted as a solution set in stone, as Darwin's seemed to be for almost 130 years. Theory is meant to be chewed, not simply swallowed. New facts and ideas spring eternal, and this age is one of the most fertile in potential for discovery and innovation.

The minds of humans are the treasure troves of evolution and should not be restricted to one or several disciplines or logics. There are new logics as well as new facts to be discovered. The complements of known facts and theories should be expressed. But there are dangers in consideration of complements. The concept of complementarity may free the human mind. Being paradoxical, it may also place the thinker in the realm of disintegration and chaos. Humans have the choice to use the tool for good, with common sense, or regress into irrationality. Yes, there are either/or choices to be made. Monistic logic is not supplanted or opposed by complementarity dynamics. They are both patterns of reasoning to be analyzed.

How much humans owe to each other in the development of their minds! No word of gratitude, reference or listing in a bibliography can express the appreciation we owe to each other, and to that spring of inspiration which remains nameless in scientific discourse!

Evolution into the 21st Century

The complementary processes of competition and cooperation function reciprocally in dynamic equilibrium. Interactional relationships occur among viruses, bacteria, cellular processes, tissues, organs, organisms and the environment.

Evolution is patterned from simple to complex, but simple is retained. Lifeforms exhibit incredible integrated variety in incredible numbers — witness the millions of existing species, the billions of organisms that are created from four DNA nucleotides, the quadrillions of human cells and numerous patterns of structure and function.

Monistic logic, dualistic logic, systems logic, simultaneity logic, reciprocal time logic, quantum logic, the logic of complementarity dynamics and complementary systems are philosophical tools to understand life patterning. Each of these logics leads to marvelous insights into life and evolution. The ability to use them all activates the mind like fireworks in a July sky.

As evolution is process, valid evolution theory is malleable, as is man's ability to understand it. The minds of humans are also evolving. But humanity is on the threshold of such a rush of understanding that it is appropriate to contemplate the emergence of a higher order of integrated intelligence than that of Homo sapiens. Humanity is witnessing the transformation to a new hominid species. It is not a biological emergence from one or a few individuals. It is a cultural transition from pre-adaptations of the whole species — a *cultural salmutation.*

The physical morphology of humans has changed very little in the last 35 thousand years, and may change very little in the next 35 thousand years, but nevertheless, cultural evolution is producing a totally new species. The new cultural species is emerging in those who are beginning to understand DNA, RNA, amino acids, proteins, genes and chromosomes; the relationships of viruses, bacteria, cells and lifeforms. It is emerging in those who understand the motions of planets and galaxies, and those who design computers, robots and spacecraft. It is arising in those who have a vision of peace and step nonviolently before troops and trains to proclaim their vision. It is arising in those who walk softly on the land, protecting other creatures. The new cultural species is arising as humans insist that world leaders assume environmental responsibility and seek new avenues of peace and goodwill.

The emergence into a new cultural species has been occurring for a long time. Socrates — one who first proclaimed that good is not an absolute — taught his students how to seek the good, three hundred years BC. The word "good" has the same root as the word "god" and perhaps the emerging cultural species is exhibiting an accelerating search — for integrity, and for spiritual growth.

Mankind has always been plagued by the opposite pole from goodness. The logic of complementarity dynamics bears witness that it is through understanding both poles, and the spectrum between them, that the dynamic equilibrium movements can be visualized.

It was necessary for humans to observe centuries of preying by animals, including themselves, before they could conceive of the concept of animal rights. It was necessary to understand the devastations of forest exploitation before generations of tree planters arose.

It was through natural spontaneous abortion, and other means of population limitation: starvation, disease, phenocide, and natural disasters, that humans conceived of the necessity for voluntary limitation of human populations and preservation of other species.

It was necessary to understand the structure and functioning of thousands of disease-causing organisms, before humans conceived the idea that viruses and bacteria also promote life and evolution. It was through the tragedies of countless wars that mankind has learned the possibilities for peace processes.

The mind of humankind has evolved to a recognition of the next cultural species into which it is entering — a species which lives to learn, to save other species, to create optimum human environments, to coexist in peace and goodwill, to decide what is "good" and do it. Humankind is evolving into *Homo deae*.

Glossary

Acausality. (New definition): Concept of interrelationships which function in non-sequential patterns. Events occurring in reciprocal time. Exemplified by self-organizing or autopoetic events.

Acquired character. Modification of habits of an organism due to environmental influences occurring during life, considered by Lamarck to be inheritable.

Adaptation. Evolutionary changes that increase the fitness of a lifeform or population to live and reproduce in its environment.

Adenine. A purine base found in RNA and DNA. With its sugar and phosphate attachment, is called a nucleotide.

Aerobic. Oxygen environment.

Algae. A taxon of plants, including photosynthetic ocean plankton, seaweeds and the filamentous fresh-water plankton.

Alive. (New definition): The state of an organism or virus that contains or is composed of DNA and-or RNA and is functioning either inherently or as part of a cell. The state of the whole and the parts of a functioning organism.

Alkaline. A basic; having a pH greater than 7.

Allantois. A saclike membrane connected to the vertebrate embryo that functions for waste storage.

Allele. One complement of a pair of genes, appearing at a given locus in a chromosome.

Allopatric. Describes populations which originate in or occupy separate geographical areas. See sympatric.

Alteration of generations. A mode of reproduction among lower animals and plants in which sexual and asexual generations alternate, exhibiting very different morphologies.

Amino acid. An organic compound containing the amino group

(NH_2) and the carboxyl group (COOH). Amino acids are building blocks of proteins.

Ammonia. NH_3, but usually present as an ammonium ion (NH_4+) at the pH levels inside animals.

Ammonites. Fossil chambered shells, with spirals similar to pearly nautilus, but having wavy partitions in complicated patterns between the chambers.

Amnion. The fluid-filled sac in which the embryo develops in reptiles, birds and mammals.

Amoeboid. Amoeba-like, flowing movement of a cell that results in continuous shape changes.

Amphibian. Member of a phylum of animals combining aquatic and terrestrial living, in which the young generally breathe by gills. Examples are toads, frogs and newts.

Anaerobic. An environment, or biological process occurring without oxygen.

Analogous. Similar in function but different in fundamental structure, such as wings of insects, pterasaurs, bats and birds.

And-or. Symbol for complementarity.

Angiosperms. Flowering plants.

Annelid. Member of a phylum of worms which includes earthworms, marine worms and leeches.

Antennae. Jointed appendages of the head of insects and crustaceans used for prehension of the environment.

Antibody. A protein produced specifically by the immune system to bind to an antigen, inactivating or destroying it.

Antigen. A foreign substance, usually a protein attached to a virus or protozoan, that stimulates the production of specific antibodies when introduced into a living organism. The antigen-antibody interaction is a complementary relationship, controlling life and health.

Aquatic. Living in water.

Arthropod. Member of phylum of animal kingdom having surface of the body divided into segments, some with jointed legs. Examples are insects and lobsters.

Artiodactyl. Mammals having an even number of toes or hooves. e.g. deer, elk, pigs.

Asexual reproduction. Any process of reproduction that does not involve the union of gametes from two sexes.

Atom. The smallest unit of matter that has the properties of a chemical element.

Atomic number. The number of protons in a particular atomic nucleus.

Autopoietic. Self-maintaining system.

Autotrophic. Photosynthetic. See Heterotrophic.

Axion. The fiber of a nerve cell that carries impulses away from the cell body.

Bacterium. A member of the kingdom Monera; a prokaryote containing one genophore (chromosome). One of the earliest of lifeforms.

Bacteriophage. A virus capable of penetrating a bacterium, co-opting the bacterial DNA, uniting its DNA with the bacterial DNA, or forming a plasmid within the bacterium.

Base. A solution with a pH of more than seven. A substance that can accept a proton (hydrogen ion) in a reaction.

Bilateral symmetry. Describes a body that has two, essentially mirror-image, sides.

Biome. A terrestrial ecosystem characterized by a climax community.

Bivalve. A mollusk that has two hinged shells.

Brachiopod. Member of phylum of marine animals with hinged, bivalve shells, and with soft fringed appendages which serve to direct food to the mouth. Example is a lamp shell.

Capillaries. The smallest blood vessels, serving to transport blood from arteries, to cells, to veins.

Carapace. The shell covering the anterior part of the body, generally of crustacians.

Carnivore. An animal that feeds on animals; contrasted with herbivore and omnivore.

Carrier. An individual which carries a recessive gene that is not expressed but rather obscured by a dominant allele.

Cartilage. A rubbery vertebral tissue, in contrast to the rigidity of bone.

Catalyst. An enzyme or other substance that accelerates a chemical reaction, without alteration to itself.

Caudal. Pertaining to the tail or posterior end of the body.

Causality. The relationship of cause and affect.

Cell. Microscopic unit of living bodies, separated from each other by membrane. One-celled organisms.

Cetacea. An order of fish-like mammals with no hind limbs, includes dolphins and whales.

Character or characteristic. One of the details of structure, form, substance or function of an individual organism.

Chimera. An organism composed of DNA from more than one species or a molecule composed of DNA segments from more than one organism.

Chlorophyll. Any of several similar green pigments that absorb light energy in photosynthesis.

Chloroplast. Also plastid. Green subcellular organelle found in plant cytoplasm, containing chlorophyll; the site of photosynthesis. Chloroplasts appear to have a different DNA structure than the DNA in the nucleus of the cells in which they occur.

Chordate. Member of phylum of animals having a notochord or vertebral structure.

Chromosomes. Nucleoprotein bodies in the cell, observable microscopically during cell division. Each species has a characteristic number of chromosomes, carrying the genes, arranged in linear order.

Chrysalis. See pupa.

Cilia. Undulipodia. Hairlike locomotor structures on certain cells or protozoans. May have evolved from spirochetes.

Cladistic affinity. Recency of common descent.

Cladogenesis. Branching of ancestral lineage into two derived taxa.

Climax community. A stage reached in an ecological succession which is in dynamic equilibrium.

Clones. Individuals descended by asexual reproduction from a single ancestor.

Codon. A three-nucleotide sequence of RNA that specifies one amino acid.

Coelenterate. Member of a phylum of invertebrate animals having tentacles and a single internal cavity serving for digestion, excretion and other functions. Examples are jellyfish, sea anemones and corals.

Coevolution. Those patterns in which different organisms such as insects and flowers; predators and prey; producers, consumers and recyclers evolved together in dynamic equilibrium.

Community. An ecological unit of all the populations of organisms living and interacting in a given area at a given time.

Competent cells. Cells capable of taking up DNA from an outside source.

Complementarity dynamics. (New definition): A logic of paradox in which two entities, an entity and a concept, or two concepts complete each other as a whole, require different criteria for analysis of their essence, and function together in dynamic equilibrium.

Compound. Subtance composed of two or more parts, elements, or ingredients.

Conjugation. A mating process in which DNA passes from one bacterial cell to another through a tube-like bridge. Joining of protozoans as part of the fertilization process; process of sexual reproduction in protist organisms.

Consumer. An organism which eats plants or animals. Does not photosynthesize.

Convergent evolution. (Classical definition): Two similar forms, believed to be descended from different ancestors, having similarity through adaptation to similar niches; e.g. fish and whales.

(New definition): Two similar forms which have received similar insertions or recombinations of DNA during the course of evolution; e.g. moths and butterflies, fish and whales. To be distinguished from coevolution.

Crossing over. A process of exchanging genes between two DNA strands. It produces genetic combinations which are different from those of the parents.

Cyanobacteria. Blue green algae.

Curie. Unit of radioactivity measured in number of disintegrations per second.

Cytology. The study of the structure and function of the cell.

Cytoplasm. The protoplasm of a cell outside the nucleus, in which are located cell organelles such as mitochondria, plastids and plasmids.

Cytoplasmic inheritance. Hereditary transmission dependent on structures in the cytoplasm rather than the nuclear genes; extrachromosomal inheritance.

Cytosine. A pyrimidine base found in RNA and DNA.

Deficiency or deletion. Absence of a segment of a chromosome involving one or more genes.

Dendrite. Slender branched extension of a nerve cell that functions as a receptor.

Dimorphism. Two different forms in a group as determined by such characteristics as sex, size or coloration. See polymorphism.

DNA. Deoxyribonucleic acid. The genetic material of all cells and many viruses which transmits coded information in two polynucleotide strands arranged in a double helix. In bacteria and viruses DNA molecules usually form a circular genophore (chromosome). "Naked" DNA may be a phage without a protein coat; a plasmid.

Dominance. Quality applied to one member of an allelic gene pair that has the ability to manifest itself wholly or largely to the exclusion of expression of the other member.

Dorsal. Pertaining to the back. Opposite of ventral.

Dualism. The concept that there are two basic substances or principles; e.g. mind and body.

Duplication. The occurrence of a segment more than once in the same chromosome or gene.

Dynamic equilibrium. (Classical definition): homeostasis or steady state; maintenance of internal stability of a system. (New definition): Concept of kaleidoscopic motion in perception of complements. Complementary combination of change and constancy. Similar to homeostasis, but involves concept of reciprocal time in perceiving interrelationships.

Echinoderm. Member of phylum of marine animals having radial symmetry and a body wall stiffened by calcium material. Examples are star fish, sea urchins and sand dollars.

Ecology. Study of interrelationships of organisms with each other and with their environment.

Ecosystem. An ecological unit composed of a community of living organisms and the physical features of their environment.

Edentate. (Xenarthra). Member of the order of mammals characterized by the absence of the middle incisor teeth in both jaws; e.g. sloths and armadillos.

Egg. A germ cell produced by a female.

Electron. A negatively charged subatomic particle.

Element. A basic chemical unit made up of a number of the same kinds of atoms.

Embryo. An organism in an early stage of development in the egg or womb; the stage following formation of the zygote from the fertilized egg.

Endemic. Within a given location.

Endosymbiotic. Mutually beneficial relationships of organelles within their cellular host.

Entropy. A measure of the disorder or randomness of a system.

Environment. The external conditions and influences surrounding an organism.

Enzyme. A protein molecule that accelerates a specific chemical reaction; a catalyst in a living system.

Epicontinental. Located on a continent.

Epidermis. Outer layer of cells of an animal embryo that originates the skin and nervous system.

Episome. An organelle which may be present or absent in different cells. It may be associated with a chromosome, or independent in the cytoplasm. It may be the equivalent of a virus or plasmid.

Eukaryote. An organism that is composed of one or more cells which have their multiple chromosomes contained within a nucleus. The group includes algae (other than blue green algae) protists, fungi, plants and animals.

Evolution. Ascent with modification.

Extrachromosomal. Structures that are not a part of the chromosomes; DNA units in the cytoplasm that control cytoplasmic inheritance.

Fauna. The totality of animals inhabiting a certain region or which have lived during a particular geological period.

Feedback mechanism. An arrangement in which a controller is itself regulated by the function that it controls. Operates in reciprocal time.

F (fertility) factor plasmid. The structure of DNA in a bacterium that controls its "sex" and may be a form of virus.

Fertilization. The fusion of a male sperm with a female egg to form a zygote.

Filter feeder. An animal that feeds on small organisms or particles strained from the water around it, or from a stream of water that passes through part of its body.

Fitness. A concept measuring success by number of offspring left by an individual compared to other individuals of the same species in competition for the means of subsistence.

Flagella. Whip-like organelles of locomotion in certain cells or in protozoa. Longer than cilia but similar in structure. May be evolved from spirochete bacteria. See Undulipodia.

Flora. The totality of plants growing in a given area or during a particular geological period.

Food chain. A sequence of producers and consumers through which energy and materials move in an ecosystem.

Foraminifera. Animals of very small size with jelly-like bodies enclosed by calcium shells, with retractable filaments for prehension of their environment.

Fungi. A kingdom of cellular plants such a mushrooms, toadstools and molds.

Gamete. A specialized reproductive cell that fuses with another gamete (usually egg and sperm) in fertilization to produce a zygote.

Gastrodermis. Inner layer of an animal embryo that originates the lining of the digestive tract and the respiratory system in vertebrates.

Gene. A sequence of DNA base pairs located in a fixed location on the chromosome that encodes information for synthesis of one species of protein. Genes determine phenotypic character.

Gene pool. The sum of all alleles of all genes in a population.

Genetics. The science of heredity and variation.

Genome. A complete set of chromosomes.

Genophore. The singular chromosome of a prokaryote.

Genotype. The gene structure of an organism, whether expressed or recessive, which may be expressed in successive generations. Contrasted to the phenotype, which is a set of observable characteristics of a organism.

Germ cell. A reproductive cell capable of being fertilized and reproducing an entire organism.

Germ plasm. The germinal material or basis of heredity; the DNA.

Gills. Projections of the body surface of aquatic animals that function in gas exchange.

Glycine. The simplest amino acid.

Guanine. A purine base found in DNA and RNA.

Gymnosperm. A plant having its seeds exposed, not enclosed in an ovary. Examples are palms, cycads, ginkgoes, pines, and firs.

Habitat. The locality in which a plant or animal naturally lives.

Haploid. The chromosome state in which each chromosome is represented once. Occurs in germ cells of eukaryotes. Prokaryotes exist only in the haploid state.

Helix. A structure with a spiral shape. DNA is structured in the form of a double helix.

Hemoglobin. Protein compound containing iron located in red blood cells of vertebrates that functions in transportation of oxygen.

Hemolymph. The fluid that circulates in an open system in the body cavity of invertebrates.

Herbivore. An animal that feeds on plants. Contrast with carnivore and omnivore.

Heredity. Transmission of traits from parents to offspring.

Hermaphrodite. An animal individual with both male and female reproductive organs; monoecious in plants.

Heterotrophic. Eats food. Does not photosynthesize.

Hominids. Humans, including those classified as *Australopithecus* and *Homo*.

Homologous. Similar in structure and shape. Homologous bones and teeth are believed, in classical evolution theory, to indicate direct, sequential lines of descent. Homologous paired chromosomes bear genes in the same loci.

Hormone. An organic product of cells in one part of the body that is transported to another part where it influences or coordinates activity.

Hypothesis. A supposition that is established by reasoning after consideration of available evidence.

Inorganic. Not organic; not a carbon compound (other than carbon dioxide).

Insectivore. Member of the order of mammals that includes shrews, moles and hedgehogs.

Intelligence, finite. (New definition): Manifestation of the ability to gather, process, utilize, distribute and exchange information. Not only the capacity for human reasoning.

Intelligence, infinite. (New definition): The intelligence manifested in and-or beyond the known universe, which is creative, coordinating, constructive and beyond human capacity to comprehend in its totality.

Interferon. A substance produced and released by virus-infected cells that protects other cells from virus infection.

Inversion. A rearrangement of the order of a group of nucleotides in the chromosome.

Invertebrate. An animal not possessing a spinal column.

Ion. An electrically charged atom or molecule, formed by the loss or gain of one or more electrons.

Kingdom. One of the six great divisions of life: Virusa (new), Monera, Protoctista, Fungi, Plantae and Animalia.

Lambda. A bacteriophage capable of specialized transduction.

Larva. An immature form of an animal issuing from an egg, usually as a caterpillar or maggot that is very different from the mature adult.

Lemurs. A group of four-handed mammals (prosimians) considered to be derived from insectivores, having a claw rather than a nail on the first finger of the posterior hands.

Leukocyte. A white blood cell.

Lichen. The combined organism resulting from the intimate symbiotic association of an alga and a fungus.

Life. (Classical definition): The condition of organisms having metabolism, growth, responsiveness and reproduction. (New definition): The animating quality of functioning organisms containing DNA/RNA.

Locus. A specific part of a chromosome that is the site of a particular gene. One of the alleles of that gene may occupy this site.

Logic. (Classical definition): The study of principles of reliable inference. A particular method of reasoning. (New definition): A systematic organization of principles and methods of reasoning. A reasoning pattern leading to one or more conclusions, a synthesis, or expanded insight into complex and-or paradoxical questions; not necessarily a system of proof.

- **Complementarity logic**. A reasoning pattern in which atomic events and other complex and-or paradoxical phenomena are viewed as fundamentals (complements) of wholeness.
- **Complementarity dynamics logic**. A reasoning pattern that entails perception of dynamic equilibrium between two entities, an entity and a concept, or two concepts.
- **Complementary systems logic**. A reasoning pattern in which interactions of several complex systems are coordinated.
- **Dualistic logic**. A reasoning pattern in which two opposing concepts are coordinated in synthesis.
- **Monistic logic**. Inductive, deductive and linear inference system of reasoning.
- **Quantum logic**. A reasoning pattern that envisions the discontinuous energy of atomic phenomena. Expressed in the relationship of position and momentum of an electron.
- **Reciprocal time logic**. A reasoning pattern in which perception oscillates instantaneously among variables. It evokes

understanding of "spontaneous" *acausal* events and "instantaneous" self-organizing systems.

Relativity logic. A reasoning pattern in which the relativity of simultaneity is envisioned.

Systems logics. Reasoning patterns that involve analysis and coordination of numerous interacting elements.

Lophophore. Feathery ring structure surrounding mouth of invertebrates used for food gathering and-or breathing.

Lysis. Destruction of bacterial or other host cell when viruses dissolve the cell wall and spread as new viruses to other cells.

Lysogenic bacteria. Bacteria harboring temperate bacteriophages, capable of lysing the cell.

Macroevolution. Processes giving rise to taxonomic categories higher than species.

Macromolecule. A large molecule. Often used to describe DNA.

Macromutation. Alteration in number and-or configuration of chromosomes. Includes mutations of plasmid genes, polyploidy, centric fusions, and centric fissions.

Macrophage. A large amoeboid phagocytic white blood cell.

Mammal. The class of animals characterized by birth of living young that are nourished after birth by milk from the mammary glands of the mother. Most mammals have a direct connection (placenta) between mother and embryo during development in the womb. Examples are hairy quadrupeds, and the primates including humans. Aplacental mammals include the marsupials and monotremes.

Marsupial. A pouched mammal, born in an immature stage of development, receiving nourishment from mammary glands only after crawling by itself into the pouch of its mother. Examples include opossums, kangaroos and wombats.

Megamutation. Genomic acquisitions. Includes transformation, transduction, sexduction, meiotic sex and symbioses.

Meiosis. The process by which the chromosome number of a reproductive cell becomes reduced to half the somatic chromosome number. Results in formation of eggs and sperm in animals and spores in plants.

Mendelian ratio. Plant or animal reproduction showing a ratio of three dominant-characteristic individuals to one recessive-characteristic individual in the first F1 generation.

Mesenchyme. The middle tissue layer that originates skeleton, musculature, circulatory system, and lining and covering layers in the body cavities of animal embryos.

Messenger RNA (mRNA). The RNA that carries the complement of the DNA code to the ribosome where it is translated into tRNA, which forms the amino acids that compose proteins.

Metabolism. The sum of chemical reactions within a cell or the sum of all cellular activities in an organism.

Metamorphosis. Change of structure, form and-or substance that transforms an immature animal (usually a larva) into an adult.

Meter (m). Metric measurement which equals 39.37 inches or 3.2808 feet.

Millimeter (mm). One thousandth of a meter.

Mitochondria. Small DNA-containing organelles in the cytoplasm of eukaryotic cells in which aerobic respiration occurs.

Mitosis. Cell division in which chromosomes replicate. They migrate to the ends of the spindle, after which the nucleus containing the chromosomes, and the cytoplasm divide into two cells, each with a full set of chromosomes, identical to the original chromosomes.

Molecule. The smallest unit of a compound, composed of two or more atoms.

Mollusk. Member of a phylum of invertebrate animals having a one, two, or several part calcium shell which encloses the soft, unsegmented body. Examples are clams, oysters, snails and pearly nautilus.

Monoecious. Separate staminate (male) and pistillate (female) flowers on the same plant.

Monotremes. Mammals that lay eggs, distinguished from those which produce live young. Found only in Australia and New Guinea. Only living examples are duck-billed platypus and spiny anteater.

Morphology. Study of the structure or form of organisms and changes over time.

Mutagen. A physical or chemical agent that is capable of inducing mutations.

Mutation. (Classical definition): Spontaneous change in genetic material.

(New definition): A "micro" recombination of DNA which involves a change in the genotype, and may or may not result in a changed phenotype in the progeny. Includes base pair changes, deletions, duplications and transpositions of DNA molecules. See Macromutation, Megamutation and Salmutation.

Mutualism. A symbiotic relationship in which both organisms benefit from the association.

Myoglobin. Protein which reversibly stores oxygen in muscle tissue; may have preceeded hemolymph and hemaglobin during evolution.

Natural selection. Competitive processes that favor individuals and groups which are better adapted and eliminate those which are unfit.

Nematode. Member of phylum of unsegmented worms known as round worms.

Nerve. A bundle of neuron fibers outside the central nervous system.

Neuron. A nerve cell.

Neutron. An electrically uncharged subatomic particle, with approximately the same mass as a proton, found in the nucleus of the atom.

Niche. Description of environment and role of a particular species in its natural community.

Non-living. Concept of substances that lack animate DNA or RNA.

Notochord. A longitudinal rod of cells within early chordates believed to give rise to the spinal column of later vertebrate animals.

Nucleic acid. Macromolecules such as DNA and RNA composed of phosphates, sugar and organic bases, that function in the genetic expression of cells.

Nucleotide. Basic unit of the DNA molecule composed of organic bases of cytosine and guanine, adenine and thymine (adenine and uracil in RNA) plus a phosphate and a sugar.

Nucleus. The spherical body that contains the chromosomes of a eukaryotic cell. The central part of an atom.

Omnivore. An animal that eats both plants and animals. Contrast with carnivore and herbivore.

Operator. The gene on a chromosome that controls the activity of one or more structural genes.

Operon. An operator and one or more structural genes which makes up a regulatory or control unit of the chromosome.

Organ. A body part specialized for a particular function or set of functions.

Organelle. Specialized part of a cell with a particular function such as the locomotor activity of the undulipodia of a protozoan.

Organic. Pertaining to living things. A carbon compound.

Organism. A single cell or a multicellular lifeform.

Oxidation and Reduction. Oxidizers are elements deficient in negatively charged electrons. They include oxygen, chlorine and nitrates. Oxidation is the reaction of oxidizers and reducers. Reducing elements are rich in electrons, including hydrogen, metals and most fuels. Reduction is the process of restoring the original elements. An example is the production of carbon by the chloroplasts in green plants.

P1. A bacteriophage capable of generalized transduction.

Paleontology. The study of fossils and the fossil record of lifeforms.

Parallel evolution. (Classical definition): Adaptations from a common stock, independently evolved in similar ecological niches. Examples are Old World monkeys and New World monkeys.

(New definition): Development of evolutionary lines of similar organisms in different locations by insertion or recombination of similar DNA; e.g. Old World monkeys and New World monkeys. To be distinguished from coevolution.

Parasite. A plant or animal living in, or upon, and at the expense of another organism.

Pathogen. A virus or other organism that causes disease.

Pelvis. The bony arch to which the posterior limbs of vertebrate animals are connected.

Peptide. A compound containing two or more amino acids which joins a carboxyl group of one to the amino group of the other.

Perissodactyl. A taxon of mammals having an even number of toes or hooves; e.g. rhinoceros, horses.

pH. Symbol for measure of acidity in solutions ranging from zero to 14; pH seven is neutral, less than seven is acidic, and more than seven is basic.

Phage. A bacterial virus.

Temperate phage. A virus that penetrates but does not destroy the host bacterial cell. It may remain dormant for varying lengths of time.

Virulent phage. A virus that uses the bacterium's own cell structure to create more viruses and lyses (dissolves) the host bacterial cell.

Phagocyte. Specialized white blood cell that engulfs and ingests foreign particles, bacteria or dead cells.

Phagocytosis. Engulfment and intake of solid particles by cells.

Phenocide. Destruction of members of a phenotype.

Phenotype. A set of observable characteristics of an organism, that identify a species, as opposed to the genotype, which is its genetic structure.

Photosynthesis. Synthesis of organic compounds from inorganic compounds using light energy absorbed by chlorophyll.

Phylum. Major classification category of lifeforms under each of the kingdoms, followed by subphylum, superclass, class, subclass, infraclass, superorder, order, suborder, infraorder, superfamily, family, subfamily, genus and species.

Placenta. An organ formed during mammalian development from tissues of both mother and embryo, within which nutrients and wastes are exchanged through a tissue barrier that separates the two circulations.

Planck's constant. The ratio of the energy of one quantum of radiation to the frequency of the radiation; usually expressed in ergs per second.

Plankton. Microscopic aquatic protists and small aquatic plant and animal organisms.

Plasma. Substance made up of electricallly charged atomic particles rather than intact atoms.

Plasmid. A microscopic, circular DNA molecule in a cell, that can replicate independently of the cell's chromosome. May be a form of virus. See F factor and R factor.

Plastid. A cytoplasmic body in cells of plants and some protozoans.

Platyhelminth. Member of the class of worms having bilateral symmetry and a flat body, such as tapeworms, trematodes, and planarians.

Proton. The positively charged subatomic particle in the nucleus of an atom.

Polarity. Having negative charge at one end and positive charge at the other.

Polymorphism. Two or more kinds of individuals in a population in which at least one form is maintained by a mechanism other than recurrent mutation — a Mendelian effect, e.g. Malaysian butterflies.

Polypeptide. A molecule of many amino acids linked together.

Polyphyletic. Originated or developed from more than one ancestral type.

Pongids. Taxon of contempory and extinct apes; e.g. orangutans, gorillas and chimpanzees.

Population. An entire interbreeding group of plants or animals that occupy a given area at a given time.

Porifera. Phylum comprising the sponges.

Precursor. A preceding developmental or ancestral link.

Predator. A free-living organism that feeds on other organisms.

Primitive. Old or resembling an ancestral condition.

Principle. A fundamental truth from which others are derived; an originating or actuating agency or force.

Probability. Likelihood of occurrence.

Proboscis. A tubular extension of the snout of an animal. A feeding or sucking tubal extremity.

Producer. An organism that synthesizes organic materials, usually by photosynthesis.

Prokaryote. A member of the group of lower organisms including bacteria and cyanobacteria that have a double-ring genophore (chromosome) but lack a well defined nucleus.

Progeny. Offspring of plants or animals from a particular fertilization or mating.

Prophage. Noninfectious stage of a temperate phage, integrated into a bacterial chromosome.

Prosimian. Early form of primates e.g. lemurs, tarsiers.

Protein. A large, complex, organic molecule in cells, made up of a chain of amino acids bonded by electrons. May be structural, functional, enzymatic or a combination of these.

Protista. A phylum of simple, single-celled organisms with gelatinous bodies and little trace of distinct organs, e.g. foraminifera, amoebae, paramecia.

Protoctista. New phylum classification including protists and other small eukaryotic organisms that are not fungi, plants or animals.

Proton. A positively charged subatomic particle in the atomic nucleus.

Protoplasm. The material within a cell, including cytoplasm, organelles and nucleus.

Pupa. A non-eating stage of development in insects occurring between larva and adult. Pupae undergo metamorphosis into the winged reproductive form. The pupal state of butterflies is the chrysalis.

Purposeful. Goal oriented result or intent inherent in function or system.

Quantum. A unit of electromagnetic energy.

R (resistance) factor plasmid. The structure of DNA in a bacterial cell that provides immunity to a chemical and is stably inherited. May be a form of virus.

Radial symmetry. Describes a body that is arranged around a central axis, such as a jellyfish.

Radula. Teeth-like feeding structure of mollusks.

Recessive. An inherited trait of one member of an allelic pair of genes, expressed in the progeny when the dominant member of the pair is absent or ineffective.

Reciprocal moment. An instant of time in which perception oscillates between two or more perspectives. The instant of time in which events occur simultaneously and-or in various sequences depending upon the perspective and-or position of the observer.

Recombinant DNA. DNA molecules of different origin joined together to form a single molecule, capable of some specific biological structure and-or function.

Reduction. A chemical reaction that adds one or more electrons to an atom or compound. Reduction of one substance occurs with oxidation of another.

Regulator gene. A gene that controls the level of production of another gene. Three types of genes are: regulator, operator and structural.

Replication. Construction of a new chain of complementary DNA or RNA from template DNA or RNA.

Resonance. A philosophical concept of electron motion based on the idea that the bond between two or more electrons, orbiting different nuclei, allows changes in orbits or energy states. This may be perceived in reciprocal time logic.

Ribosome. Cytoplasmic structure on which mRNA replicates tRNA to produce the amino acids which form proteins.

RNA. Ribonucleic acid. The information carrying strand, complementary to DNA, which carries the code for specific proteins from DNA as mRNA, to transcribe as tRNA, to form amino acids that create proteins. Cells contain both DNA and RNA. Viruses may have DNA or RNA as their genetic material. RNA has the sugar ribose instead of deoxyribose and has the base uracil in place of the base thymine of DNA. RNA is usually single-stranded.

Rodents. The gnawing mammals especially characterized by a single pair of chisel-like cutting teeth in each jaw, with a gap between this pair and the grinding teeth. Examples are rats, mice, beavers and squirrels.

Salmutation. (New word.) Macromutational and-or megamutational process of DNA reorganization forming a new taxon line. May be virus or microorganism mediated and is stably inherited.

Saltation. A jump in the fossil record.

Sedimentary formations. Rocks deposited as sediments from water.

Segregation. The separation of chromosomes at meiosis.

Selection. The competitive process favoring the survival and propagation of the most fit individuals or species.

Sex chromosomes. Chromosomes which determine the sex of the individual. See X and Y chromosome.

Sexduction. The process in which ring-form DNA is carried from one bacterium to another. An example is the functioning of fertility factor F.

Sex pilus. A structure emanating from a F+ bacterium which unites with an F- bacterium during conjugation.

Sexual reproduction. Formation of a mature germ cell from the union of an egg and a sperm.

Siphon. A tubular structure used to draw in or expel fluids.

Somatic cells. Body cells having one set of chromosomes from each parent, as contrasted with germ cells with half the number of somatic chromosomes, capable of reproduction of the organism.

Somatoplasm. The nonreproductive tissue of the organism, in contrast to germ plasm (gametes).

Speciation. (Classical definition): Formation of a new species. (New definition): The process of formation of new taxon lines by means of mutations or salmutations.

Species. The basic, but arbitrary, category of classification of plants and animals between the genus level and the variety level, expressed as the second name following the capitalized genus name. A population of organisms capable of interbreeding. A group of microorganisms which share many common characteristics and are classified together.

Sperm. A mature male germ cell, a gamete.

Spindle. Set of microtubules which coordinate movement of chromosomes during mitosis or meiosis.

Spirochete. Any of several spiral-shaped bacteria.

Spore. A unit of plant protoplasm which develops into a new individual. A specialized encapsulated bacterial cell which can survive adverse conditions and germinate into a new cell.

Statistic. A value based on a sample of a population.

Stratosphere. Region of the atmosphere extending upward from about 12 to 15 miles above the earth; the site of the ozone layer.

Structural gene. A gene composed of DNA which controls protein production by coding the amino acid sequence.

Succession. Progressive change in kinds of species in a biome under changed conditions.

Suctorial. Adapted for sucking.

Symbiosis. (Classical definition): Association of two organisms of different species living together in an intimate relationship. (New definition): Intimate, cooperative relationships among organisms or among organelles and the nucleus within a cell. The cooperative relationships within cells are endosymbioses.

Sympatric. Originating in, or occupying the same geographical area. See Allopatric.

Synapse. The chemical interface between the axon of one neuron (nerve cell) and the dendrite or cell body of another.

Synapsis. The pairing of homologous chromosomes.

T4. A phage. Inhabits the bacteria (*Escherichia coli*) of the human intestinal tract. Was studied in the process of determining the structure of DNA.

Taxon. Classification group.

Taxonomy. Classification based on analysis of structural similarities and projected evolutionary relationships.

Template. A pattern or model. DNA stores information and acts as a model or mold from which information is transported by complementary mRNA.

Terrestrial. Pertaining to land, land-dwelling.

Theca. Main truck of echinoderms.

Theory. A scientific concept or set of concepts supported by significant data.

Thymine. A pyrimidine base of DNA. Adenine, cytosine and guanine are found in both DNA and RNA, but in RNA, thymine is replaced by uracil.

Transcription. The process of production of a strand of tRNA, complementary to a segment of mRNA.

Transduction. Transfer of genetic material from one bacterium to another by a virus.

Generalized transduction. The moving of any gene from one bacterium to another by a phage such as P1.

Specialized transduction. The moving of only one or two genes adjacent to the phage on the bacterial genophore into a different bacterium by a phage such as Lambda.

Transfer RNA (tRNA). The RNA that transfers the genetic code in a complementary strand from messenger RNA into the amino acids which compose proteins.

Transformation. Genetic recombination in a cell brought about by penetration and incorporation of foreign DNA into the cell body, often by virus action.

Translation. The interaction of the ribosome, mRNA, and tRNA, in the cytoplasm of the cell that results in synthesis of a protein.

Translocation. Change in position of a segment of a chromosome to a different chromosome, or to another part of the same chromosome.

Transpiration. Evaporation of water from plants into the surrounding environment.

Transposons. Insertion sequence of genes which appear to interchange as plasmids and-or as viruses.

Triplet. A string of three nucleotides of RNA, a codon, which forms one amino acid, and-or may code a start or stop signal.

Uncertainty principle. Definite values cannot be given simultaneously to positional coordinates and momentum components of atomic particles.

Undulipodia. Microtubule structure (9+2) found in cilia, flagella, spindles, axions and dendrites.

Ungulates. Four-legged mammals with fused digits that are generally hoofed.

Unicellular. Consisting of a single cell.

Uracil. A pyrimidine base found in RNA but not in DNA. In DNA uracil is replaced by thymine.

Variation. Differences among individuals of the same species

Vector. An organism, plasmid or viral DNA molecule into which another DNA molecule is inserted without disruption of the ability to replicate.

Ventral. The front part. Opposite of dorsal.

Vertebrate. Member of the taxon possessing a vertebral column, a series of skeletal elements that enclose the nerve cord. Includes fish, amphibians, reptiles, mammal-like reptiles, birds and mammals.

Virion. A free virus particle. A reproductively inactive form of virus existing outside of a host cell.

Viroid. An virus particle consisting of a short strand of RNA without a protein coat, much smaller that the smallest virus.

Virology. Study of viruses.

Virus. A living microscopic organism composed of a nucleic acid core and a protein coat, capable of transforming the DNA or RNA within cells of living organisms. Note: Not considered to be a living organism by many scientists.

Virusa. New kingdom of viruses and virus particles proposed here as one of six kingdoms of living organisms.

Viviparous. Capable of bearing live young born at a relatively advanced stage after an extended period of development in the female body, where nutrients and wastes are exchanged between the embryo and female.

Wild type. The normal phenotype or standard for comparison.

X chromosome. A chromosome which determines sex. In most animals the male has one X and the female has two X. In birds, the chromosome arrangement is the opposite of mammals; the male has two X and the female has one X and one Y.

Y chromosome. The mate to the X chromosome in the male of most animal species. It is frequently the smallest chromosome.

Zooids. Individuals of colonial animals; e.g. bryozoans.

Zygote. The cell produced by the union of egg and sperm or the individual developing from this cell.

Bibliography

Agazzi, Evandro (ed.) (1981). *Modern Logic — a Survey*. London: R. Reidel Publishing Co.

Ahmadjian, Vernon (1967). *The Lichen Symbiosis*. Waltham, MA: Blaisdell Publishing Co.

Allee, Warder Clyde (1951). *Cooperation Among Animals, with Human Implications*. NY: Henry Schuman. (1st. ed. 1938).

Anati, Emmanuel (1963). *Palestine Before the Hebrews; A History from the Earliest Arrival of Man to the Conquest of Canaan*. NY: Alfred A. Knopf.

Andrewartha, H. G. and L. C. Birch (1954). *The Distribution and Abundance of Animals*. Chicago, IL: University of Chicago Press.

Andrews, Henry N. Jr. (1964). *Ancient Plants; and the World They Lived In*. NY: Cornell University Press.

Aristotle (1966). *Metaphysics*. Grinnell, Iowa: The Peripatetic Press. 58-70. (tr. Hippocrates G. Apostle, *Aristotle's Metaphysics)*.

Ashby, W. R. (1958). General Systems Theory as a new discipline. *General Systems* 3: 1-6.

Asimov, Isaac (1989). Fundamental particles of universe. *Los Angeles Times*, 9-15-89. CA: Times Mirror Company.

Augros, Robert & George Stanciu (1987). *The New Biology; Discovering the Wisdom in Nature*. Boston, MA: Shambhala, New Science Library.

Augusta, Josef (1962). *A Book of Mammoths*. London: Paul Hamlyn, Ltd.

____. (1967). *Prehistoric Animals*. London: Paul Hamlyn. (tr. Greta Hort).

Axelrod, Daniel I. (1981). Role of volcanism in climate and evolution. Special Paper 185. CO: *The Geological Society of America*.

Axelrod, Robert (1984). *The Evolution of Cooperation*. NY: Basic Books.

Ayala, F. J. (1970). Invalidation of principle of competitive exclusion defended. *Nature* 227: 89.

Bachofen, Johann J. (1967), *Selected Writings; Myth, Religion and Mother Right*. NJ: Princeton University Press. (tr. Ralph Manheim).

Bainbridge, Brian (1980). *The Genetics of Microbes*. NY: Halsted Press, John Wiley & Sons

Bakker, R. T. (1977). Tetrapod mass extinctions; a model of the regulation of speciation rates and immigration by cycles of topographic diversity. *Patterns of Evolution as Illustrated by the Fossil Record,* NY: Elservier Scientific Publishing Co. pp 439-468. (ed. A. A. Hallam).

Barghoorn, Elso S. (1981). Aspects of Precambrian paleobiology: the early Precambrian. *Paleobotany, Paleoecology and Evolution.* NY: Praeger pp 1-54. (ed. Karl J. Niklas).

____. (1982). The oldest fossils. *Evolution and the Fossil Record.* San Francisco, CA: W. H. Freeman & Co. 44-56. (ed. Leo F. Laporte).

Barrow, John D. (1991). *Theories of Everything; The Quest for Ultimate Explanation.* Oxford: Clarendon Press.

Barthelemy-Madaule, Madeleine (1982). *Lamarck, The Mythical Precurson: A Study of the Relations between Science and Ideology.* MA: MIT Press. (English tr. M. H. Shank).

Bartlett, John R. (1982). *Jericho.* MI: William B. Eerdman Pub. Co.

Bates, Marston (1961). *Man in Nature.* NJ: Prentice Hall.

Beadle, George W. (1977). The origins of Zea mays. *Origins of Agriculture.* Paris: Mouton Publishers. pp 616-635. (ed. Charles A. Reed).

Beckwith, Jon (ed.) (1983). *Gene Function in Prokaryotes.* NY: Cold Spring Harbor Laboratory.

Bender, Barbara (1975). *Farming in Prehistory; From Hunter-gatherer to Food-producer.* NY: St. Martin's Press.

Bergson, Henri (1944). *Creative Evolution.* NY: Random House. (1st. ed. 1911).

Bertalanffy, Ludwig von (1969). *General Systems Theory.* NY: George Braziller, Inc. Copyright 1969 by George Braziller, Inc.

Bignot, Gerard (1985). *Elements of Micropalaeontology: Microfossils — Their Geological and Palaeobiological Applications.* London: Graham and Trotman Ltd.

Birdsell, J. B. (1981). *Human Evolution.* NY: Houghton Mifflin Co. (3rd. ed.).

Birx, H. James (1984). *Theories of Evolution.* Springfield, IL: Charles C. Thomas Publisher.

Black, Craig C. & Mary R. Dawson (ed.) (1989). *Papers on Fossil Rodents, In Honor of Albert Elmer Wood.* # 33 Science Series, CA: Natural History Museum of Los Angeles County.

Boardman, Richard S., Alan H. Cheetham, & Albert J. Rowell (eds.) (1987). *Fossil Invertebrates.* Palo Alto, CA: Blackwell Scientific Publications.

Boberg, Dorothy Kurth (1970). *Legacy: The Land* (unpublished).

Bohm, David (1980). *Wholeness and the Implicate Order.* London: Routledge & Kegan Press.

Bohr, Niels (1958). *Atomic Physics and Human Knowledge.* NY: John Wiley & Sons.

____. (1963). *Essays 1958-1962 on Atomic Physics and Human Knowledge.* NY: Interscience Publishers, A Division of John Wiley & Sons.

Bondi, Herman (1961). *Cosmology.* Cambridge: Cambridge University Press. (1st ed. 1952).

Boole, George (1952). *An Investigation of the Laws of Thought.* Cambridge: Macmillan & Co. (1st ed. 1854).

____. (1952). *Collected Logical Works, Laws of Thought*. LaSalle, IL: The Open Court Publishing Co.

Bowler, Peter J. (1988). *The Non-Darwinian Revolution.* Baltimore, MD: The Johns Hopkins University Press.

Brace, C. Loring (1979). *The Stages of Human Evolution; Human and Cultural Origins*. NJ: Prentice-Hall, Inc.

Brasier, M. D. (1980). *Microfossils.* London: George Allen and Unwin.

Briffault, R. (1959). *The Mothers: A Study of the Origins of Sentiments and Institutions.* Abridged: *The Mothers.* (Gordon Rattray Taylor). London: George Allen and Unwin Ltd. (1st ed. 1927, 3 vols.)

British Museum of Natural History, (1979). *Dinosaurs and Their Living Relatives.* London: British Museum.

____. (1980). *Man's Place in Evolution.* London: British Museum.

Broughey, Arthur (1973). *Ecology of Populations.* NY: Macmillan Publishing Co., Inc. p 167. (2nd. ed.).

Brown, Michael H. (1990). *The Search for Eve.* NY: Harper and Row

Buffetaut, Eric (1987). *A Short History of Vertebrate Palaeontology.* London: Croom Helm.

Burt, William H., Richard P. Grossenheider, 1976. *Mammals.* Peterson Field Guides. MA: Houghton Mifflin Co.

Butler, Samuel (1878). *Life and Habit.*

Cairns-Smith, A. G. (1985). *Seven Clues to the Origin of Life: A Scientific Detective Story.* Cambridge: Cambridge University Press.

Campbell, Allan M. (1969). *Episomes.* NY: Harper and Row.

____. (December, 1976:102). How viruses insert their DNA into the DNA of the host cell. *Scientific American.*

____. (1983). Transposons and their evolutionary significance. *Evolution of Genes and Proteins.* Masalosshi Nei and Richard K. Koehn (ed). MS: Sinauer Associates, Inc., Publishers.

Campbell, Bernard G. (1976). *Humankind Emerging.* Boston: Little, Brown and Co.

Campbell, Joseph (1988). *The Power of the Myth.* NY: Doubleday.

Cann, Rebecca L., Mark Stoneking, Allan C. Wilson (1987). Mitochondria DNA and human evolution. *Nature* 1-1-87: 31.

Cannon, H. Graham, FRS (1959). *Lamarck and Modern Genetics.* Springfield, IL: Charles C. Thomas, Publisher.

Capra, Fritjof (1982). *The Turning Point; Science, Society and the Rising Culture.* NY: Simon & Schuster.

____. (1983). *The Tao of Physics.* NY: Bantam Books.

Carmody, Denise Lardner (1981). *The Oldest God; Archaic Religion Yesterday and Today.* TN: Abingdon/Nashville.

____. (1989) *Women and World Religions.* NJ: Prentice Hall. (1st edition 1979).

Carroll, Robert L. (1977). Patterns of amphibian evolution; an extended example of the incompleteness of the fossil record. *Patterns of Evolution as Illustrated by the Fossil Record.* NY: Elservier Scientific Pub. Co. 405-438.

____. (1988). *Vertebrate Paleontology and Evolution.* NY: W. H. Freeman and Co.

Casson, Stanley (1940). *The Discovery of Man; The Story of the Inquiry into Human Origins.* London: Star and Gazette Ltd.

Charig, Alan (1979). *A New Look at the Dinosaurs.* NY: Mayflower Books.

Cheng, Thomas (1970). *Symbiosis.* NY: Pegasus.

Christ, Carol and Judith Plaskow (1979). *Womenspirit Rising; A Feminist Reader in Religion.* SF: Harper and Row.

Cifelli, Richard (1985). South American ungulate evolution and extinction. *The Great American Interchange.* Francis Stelli & David S. Webb (ed.) NY: Plenum Press. pp 249-266.

Ciochon, Russell L. & John G. Fleagle (eds.) (1987). *Primate Evolution and Human Origins.* NY: Aldine De Gruyter.

Civetta, L., P. Gasparinis, G. Luongo, A. Papolla (ed.) (1974). *Physical Volcanology.* Developments in Solid Earth Geophysics 6. NY: Elsevier Scientific Publishing Co.

Clark, Ronald W. (1984). *The Survival of Charles Darwin.* NY: Random House.

Clark, W. E. Le Gros (1965). *History of the Primates; An Introduction to the Study of Fossil Man.* IL: University of Chicago Press.

Clarkson, E. N. K. (1966). Schizochroal eyes and vision of some Silurian acastid trilobites. *Palaeontology* 9: 1-29.

____. (1986). *Invertebrate Palaeontology and Evolution.* London: Allen & Unwin. (1st ed., 1979).

Cohen, I. L. (1984). *Darwin Was Wrong: A Study in Probabilities.* NY: New Research Publications, Inc.

Cohen, Robert S. & Marx W. Wartofsky (ed.) (1984). *Hegel and the Sciences.* Dordrecht/Boston/Lancaster: D. Reidel Publishing Co.

Colbert, Edwin H. (March, 1949:42). The ancestors of Mammals. *Scientific American.*

____. (1965). *The Age of Reptiles.* NY: W. W. Norton & Co., Inc.

____. (1968). *Dinosaurs, Their Discovery and Their World.* NY: E. P. Dutton & Co., Inc.

____. (1973). *Wandering Lands and Animals, The Story of Continental Drift and Animal Populations.* NY: Dover.

____. (1980). *Evolution of the Vertebrates, A History of the Backboned Animals Through Time.* NY: John Wiley and Sons.

____. (1983). *Dinosaurs, An Illustrated History.* NJ: A Dembner Book, Hammond Incorporated.

Conroy, Glenn C. (ed). (1976). Primate postcranial remains from the Oligocene of Egypt. *Contributions to Primatology,* Vol. 8. Ed: F. S. Szalay. NY: S.Karger.

Cope, E. D. (1884). The vertebrata of the Tertiary formations of the West. Report, *U. S. Geological Survey Territ.* Vol 3: 1044.

Cowen, Richard (1976). *History of Life*. NY: Mcgraw-Hill Book Co., Inc.
Cowen, Robert C. (1988). Regulating biotech: How much is enough? *The Christian Science Monitor* 6-3-1988, pp. 16-17.
Cowles, H. C. (1904). The work of the year 1903 in ecology. *Science*, 19: 879-885.
Cox, Allen (1969). Geomagnetic reversal. *Science* 163: 237-245.
Cuvier, Georges (1969). *The Animal Kingdom*. NY: Kraus Reprint.
Darrah, William C. (1960). *Principles of Paleobotany*. NY: The Ronald Press Co.
Darwin, Charles (1872). *The Origin of Species*. NY: The Modern Library. (6th. ed.)
____. (1885). *The Origin of Species*. NY: D. Appleton & Co. (6th. ed.)
____. (1964). *The Origin of Species*. Cambridge, MA: Harvard University Press. (A facsimile of the 1st. ed., 1859).
____. (1971). *The Origin of Species*. London: J. M. Dent and Sons, Ltd.
____. (1979). *The Origin of Species*. London: Murray. (6th. ed. republished from 1872 ed.)
Davidson, Mark (1983). *Uncommon Sense, The Life and Thought of Ludwig von Bertalanffy (1901-1972), Father of General Systems Theory*. Los Angeles, CA: J. P. Tarcher, Inc.
Dawkins, Richard (1986). *The Blind Watchmaker*. Harlow: Longman Scientific and Technical.
Delbruck, Max (1986). *Mind from Matter?* Palo Alto, CA: Blackwell Scientific Publications.
Delson, Eric (ed.) (1985). *Ancestors; the Hard Evidence*. NY: A. R. Liss
Denton, Michael (1986). *Evolution: A Theory in Crisis*. NY: Adler & Adler.
Diamond, J. M. (1978). Niche shifts and the rediscovery of competition. *American Scientist* 66: 322-331.
Dobzhansky, Theodosius, and Carl Epling (1944). Contributions to the genetics, taxonomy, and ecology of Drosophila pseudoobscura and its relatives. *Carnegie Institution of Washington Publication*. p 554. MD: The Lord Baltimore Press.
Dobzhansky, Theodosius (1964). *Heredity and the Nature of Man*. NY: Harcourt, Brace & World.
Dobzhansky, Theodosius, Francisco J. Ayala, Ledyard G. Stebbins, James W. Valentine (1977). *Evolution*. San Francisco, CA: W.H. Freeman & Co.
Dobzhansky, Theodosius and Ernest Boesiger (1983). *Human Culture: A Moment in Evolution*. NY: Columbia University Press. (edited and completed by Bruce Wallace).
Donnelly, Thomas W. (1985). Mesozoic and Cenozoic plate evolution of the Caribbean region. *The Great American Interchange*. (ed. Francis Stelli & S. David Webb) NY: Plenum Press. pp 89-121.
Dowdeswell, W. H. (1984). *Evolution: A Modern Synthesis*. London: Heinemann Educational Books.
Downs, Theodore (1968) *Fossil Vertebrates of Southern California*. Berkeley:

University of California Press.
Eddington, Arthur (1962). *The Expanding Universe*. MI: University of Michigan Press.
Edey, Maitland A. & Donald C. Johanson (1989). *Blueprints: Solving the Mystery of Evolution*. Boston: Little Brown and Co. Copyright 1989 by Maitland A. Edey. By permission of Little, Brown and Company.
Ehrlich, Paul P. & Richard W. Holm (1963). *The Process of Evolution*. NY: McGraw Hill Book Co., Inc.
Ehrlich, Paul P. & Peter Raven (1965). Butterflies and plants: A study in coevolution. *Evolution*, 18: 586-608.
Einstein, Albert (1961). *Relativity; The Special and General Theory*. NY: Crown Publishers. pp 25-26.
Eldridge, Niles (1977). Trilobites and evolutionary patterns. *Patterns of Evolution As Illustrated by the Fossil Record*. NY: Elservier Scientific Pub. Co. pp 305-332.
____. (1982). *The Monkey Business: A Scientist Looks at Creationism*. NY: Washington Square Press.
____. (1985a). *Time Frames: The Rethinking of Darwinian Evolution and the Theory of Punctuated Equilibria*. NY: Simon and Schuster.
____. (1985b). *Unfinished Synthesis: Biological Hierarchies and Modern Evolutionary Thought*. NY: Oxford University Press.
____. (1987). *Life Pulse*. NY: Facts on File Publications.
Eldridge, Niles & Ian Tattersall (1982). *The Myths of Human Evolution*. NY: Columbia University Press.
Elton, C. (1927). *Animal Ecology*. London: Sedgwick & Jackson.
Emmel, Thomas E. (1975). *Butterflies, Their World, Their Life Cycle, Their Behavior*. NY: Knopf.
Ephrussi, Boris (1953). *Nucleo-Cytoplasmic Relations in Microorganisms, Their Bearing in Cell Heredity and Differentiation*. Oxford: Clarendon Press.
Eron, Carol (1981). *The Virus that Ate Cannibals*. NY: Macmillan Publishing Co., Inc.
Evans, J. D. G. (1977). *Aristotle's Concept of Dialectic*. NY: Cambridge University Press.
Fenton, Carroll Lane (1966). *Tales Told By Fossils*. NY: Doubleday and Co. Inc.
Festinger, Leon (1983). *The Human Legacy*. NY: Columbia University Press.
Fields, Bernard, Malcolm A. Martin, & Daphne Kamely (ed.) (1985). *Genetically Altered Viruses and the Environment*. NY: Cold Springs Harbor Laboratory.
Fisher, Ernest P. and Carol Lipson (1988). *Thinking About Science: Max Delbruck and the Origins of Molecular Biology*. NY: W. W. Norton and Company.
Fleagle, J. G. & R. F. Kay (1987). New interpretations of the phyletic position of Oligocene hominoids. *Primate Evolution and Human Origins*, (ed. Russell L. Ciochon and John G. Fleagle). NY: Aldine De Gruyter. pp 157-172.

Florkin, Marcel (1944). *Biochemical Evolution*. NY: Academic Press

Forbes, Stephen Alfred (1887). The lake as a microcosm. *Bulletin, Science Association of Peoria, Illinois.* pp 77-87.

____. (1920). *The Fishes of Illinois.* IL: Illinois State Journal Co., State Printer.

Foster, David Blythe (1975). *The Intelligent Universe; A Cybernetic Philosophy.* NY: G. P. Putnam's Sons.

Fox, Sidney (1988). *The Emergence of Life: Darwinian Evolution from the Inside.* NY: Basic Books Inc., Publisher.

Fraenkel-Conrat, Heintz (1976). *Regulation and Genetics: Bacterial DNA Viruses.* Vol. 8. NY: Plenum Press.

Frakes, L. A. (1979). *Climates Throughout Geologic Time.* NY: Elsevier Scientific Publishing Co.

Fraser, Dean (1967). *Viruses and Molecular Biology.* NY: The Macmillan Co.

Friedl, Ernestine (1975). *Women and Men: An Anthropologist's View.* NY: Holt, Rinehart and Winston.

Gadon, Elinor W. (1989). *The Once and Future Goddess: A Symbol for our Time.* S.F.: Harper and Row, Pub., Inc.

Galinat, Walton C. (1971). The origins of maize. *Annual Review of Genetics* Vol. 5. (ed. Hershel L. Roman, Laurence M. Sandler, and Allan Campbell). Palo Alto, CA: Annual Reviews Inc. pp 447-478.

Gallagher, Idella J. (1970). *Morality in Evolution, The Moral Philosophy of Henri Bergson.* The Hague, Neatherlands: Martinus Myhoff.

Gamow, George (1952). *The Creation of the Universe.* NY: Viking Press.

Gardner, A. L., and J. L. Patton (1976). Karyotypic variation in oryzomyine rodents (Cricetinae) with comments on chromosomal evolution in the neotropical cricetine complex. *Occasional Papers, Museum Zoology.* LA: Louisiana State University. 49: 1-48.

Gardner, Eldon John (1984). *Principles of Genetics.* NY: John Wiley.

Gilson, Etienne (1984). *From Aristotle to Darwin and Back Again; A Journey in Final Causality, Species, and Evolution.* (tr. John Lyon) France: University of Notre Dame Press.

Gimbutas, Marija (1976). *The Goddesses and Gods of Old Europe: 7000-3500 BC: Myths, Legends and Cult Images.* Berkeley: University of California Press.

____. (1989). *The Language of the Goddess.* SF: Harper and Row.

Gingerich, Philip D. (1977). Patterns of evolution in the mammalian fossil record. *Patterns of Evolution as Illustrated by the Fossil Record.* (ed. A. A. Hallam.) NY: Elservier Scientific Pub. Co. pp 469-500.

____. (1985). South American mammals in the Paleocene of North America. *The Great American Interchange.* (ed. Francis Stelli & S. David Webb). NY: Plenum Press. pp 123-137.

Goodrich, L. Carrington (1963). *A Short History of the Chinese People.* NY: Harper and Row. (3rd ed., 1959).

Gose, Wulf A. (1985). Caribbean tectonics from a paleomagnetic perspective. In: *The Great American Interchange* (ed. Francis Stelli & S. David

Webb). NY: Plenum Press. pp 285-301.
Gould, Stephen Jay (1977). Eternal metaphors of palaeontology. In: *Patterns of Evolution as Illustrated by the Fossil Record*. Developments in Palaeontology and Stratigraphy, Vol. 5. (ed. A. A. Hallam) NY: Elservier Scientific Pub. Co. pp 1-26.
____. (1980a). *The Panda's Thumb*. NY: W. W. Norton & Co.
____. (1980b). Is a new and general theory of evolution emerging? *Paleobiology* 6: 119-130.
____. (1983). *Hen's Teeth and Horse's Toes*. NY: W. W. Norton & Co.
____. (1987). *Time's Arrow, Time's Cycle, Myth and Metaphor in the Discovery of Geological Time*. Cambridge, MS: Harvard University Press.
____. (1989). *Wonderful Life, The Burgess Shale and the Nature of History*. NY: W. W. Norton and Co., Inc.
Grant, Verne (1985). *The Evolutionary Process: A Critical Review of Evolutionary Theory*. NY: Columbia University Press.
Grasse, P. P. (1977). *Evolution of Living Organisms*. NY: Academic Press.
Greenwood, Paul J., P. H. Harvey, & M. Slatkin (1985). *Evolution: Essays in Honor of John Maynard Smith*. NY: Cambridge University Press.
Gregory, J. T. (1951). Convergent evolution. *Evolution*. 5 (4): 345-354.
Gregory, William K. (1951). *Evolution Emerging; A Survey of Changing Patterns from Primeval Life to Man*. NY: Macmillan Co.
Gregory, William K. (1963). *Our Face from Fish to Man, A Portrait Gallery of Our Ancient Ancestors and Kinsfolk together with a Concise History of Our Best Features*. NY: Hafner Publishing Co. (1st. edition, 1929).
Gribbin, John (1985). *In Search of the Double Helix; Darwin, DNA, and Beyond*. England: Wildwood House Ltd., Gower House, Croft Rd. Aldershot, Hants GU11 3 HR.
Gurney, O. R. (1964). *The Hittites*. MD: Penguin Books. (1st ed. 1952).
Haack, Susan (1974). *Deviant Logic; Some Philosophical Issues*. NY: Cambridge University Press.
Haeckel, Ernst (1866). *Generelle Morphologie der Organismen*. Berlin: Reimer.
Haeckel, Ernst, J. A. Thomson, A. Weismann, et al (1909). *Evolution in Modern Thought*. NY: Boni & Liveright, Inc.
Haldane, J. B. S., (1966). *The Causes of Evolution*. Ithaca, NY: Cornell University Press.
Hallam, A. A. (ed.) (1973) *Atlas of Palaeobiologeography*. Amsterdam, London, NY: Elsevier Scientific Pub. Co.
____. (ed.) (1977). *Patterns of Evolution as Illustrated by the Fossil Record*. Developments in Paleontology and Stratigraphy, Vol. 5, NY: Elservier Scientific Publishing Co.
Halstead, L. B. (1968). *The Pattern of Vertebrate Evolution*. CA: W. H. Freeman & Co.
____. (1982). *The Search for the Past; Fossils, Rocks, Tracks, and Trails; The Search for the Origin of Life*. NY: Doubleday and Co. Inc.
Hamilton, Edith (1961). *The Greek Way to Western Civilization*. NY: Mentor

Book, W. W. Norton & Co., Inc.

Harlan,Jack R. (1976). The races of sorghum in Africa. *Origins of African Plant Domestication.* Paris: Mouton Publishers. pp 465-478.

Harlan,Jack R., Jan M. J. De Wet, Ann B. L. Stemler (ed.) (1976). *Origins of African Plant Domestication. World Anthropology.* Paris: Mouton Publishers. pp 465-478.

Harris, Errol E. (1983). *An Interpretation of the Logic of Hegel.* Lanham: University Press of America.

Harris, Marvin (1977). *Cannibals and Kings; The Origins of Cultures.* NY: Random House.

Harris, Marvin and Eric B. Ross (ed.) 1987. *Food and Evolution: Toward a Theory of Human Food Habits.* PA: Temple University Press.

Hartman, C. G. (1952). *Opossums.* TX: University of Texas Press.

Hawksworth, D. L. (1984). *The Lichen-Farming Fungi.* Glasco: Blackie.

Hegel, Georg (1963). *The Logic of Hegel.* Oxford: Oxford University Press. (tr. William Wallace, from *Encyclopedia of the Philosophic Sciences,* from 2nd ed. of 1892, rev. and augmented.)

Hegel, Georg (1975). *Logic.* (tr. W. Wallace). Oxford: Clarendon Press.

Heilman, G., (1926). *The Origin of Birds.* London: Appleton-Century Crofts, Inc.

Heisenberg, Werner (1958). *Physics and Philosophy; The Revolution in Modern Science.* NY: Harper & Row, Publishers.

____. (1973). *Across the Frontiers.* (tr. from German, Peter Heath) NY: Harper & Row, Publishers.

____. (1977). *The Uncertainty Principle and the Foundations of Quantum Mechanics, A Fifty Year Survey.* NY: Wiley.

____. (1979). *Philosophical Problems of Quantum Physics.* CT: Ox Bow Press.

____. (1983). *Tradition in Science.* (tr. Peter Heath). NY: Seabury Press. 116-117.

Hoffstetter, R. (1987). Relationships, origins, and history of the ceboid monkeys and caviomorph rodents: A modern interpretation. *Primate Evolution and Human Origins.* (ed. Ciochon, Russell L. & John G. Fleagle) NY: Aldine De Gruyter. pp 101-109.

Holland, W. J. (1916). *The Butterfly Book; A Popular Guide to Knowledge of Butterflies of North America.* NY: Doubleday, Page and Co.

Holloway, Ralph L. (July, 1974: 109-110). The casts of fossil hominid brains. *Scientific American.*

Hood, Sinclair (1967). *The Home of the Heroes; The Aegeans Before the Greeks.* NY: McGraw-Hill.

Hopkins, Colin R. (1978). *Structure and Function of Cells.* London; Philadelphia: Saunders.

Hotton, Nicholas III, et al, eds. (1986). *The Ecology and Biology of Mammal-like Reptiles.* Washington, DC: Smithsonian Institution Press.

House, M. R. (ed.) (1979). *The Origin of Major Invertebrate Groups.* Published for the Systematic Association, NY: Academic Press.

Howard, George Elliott (1964). *A History of Matrimonial Institutions.*

Vol I (of 3). NY: Humanities Press. (1st ed. 1904).
Howells, William W. (1967). *Mankind in the Making.* NY: Doubleday & Co.
____. (1982). The distribution of man. *Evolution and the Fossil Record. Readings from Scientific American.* (ed. Leo F. Laporte). San Francisco, CA: W. F. Freeman & Co.
Hoyle, Fred (1962). *Astronomy.* NY: Crescent Books.
Hubble, Edwin (1982). *The Realm of the Nebulae.* New Haven CT: Yale University Press.
Hughes, Sally Smith (1977). *The Virus: A History of the Concept.* NY: Heinemann Educational Books.
Hull, David L. (1988). *Science as a Process; An Evolutionary Account of the Social and Conceptual Development of Science.* IL: The University of Chicago Press.
Hutchinson, H. N., J. W. Gregory, R. Lydekker (ed). *The Living Races of Mankind.* Vol I. London: Hutchinson and Co.
Huxley, Julian (1943). *Evolution, the Modern Synthesis.* NY: Harper and Brothers.
Hyman, L. H. (1940). *The Invertebrates.* NY: McGraw-Hill Book Co.
Jacob, Francois (1974). *The Logic of Life: A History of Heredity.* NY: Pantheon Books. (tr. Betty E. Spillman).
Jantsch, Erich (1980). *The Self-organizing Universe: Scientific and Human Implications of the Emerging Paradigm of Evolution.* NY: Pergamon Press.
Jarvik, Erik (1980). *Basic Structure and Evolution of Vertebrates.* 2 Vol., V 2. NY: Academic Press.
Jeans, James (1958). *Physics and Philosophy.* MI: Ann Arbor Paperbacks, University of Michigan Press.
Jefferies, R. P. S. (1967). Some fossil chordates with echinoderm affinities. *Symposium, London Zoological Society.* 20: 163-208.
____. (1975). Fossil evidence concerning the origin of the chordates. *Symposium,* London Zoological Society. 36: 253-318.
____. (1979). On origin of the chordates — a methodological essay. *The Origin of Major Invertebrate Groups,* (ed. M. R. House) NY: Academic Press. pp 443-477.
____. (1986). *The Ancestry of the Vertebrates.* London: British Museum (Natural History).
Johanson, Donald and Maitland Edey (1981). *Lucy: The Beginnings of Humankind.* NY: Warner Books.
John, Brian S. (1977). *The Ice Age; Past and Present.* London: William Collins Sons and Co. Ltd.
Johnson, Leland G. (1983). *Biology.* IA: Wm. C. Brown Co., Publishers.
Joseph, Lawrence E. (1990). *Gaia: The Growth of an Idea.* NY: St. Martin's Press.
Keast, Allen, Frank C. Erk and Bentley Glass, eds. (1972). *Evolution, Mammals and the Southern Continents.* Albany, NY: State University Press.

Keller, Evelyn Fox (1983). *A Feeling for the Organism; The Life and Work of Barbara McClintock*. SF: W. H. Freeman and Co.

Kemp, T. S. (1982). *Mammal-like Reptiles and the Origin of Mammals.* NY: Academic Press.

Kenyon, Kathleen Mary (1957). *Digging Up Jericho.* NY: Praeger.

Kerr, Richard (1987). A better fit for the plate tectonic puzzle. *Science,* 238: 30.

Kitching, R. L. (1983). *Systems Ecology: An Introduction to Ecological Modelling.* St. Lucia: University of Queensland Press.

Knight, Clifford B. (1965). *Basic Concepts of Ecology.* NY: Macmillan Co.

Koestler, Arthur (1978). *Janus: A Summing Up.* NY: Random House.

Konigsson, Lars-Konig (ed). (1980). *Current Argument on Early Man; Report from a Nobel Symposium.* Oxford: Pergamon Press.

Kopchick, John J. and Wen Y. Chen (July, 1990). *Proceedings of the National Academy of Sciences.* Vol 87, No. 13: 5061-5065.

Kramer, Samuel Noah and Editors of Time-Life Books. (1967). *Cradle of Civilization.* NY: Time Inc.

Kroeber, A. L. & T. T. Waterman (ed.) (1920). *Source Book in Anthropology.* Berkeley, University of California Press.

Kropotkin, Petr (1955). *Mutual Aid: A Factor of Evolution.* (1st ed. 1888). Boston: Extending Horizons Book.

Kuhn, Thomas S. (1962). *The Structure of Scientific Revolutions.* IL: University of Chicago Press.

Kummel, Bernhard (1961). *History of Earth; An Introduction to Historical Geology.* S. F.: W. H. Freeman and Co.

Kurtin, Bjorn (1971). *The Age of Mammals.* London: Weidenfeld and Nicolson.

Kurtin,Bjorn (1982). Continental drift and evolution. *Evolution and the Fossil Record. Readings from Scientific American.* (ed. Leo F. Laporte). San Francisco, CA: Freeman. pp 203-212.

Kurstak, Edouard, and Karl Maramorosch (ed.) (1974). *Viruses, Evolution, and Cancer; Basic Considerations.* NY: Academic Press.

Lamanna, Carl, M. Frank Mallette, and Leonard N. Zimmerman (1973). *Basic Bacteriology: Its Chemical Background.* Baltimore: Ballimore, Williams, and Wilkins.

Lamarck, J. B. (1914). *Zoological Philosophy.* (Reprinted 1963, from 1914 ed., 1st ed. 1829) NY: Hafner Publishing Co., Inc.

Lambert, David (1978). *Dinosaurs.* NY: Crown Publishers, Inc.

____. The Diagram Group (1983). *A Field Guide to Dinosaurs.* NY: Avon.

Lanham, Url (1973). *The Bone Hunters.* NY: Columbia University Press.

Laporte, Leo F. (ed.) (1982). The fossil record and evolution. *Evolution and the Fossil Record. Readings from Scientific American.* San Francisco, CA: W. A. Freeman.

Leake, Chauncey D. and Patrick Romanell (1950). *Can We Agree? A Scientist and Philosopher Argue about Ethics.* TX: University of Texas Press.

Leakey, Richard (1981). *The Making of Mankind.* NY: E. P. Dutton.

Leakey, Richard (1981). *The Illustrated Origin of Species by Charles Darwin.* NY:

Hill and Wang.
Ledoux, L. (ed.) (1972). *Uptake of Information Molecules by Living Cells.* Amsterdam: North-Holland Publishing Co. NY: American Elservier Publishing Co., Inc.
Lehrman, Robert L. (1961). *The Long Road to Man.* NY: Basic Books, Inc.
Leifson, Einar (1960). *Atlas of Bacterial Flagellation.* NY: Academic Press.
Lerner, Eric J. (1991). *The Big Bang Never Happened: A Startling Refutation of the Dominant Theory of the Origin of the Universe.* NY: Random House, Inc.
Lerner, Gerda (1986). *The Creation of Patriarchy.* NY: Oxford University Press.
Levi-Setti, Riccardo (1975). *Trilobites: A Photographic Atlas.* Chicago: University of Chicago Press.
Lewin, Roger (1987). My close cousin the chimpanzee. *Science,* 238: 273-275.
Lewis, Gilbert Newton (1966). *Valence and the Structure of Atoms and Molecules,* NY: Dover Publications, Inc. (first ed., 1923).
Lewis, Harlan and Margaret Ensigh Lewis (1955). *The Genus Clarkia.* University of California Publications in Botany. Berkeley: University of California Press. Vol 20, #4: 241-392.
Lillegraven, Jason A., Zofia Kielan-Jaworowska, & William A. Clemens, eds., (1979). *Mesozoic Mammals; The First Two-Thirds of Mammalian History.* Berkeley, CA: University of California Press.
Lorenz, Konrad (1965). *Evolution and Modification of Behavior.* IL: University of Chicago Press.
Louis, David (1983). 2201 *Fascinating Facts.* NY: Greenwich House.
Lovelock, J. E. (1979). *Gaia; A New Look at Life on Earth.* Oxford: Oxford University Press.
____. 1992). Personal communication.
Lovtrup, Soren (1987). *Darwinism: The Refutation of a Myth.* Kent: Croom Helm Ltd.
Lull, R. S. and Wright, N. E. 1942. Hadrosaurian dinosaurs of North America. *Geological Society America, Special Papers* No. 40.
Lyell, Charles (1853). *Principles of Geology.* Boston: Little Brown. (9th ed.)
Maglio, Vincent J. (1973). *Origin and Evolution of the Elephantidae.* Transactions of the American Philosophical Society. New Series Vol. 63: 1-149.
Mangelsdorf, P. C. & R. G. Reeves (1938, 1939). The origin of maize. *Proceedings of the National Academy of Science.* 24: 303-312.
Manton, S. M. and D. T. Anderson (1979). Polyphyly and the evolution of arthropods. *The Origin of Major Invertebrate Groups.* Systematics Association Special Vol B, (ed. M. R. House) NY: Academic Press. pp 269-321.
Mares, Michael (1985). Mammal faunas of xeric habitats and the Great American Interchange. *The Great American Interchange.* (ed. Francis Stelli & S. David Webb). NY: Plenum Press. pp 489-520.
Margulis, Lynn (1970). *Origin of Eukaryotic Cells; Evidence and Research Implications for a Theory of the Origin and Evolution of Microbial Plant*

and Animal Cells on the Precambrian Earth. New Haven, CT: Yale University Press.

____. (1978). Microtubules in Prokaryotes. NY: *Science,* 200: 1118-1124.

____. (1981). *Symbiosis in Cell Evolution.* CA: W. H. Freeman.

____. (1993). *Symbiosis in Cell Evolution; Microbial Communities in the Archean and Proterozoic Eons.* NY: W. H. Freeman and Company.

Margulis, Lynn and Rene Fester (ed.)(1991). *Symbiosis as a Source of Evolutionary Innovation: Speciation and Morphogenesis.* MA: The MIT Press.

Margulis, Lynn and Lorraine Olendzenski (ed.) (1992). *Environmental Evolution: Effects of the Origin and Evolution of Life on Planet Earth.* Cambridge MA: The MIT Press.

Margulis, Lynn and Dorion Sagan (1986a). *Microcosmos; Four Billion Years of Evolution from Our Microbial Ancestors.* Summit Books, Division of Simon & Schuster, Inc.

____. (1986b). *Origins of Sex; Three Billion Years of Genetic Recombination.* New Haven and London: Yale University Press.

____. (1991). *Mystery Dance; On the Evolution of Human Sexuality.* NY: Summit Books.

Maringer, Johannes (1960). *The Gods of Prehistoric Man.* NY: Alfred A. Knopf, Inc. (ed. and tr. from German, Mary Ilford).

Marsh, George (1965). *Man and Nature.* (1st ed. 1864) NY: Belknap Press of Harvard University Press.

Matthews, William E. (1909). *The Carnivora and Insectivora of the Bridger Basin, Mid Eocene.* MA: E. W. Wheeler, Printer.

Matthews, William H. III (1962). *Fossils; An Introduction to Prehistoric Life.* NY: Barnes and Noble, Inc.

Mayo, M. A. and K. A. Harrap (ed). (1984). *Vectors in Virus Biology.* Society for General Microbiology. NY: Academic Press.

Mayo, Oliver (1983). *Natural Selection and Its Constraints.* NY: Academic Press.

Mayr, Ernst, & W. B. Provine (1980). *The Evolutionary Synthesis: Perspectives on the Unification of Biology.* Cambridge: Harvard University Press.

McClintock, Barbara, Takeo Angel Kato, Almiro Blumenschein (1981). *Chromosome Constitution of Races of Maize.* Mexico: Colegio de Postgraduados.

McClintock, Barbara (1984). The significance of the responses of the genome to challenge. *Science* 226: 800.

____. (1987). The discovery and characterization of transposable elements. *Genes, Cells and Organisms.* (The Collected Papers of Barbara McClintock, ed. John A. Moore). NY: Garland Publishing. Inc.

McGinnis, Helen J. (1982). *Carnegie's Dinosaurs.* PA: Board of Trustees, Carnegie Institute.

McIntosh, Robert P. (1985). *The Background of Ecology; Concept and Theory.* Cambridge, MA: Cambridge University Press.

McKenna, Malcolm C., Chow Minchen, Ting Suyin, and Luo Zhexi. (1989).

Radinskya yupingae, a perissodactyl-like mammal from the late Paleocene of southern China. *Oxford Monographs on Geology and Geophysics # 15*. (ed. Prothero, Donald and Robert M. Schoch). NY: Clarendon Press. pp 24-36.

McLoughlin, John C. (1980). *Synapsida; A New Look into the Origins of Mammals.* NY: The Viking Press.

Mellaart, James (1967). *Catal Huyuk: A Neolithic Town in Anatolia.* NY: McGraw-Hill.

Mellett, James S. (1977). *Paleobiology of North American Hyaenodon (Mammalia Creodonta)*. Basel, Switzerland: S. Karger.

Mendel, Gregor (1965). *Experiments in Plant Hybridisation.* Edinburgh: Oliver & Boyd. (tr. Ronald A. Fisher, 1866 edition).

Messel, H. (ed.) (1981). *The Biological Manipulation of Life.* Sydney: Pergamon Press.

Miller, Malcolm (1980). *Anatomy of the Dog.* PA: Saunders.

Minelli, Guiseppe (1985). *The Evolution of Life.* NY: Facts on File Publications.

Monastersky, R. (1986). Reining in a Runaway Theory. *Science News,* 130: 374.

Montagu, Ashley (1952). *Darwin, Competition and Cooperation.* NY: H. Schuman.

Moore, Patrick (1969). *The Development of Astronomical Thought.* Edinburgh: Oliver & Boyd.

Moore, Raymond E., Cecil G. Lalicker, & Alfred G. Fisher (1952). *Invertebrate Fossils.* NY: McGraw-Hill Book Co., Inc.

Morgan, Elaine (1972). *The Descent of Woman.* NY: Stern & Day.

____. (1982). *The Aquatic Ape.* NY: Stein and Day.

Munn, Norman (1971). *The Evolution of the Human Mind.* Boston, MA: Houghton Mifflin Co.

Newell, Norman D. (1982). Crises in the history of life. *Evolution and the Fossil Record. Readings from Scientific American.* (ed. Leo F. Laporte). San Francisco, CA: Freeman.

Nicholson, A. J. (1933). The balance of animal populations. *Journal of Animal Ecology,* 2: 132-178.

Nicholson, Max (1970). *The Environmental Revolution: A Guide for the New Masters of the World.* NY: McGraw Hill.

____. (1973). *The Big Change: After the Environmental Revolution.* NY: McGraw Hill.

Niklas, Karl J., ed. (1981). *Paleobotany, Paleoecology, and Evolution.* NY: Praeger Publishers.

NSF Mosaic Reader (1983). *Human Evolution.* NJ: Avery Publishing Group, Inc.

Oakley, K. D. (1964). *Man the Toolmaker.* Chicago, IL: University of Chicago Press.

Odum, Eugene P. (1963). *Ecology.* NY: Holt, Rinehart & Winston.

Olsen, Stanley J. (1985). *Origins of the Domestic Dog; The Fossil Record.* AZ: The University of Arizona Press. Copyright 1985.

Ornstein, Robert, & Paul Ehrlich (1989). *New World New Mind: Moving toward Conscious Evolution.* NY: Doubleday.

Osborn, Henry Fairfield (1910). *The Age of Mammals in Europe, Asia and North America.* NY: The Macmillan Co.

____. (1942). *Proboscidea; A Monograph of the Discovery, Evolution, Migration, and Extinction of the Mastodonts and Elephants of the World.* Vol 1, 2. Trustees of American Museum of Natural History. NY: American Museum Press.

Pagels, Heinz R. (1982). *The Cosmic Code: Quantum Physics as the Language of Nature.* NY: Bantam Books.

Pascual, Rosendo, Maria G. Vucetich, Gustavo J. Scillato-Yane, and Mariano Bond (1985). Main pathways of mammalian diversification in South America. *The Great American Interchange.* (ed. Francis Stelli & S. David Webb). NY: Plenum Press. pp 219-247.

Patai, Raphael (1967). *The Hebrew Goddess.* KTAV Publishing House, Inc.

Patten, Bernard C. (ed.) (1976). *Systems Analysis and Simulation in Ecology.* NY: Academic Press.

Patten, William (1920). *The Grand Strategy of Evolution; The Social Philosophy of a Biologist.* Boston MA: R. G. Badger.

Pauling, Linus (1967). *The Chemical Bond: a Brief Introduction to Modern Structural Chemistry.* Ithaca, NY: Cornell University Press.

Pauling, Linus, and Peter Pauling (1975). *Chemistry.* San Francisco, CA: W. H. Freeman. p 443.

Perelman, Lewis J. (1976). *The Global Mind: Beyond the Limits to Growth.* NY: Mason/Charter.

Pessagno, Emile A. Jr., Charles D. Blome, and Jose F. Gongoria (1984). A Revised radiolarian zonation for the Upper Jurassic of western North America. *Bulletins of American Paleontology.* December 14, 1984. Vol. 87, No. 320.

Peters, David (1986). *Giants of Land, Sea and Air; Past and Present.* Sierra Club Books. NY: Alfred A. Knopf.

Pfeiffer, John (1972). *The Cell.*NY: Time-Life Books.

Pilbeam, David (1970). *The Evolution of Man.* First American edition published by arrangement with Thames and Hudson International Ltd, London. NY: Funk and Wagnalls.

____. (1972). *The Ascent of Man; An Introduction to Human Evolution.* NY: The Macmillan Co.

____. (1987). Rethinking human origins. *Primate Evolution and Human Origins.* (ed. Russell L. Ciochon and John G. Fleagle) NY: Aldine de Gruyter.

Planck, Max (1932). *Where is Science Going?* NY: W. W. Norton & Co.

____. (1936). *The Philosophy of Physics.* NY: W. W. Norton & Co.

____. (1960). *A Survey of Physical Theory.* NY: Dover Publisher

Ponten, Jan (1971). *Spontaneous and Virus Induced Transformation in Cell Structure.* Virology Monography Vol. 8. NY: Springer-Verlag

Prothero, Donald and Robert M. Schoch, (eds) (1989). *The Evolution of the Perissodactyls.* Oxford Monographs on Geology and Geophysics # 15. NY: Clarendon Press.

Protsch, Reiner (1975). The absolute dating of Upper Pleistocene sub-Saharan fossil hominids and their place in human evolution. *Journal of Human Evolution* 4: 297-322.

Radetsky, Peter (1991). *The Invisible Invaders; The Story of the Emerging Age of Viruses.* MA: Little Brown and Co.

Random House Dictionary of the English Language (1987). Second Edition, Unabridged. Copyright © 1987 by Random House, Inc. Reprinted by permission.

Ratkevich, Ronald Paul (1976), *Dinosaurs of the Southwest.* Albuquerque, NM: University of New Mexico Press.

Raup, David (1977). Stochastic models in evolutionary palaeontology. *Patterns of Evolution as Illustrated by the Fossil Record.* (ed. A. A. Hallam). Developments in Paleontology and Stratigraphy, Vol. 5., NY: Elservier Scientific Publishing Co. pp 59-78.

Reed, Charles A. (ed.) (1977). *Orgins of Agriculture.* Paris: Mouton Publishers.

Ridley, Mark (1985). *The Problems of Evolution.* Oxford: Oxford University Press.

Roberts, Leslie (1987) New sequence to take on the genome. *Science* 238: 271-172.

Rockham, Harris (1944). Introduction. *Aristotle's Ethics for English Readers.* Oxford: Basil Blackwell.

Roff, Rudolf A. & Thomas C. Kaufman (1983). *Embryos, Genes and Evolution.* NY: Macmillan Publishing Co., Inc.

Romer, A. S. (1966). *Vertebrate Paleontology.* (3rd ed.) (1st ed. 1933). Il: University of Chicago Press.

____. (1968). *Notes and Comments on Vertebrate Paleontology.* IL: The University of Chicago Press.

Rose, K. D., & J. G. Fleagle (1987). The third radiation - higher primates. *Primate Evolution and Human Origins.* (ed. Russell L. Ciochon and John G. Fleagle) NY: Aldine De Gruyter. pp 124-132.

Rosensweig, M. L. (1977). Coexistence and diversity in heteromyid rodents. *Evolution and Ecology.* (ed. Bernard Stonehouse and Christopher Perrins) Baltimore, MD: University Park Press.

Rosensweig, M. L. (1992). Personal communication.

Ross, Herbert H. (1966). *Understanding Evolution.* Englewood Cliffs, NJ: Prentice-Hall.

Roth, Leon (1937). *Descartes' Discourse on Method.* Oxford: Clarendon Press.

Runcorn, S. K. and P. A. Davies (ed.) (1980). *Mechanisms of Continental Drift and Plate Tectonics.* NY: Academic Press.

Ruse, Michael (1986). *Taking Darwin Seriously; A Naturalistic Approach to Philosophy.* NY: Basil Blackwell, Inc.

Russell, D. A. (1979). The enigma of the extinction of the dinosaurs. *Annual*

Review of Earth and Planetary Science. Vol. 7: 163-182.

Russell, W. M. S. (1969). *Man, Nature and History; Controlling the Environment.* NY: The American Natural History Press.

Sachs, Mendel (1988). *Einstein Versus Bohr; The Continuing Controversies in Physics.* IL: Open Court Publishing Co.

Sagan, Elli (1974). *Cannibalism; Human Aggression and Cultural Form.* NY: Harper and Row, Publishers.

Sager, Ruth and Francis J. Ryan (1967). *Cell Heredity.* NY: John Wiley & Son, Inc. (1st ed. 1961).

Sager, Ruth (1972). *Cytoplasmic Genes and Organelles.* NY: Academic Press.

Sahakian, William S. and Mabel Lewis Sahakian (1966). *Ideas of the Great Philosophers.* NY: Barnes & Noble.

Sapp, Jan (1987). *Beyond the Gene: Cytoplasmic Inheritance and the Struggle for Authority in Genetics.* NY: Oxford University Press.

Savin, Samuel M. and Robert G. Douglas (1985). Sea level, climate and the Central American bridge. *The Great American Interchange.* (ed. Francis Stelli & S. David Webb). NY: Plenum Press. pp 303-324.

Scagel, Robert F. et al (1966). *An Evolutionary Survey of the Plant Kingdom.* Belmont CA: Wadsworth Publishing Co.

Schieve, William C. & Peter M. Allen (ed.). (1982). *Self-Organization and Dissipative Structures; Applications in the Physical and Social Sciences.* Austin: University of Texas Press.

Schlosser, Max. See Zittel.

Schmalhausen, J. J. (1968). *The Origin of Terrestrial Vertebrates.* NY: Academic Press.

Schmucker, Samuel Christian (1913). *The Meaning of Evolution.* NY: The Chautauqua Press.

Schneider, Stephen H. and Randi Londer (1984). *The Coevolution of Climate and Life.* San Francisco, CA: Sierra Club Books.

Schopf, J. William (ed.) (1992). *Major Events in the History of Life.* Boston: Jones and Bartlett Publishers.

Schopf, Thomas J. M. (1977). Patterns of evolution: a summary and discussion. *Patterns of Evolution as Illustrated by the Fossil Record.* (ed. A. A. Hallam). Developments in Paleontology and Stratigraphy, Vol. 5., NY: Elservier Scientific Publishing Co. pp 547-562.

____. (1981). Evidence from findings of molecular biology with regard to the rapidity of genomic change; implications for species durations. *Paleobotany, Paleoecology, and Evolution.* Vol 1. NY: Praeger Publishers. (ed. Karl J. Niklas.) pp 135-192.

Schrodinger, Erwin (1967). *What is Life? The Physical Aspects of the Living Cell.* (and) *Mind and Matter.* Cambridge: Cambridge University Press.

Schultz, Gwen (1963). *Glaciers and the Ice Age; Earth and Its Inhabitants During the Pleistocene.* NY: Holt, Rinehart and Winston.

____. (1974). *Ice Age Lost.* NY: Anchor Press/Doubleday.

Scientific American, Readings From (1979). *Human Ancestors.* S F: W. H.

Freeman and Co.
Scientific American, Readings From (1980). *Birds.* SF: W. H. Freeman & Co.
Scott, Andrew (1985). *Pirates of the Cell: The Story of Viruses from Molecule to Microbe.* NY: Basil Blackwell.
Scott, W. B. & G. L. Jepsen (eds). (1940). The mammalian fauna of the White River Oligocene. Part IV, Artiodactyla. *Transactions of the Americal Philosophical Society.* New Ser. 28: 363-746.
Selsam, Millicent E. (1977). *Sea Monsters of Long Ago.* NY: Four Winds Press.
Shapiro, James A. (ed). (1983). *Mobile Genetic Elements.* NY: Academic Press.
Shapiro, Robert (1986). *Origins: A Skeptic's Guide to the Creation of Life on Earth.* NY: Academic Press.
Sharov, A. G. (1966). *Basic Arthropodan Stock with Special Reference to Insects.* Oxford: Pergamon Press.
Sharp, Robert P. (1988). *Living Ice: Understanding Glaciers and Glaciation.* Cambridge & NY: Cambridge University Press.
Sheldrake, Rupert (1988). *The Presence of the Past.* NY: Times Books.
Shields, John Potter (1967). *Projects in Basic Magnetism.* Photofact Publication 20556. IN: Howard W. Sams & Co., Inc.
Shimer, Hervey Woodburn & Robert Rakes Shrock (ed.). (1965) *Index Fossils of North America.* Cambridge : MIT Press.
Simons, Elwyn L. (1972). *Primate Evolution: An Introduction to Man's Place in Nature.* NY: The Macmillan Co.
____. (1987). New fossil apes from Egypt and the initial differentiation of Hominoidea. *Primate Evolution and Human Origins.* (ed. Russell L. Ciochon & John G. Fleagle). NY: Aldine De Gruyter. pp. 145-150.
Simpson, George Gaylord (1951). *Horses: The Story of the Horse Family in the Modern World and Through Sixty Million Years of History.* NY: Oxford University Press.
____. (1964). *This View of Life.* Harcourt, Brace, and World.
____. (1965). *Tempo and Mode in Evolution.* NY: Hofner Publishing Co.
____. (1967). *The Meaning of Evolution: A Study of the History of Life and Its Significance for Man.* New Haven: Yale University Press. (1st. edition, 1949).
____. (1978). *Concession to the Improbable: An Unconventional Autobiography.* New Haven: Yale University Press.
____. (1980). *Splendid Isolation: The Curious History of South American Mammals.* New Haven: Yale University Press.
____. (1983). *Fossils and the History of Life.* NY: Scientific American Library.
____. (1984). *Discoverers of the Lost World: An Account of Some of Those Who Brought Back to Life South American Mammals Long Buried in the Abyss of Time.* New Haven: Yale University Press.
Singleton, Paul and Diana Sainsbury (1981). *Introduction to Bacteria.* NY: John Wiley & Sons.
Sinnott, Edmund (1961). *Cell and Psyche: The Biology of Purpose.* NY: Harper and Row. p 15.

____. (1966). *The Bridge of Life; From Matter to Spirit.* NY: Simon & Schuster.

Slaatte, Howard A. (1968). *The Pertinence of the Paradox.* NY: Humanities Press.

Sleigh, M. A. (1979). Radiation of the eukaryote Protista. *The Origin of the Major Invertebrate Groups.* (ed. M. R. House). NY: Academic Press. pp 23-53.

Sloan, R. E. (1987). Paleocene and latest Cretaceous mammal ages, biozones, magnetozones, rates of sedimentation, and evolution. *Geological Society of America.* Special Paper 209: 165-200.

Smith, Douglas L. (1985). Caribbean plate relative motions. *The Great American Interchange.* (ed. Francis Stelli & S. David Webb). NY: Plenum Press. pp 17-46.

Smith, H. H. (et al, ed.) (1972) *Evolution of Genetic Systems.* NY: Gordon & Breach Scientific Publications, Inc.

Smith, Robert L. & Robert G. Leudke (1984). Potentially active volcanic lineaments and loci in western conterminous United States. *Explosive Volcanism: Inception, Evolution and Hazards.* Washington, D.C.: National Academy Press.

Society of General Microbiology (1963, # 13). *Symbiotic Associations.* London: Cambridge University Press.

____. (1974, # 24). *Evolution in the Microbial World.* London: Cambridge University Press.

Squires, Richard (1986). The first horse, a Mexican import. *Horse Country Gazette.* Vol 2 #1. pp 12-15.

Stanley, Steven M. (1977). Trends, roles, and patterns of evolution in the bivalvia. *Patterns of Evolution As Illustrated by the Fossil Record.* (ed. A. A. Hallam). Developments in Paleontology and Stratigraphy, Vol. 5., NY: Elservier Scientific Publishing Co. pp 209-250.

____. (1979). *Macroevolution: Pattern and Process.* S F: W. H. Freeman and Company.

____. (1981). *The New Evolutionary Timetable: Fossils, Genes, and the Origin of Species.* NY: Basic Books, Inc. Publishers.

Stearn, Colin W. and Robert L. Carroll (1987). *Paleontology: The Record of Life.* NY: John Wiley and Sons, Inc.

Stebbins, G. Ledyard (1982). *Darwin to DNA, Molecules to Humanity.* San Francisco, CA: W. H. Freeman.

Stein, Jess (ed.) (1966). *Random House Dictionary of the English Language.* NY: Random House.

Stelli, Francis, and S. David Webb (1985). A kaleidoscope of plates, faunal and floral dispersals, and sea level changes. *The Great American Interchange.* NY: Plenum Press.

Stephenson, June (1989). *Women's Roots.* CA: Diemer, Smith Publishing Co., Inc.

Stewart, Wilson N. (1983). *Paleobotany and the Evolution of Plants.* Cambridge: Cambridge University Press.

Stonehouse, Bernard, and Christopher Perrins (ed.). (1977). *Evolutionary Ecology*. Baltimore, MD: University Park Press.
Szalay, F. S. (ed.) (1975). *Approaches to Primate Paleobiology*. Contributions to Primatology. Vol. 5. Basel, Switzerland: S. Karger A. G.
Taylor, Gordon Rattray (1983). *The Great Evolution Mystery*. NY: Harper & Row.
Taylor, Thomas N. (1981). *Paleobotany: An Introduction to Plant Biology*. NY: McGraw-Hill Book Co.
Terzaghi, E. A., A. S. Wilkins, & D. Penny (ed.) (1984). *Molecular Evolution*. Boston: Jones & Bartlett Pub., Inc.
Thomas, Lewis (1974). *The Lives of a Cell: Notes of a Biology Watcher*. NY: Bantam Books.
Todd, Ian A. (1976). *Catal Huyuk in Perspective*. CA: Cummings Publishing Co.
Tyler, S. A. and E. S. Barghoorn (1954). Occurrence of structurally preserved plants in Pre-Cambrian rocks of the Canadian Shield. *Science* (119: 606-608)
Tylor, Edward Burnett (1958). *The Origins of Culture*. (1st published 1871, *Primitive Culture*) NY: Harper and Brothers Publishers.
Ulrich, E. O. & G. A. Cooper (1938). Ozarkian and Canadian Brachiopoda. *Geological Society of America*. Special Paper Vol. 13, pp 1-323.
Urey, Harold Clayton (1952). *The Planets, their Origin and Development*. New Haven: Yale University Press.
Urey, Harold Clayton, & Arthur Edward Ruark (1964) *Atoms, Molecules, and Quanta*. NY: Dover Publications.
Valentine, J. W. (1973). *Evolutionary Paleontology of the Marine Biosphere*. NJ: Prentice-Hall.
____. (1977). General patterns of metazoan evolution. *Patterns of Evolution as Illustrated by the Fossil Record*. (ed. A. A. Hallam). Developments in Paleontology and Stratigraphy, Vol. 5, NY: Elservier Scientific Publishing Co. pp 27-58.
Valentine, James W. and Eldridge M. Moores (1982). Plate tectonics and the history of life in the oceans. *Evolution and the Fossil Record. Readings from Scientific American*. (ed. Leo F. Laporte). San Francisco, CA: Freeman. pp 193-212.
Waddington, Conrad H. (1975). *The Evolution of an Evolutionist*. Ithaca, NY: Cornell University Press.
Walcott, C. D. (1911). Middle Cambrian Holothurian and Medusae. *Smithsonian Miscellaneous Collection*. 57: 42-68.
Wallace, Alfred Russell (1891). *Natural Selection and Tropical Natures, Essays on Descriptive and Theoretical Biology*. NY: McMillan.
Washburn, S. L. (1963). *Classification and Human Evolution*. Chicago: Aldine De Gruyter.
Watson, James D. (1968). *The Double Helix*. NY: A Mentor Book, New American Library, Times, Mirror.
Watt, Kenneth E. F. (ed.) (1966). *Systems Analysis in Ecology*. NY: Academic

Press.
Webb, S. David (1985) Main pathways of mammalian diversification in North America. pp 201-217. Late cenozoic mammal dispersals between the Americas. pp 357-386. *The Great American Interchange.* (ed. Francis Stelli & S. David Webb) NY: Plenum Press.
Wegener, Alfred Lothar (1966). *The Origin of Continents and Oceans.* (tr. John Biram). NY: Dover Publisher.
Weiner, Norbert (1961). *Cybernetics or Control and Communication in the Animal and the Machine.* NY: The MIT Press and John Wiley and Sons, Inc.
Weismann, J. A. See Haeckel et al (1909, 1892).
Wesley, James Paul (1974). *Ecophysics, The Application of Physics to Ecology.* IL: Charles C. Thomas, Publisher.
Weyl, P. K. (1968). The role of the oceans in climatic change: a theory of the ice ages. *American Meteorological Society Meteorological Monographs.* 8: 37-62.
Whitehead, Alfred North (1925). *Science and the Modern World.* NY: The Macmillan Company.
____. (1934). *Nature and Life.* London: Cambridge University Press.
Whittaker, R. H. (1953). A consideration of climax theory: the climax as a population and pattern. *Ecology Monographs* 23: 41-78.
____. (1969). New concepts of the kingdoms of organisms. NY: *Science* 163: 150-159.
Wilkes, H. C. (1972). Maize and its wild relatives. *Science* 177: 1071-1077.
Wills, Christopher (1989). *The Wisdom of the Genes; New Pathways in Evolution.* NY: Basic Books, Inc., Publishers.
Wilner, Burton I. (1965). *A Classification of the Major Groups of Human and Other Animal Viruses.* Minneapolis MN: Burgess Publishing Co.
Wolf, Fred Alan (1981). *Taking the Quantum Leap: The New Physics for Nonscientists.* NY: Harper & Row, Publishers.
Wolkstein, Diane and Samuel Noah Kramer (1983). *Inanna: Queen of Heaven and Earth; Her Stories and Hymns from Sumer.* NY: Harper and Row.
Wolstenholme, G.E.W. and Maeve O'Connor (1971). *Strategy of the Viral Genome.* Edinburgh: Churchill Livingstone.
Wood, A. C. (1957). What is a rabbit? *Evolution* 11: 417-425.
Wood, Albert E. (1980). The Oligocene rodents of North America. PA: *Transactions of the American Philosophical Society.* Vol. 70 Pt 5, 9-24-1980.
____. (1985). Northern waif Primates and rodents. *The Great American Interchange.* (ed. Francis Stelli & S. David Webb). NY: Plenum Press. pp 267-282.
Woolley, C. Leonard (1930). *Ur of the Chaldees.* London: Ernest Benn Ltd.
____. (1954). *Excavations at Ur.* London: Ernest Benn Ltd.
Wynne-Edwards, V. C. (1962). *Animal Dispersion in Relation to Sociel Behavior.* Edinburgh: Alwer & Boyd.
Yates, F. Eugene, et al (ed.) (1987). *Self Organizing Systems: The Emergence of*

Order. Life Science Monographs. NY: Plenum Press.
Zambryski, Patricia, Howard M. Goodman, Marc Van Montagu, Jeff Schell (1983). Agrobacterium tumor induction. *Mobile Genetic Elements.* NY: Academic Press. (ed. James A. Shapiro).
Zittel, Karl A. von (1925). *Text-Book of Palaeontology.* London: Macmillan & Co., Ltd.
Zohary, Daniel, and Maria Hopf (1988). *Domestication of Plants in the Old World; The Origin and Spread of Cultivated Plants in West Asia, Europe and the Nile Valley.* Oxford: Clarendon Press.
Zukav, Gary (1980). *The Dancing Wu Li Masters; An Overview of the New Physics.* NY: Bantam Books.

Taxonomy of Selected Genera

This taxonomy is a simplified classification of lifeforms which are illustrated in the text. It is a composite of the classifications of many scientists and is not to be considered definitive.

The kingdom of Virusa has been added to the five kingdom taxonomies referenced from other authors because I conclude that viruses are lifeforms, having DNA or RNA that interacts with the DNA/RNA of other organisms (even though viruses cannot live outside a cell of another organism). More than 6,000 viruses have been identified in several classification systems. The Virusa kingdom classification includes here only major groups of animal and human viruses.

The other kingdoms include classical taxonomy of genera illustrated in the text. See Margulis (1993) for a new taxonomy including 16 phyla of Monera and 37 phyla of Protoctista. It is becoming apparent that the interactions and recombinations of DNA/RNA from a variety of organisms will soon require new classifications in conjunction with morphologies of living and fossil lifeforms.

KINGDOM I: VIRUSA: all viruses and viral elements

- RNA Viruses:
    - Arboviruses
    - Myxoviruses
    - Picornaviruses
    - Reoviruses
    - RNA — Enveloped Viruses
- DNA Viruses:
    - Adenoviruses
    - Cytoplasmic Viruses
    - Herpesviruses

KINGDOM I: Virusa (continued)

Papovaviruses
Parvoviruses
Poxviruses
Retroviruses
Myoviruses
FAMILY: Myoviridae
GENUS: T4 (Fig. 11.25)

KINGDOM II: MONERA: Prokaryotic cells with ring-form genophore, no nucleous

CLASS: Thermoplasma: fermenting bacteria
CLASS: Bacteria: Includes mitochondrial respiratory type (Fig. 13.1 a)
CLASS: Cyanobacteria: (blue green algae) photosynthesizing bacteria (Fig. 13.1 b)
CLASS: Spirocheta: motile bacteria (Fig. 13.3 a)

KINGDOM III: PROTOCTISTA: Eukaryotes, with nucleous (Classification Protista/Protoctista is in process of reorganization)

SOFT BODY:
ORDER: Chlorophyta, green algae
ORDER: Amoeba
GENUS: *Amoeba* (Fig. 13.5)
ORDER: Dinoflagellata (trace acritarchs)
ORDER: Collar flagellata
ORDER: Ciliata
GENUS: *Paramecium* (Fig. 13.6)
ORDER: Sporozoa
ORDER: Euglenida, yeasts
GENUS: *Euglena* (Fig. 13.2)

HARD BODY:
ORDER: Haplophyta (Coccolithophores, Fig. 10.2 a)
ORDER: Bacillariophyta (Diatoms Fig. 10.2 b)
ORDER: Foraminifera (Fig. 12.1 a)
ORDER: Radiolaria (Fig. 12.1 b)

KINGDOM IV: FUNGI: Multicelled, non-photosythesizing plants (Plate 12.3)

KINGDOM V: PLANTAE: Multicelled, photosynthesizing lifeforms (Plate 12.3)

KINGDOM VI: ANIMALIA: Multicelled, non-photosynthesizing lifeforms

SUBKINGDOM: INVERTEBRATA

PHYLUM: PORIFERA

SUBPHYLUM: Gelantinosa

CLASS: Demospongea, Spicular rays, diverge at 60 or 120 degrees (Fig. 14.1 c)

SUBPHYLUM: Nuda

CLASS: Hexactinellida (Hyalospongea) Spicular rays with 90 degree divergence.

GENUS: *Chancelloria* (Fig. 14.1 a, b)

PHYLUM: COELENTERATA (CNIDARIA)

CLASS: uncertain

GENERA: *Charnia, Rangea, Arborea* (Fig. 13.4)

CLASS: Scyphozoa, true jellyfish

GENUS: *Kimberella* (Fig. 13.8 a)

GENUS: *Brooksella* (Fig. 14.13 a)

CLASS: Anthozoa

SUBCLASS: Zoantharia

ORDER: Rugosa, tetracorals of bilateral symmetry.

GENUS: *Cothonia* (Fig. 13.9 a)

PHYLUM: BRYOZOA

GENUS: *Floridinella* (Figure 15.2 a)

GENUS: *Bugula* (Figure 15.2 b)

PHYLUM: BRACHIOPODA

CLASS: Inarticulata

CLASS: Articulata

GENUS: Billingsella (Fig. 14.6 a, b)

PHYLUM: MOLLUSCA

CLASS: Monoplacophora

GENUS: *Pilima* (Fig. 14.7 a)

GENUS: *Neopilima* (Fig. 14.7 b, c)

CLASS: Amphineura

GENUS: *Chiton* (Fig. 14.8)

CLASS: Gastropoda, snails

GENUS: *Aldanella* (Fig. 14.10 a)

GENUS: *Buccinum* (Fig. 14.10 b, c)

PHYLUM: MOLLUSCA (continued)

CLASS: Cephalopoda
GENUS: *Plectronoceras* (Fig. 14.11 a)
GENUS: *Loligo,* squid (Fig. 14.11 c)
GENUS: *Cycloplectoceras* (Fig. 15.1 a)
GENUS: *Nautilus* (Fig. 15.1 b, c)
GENUS: *Ammonite* (Fig. 17.6)
CLASS: Pelecypoda, bivalve clams
GENUS: *Fordilla* (Fig. 14.9 a)
GENUS: *Pterinea,* scallop (Fig. 14.9 b)
PHYLUM: ECHINODERMATA
CLASS: uncertain
GENUS: *Tribrachedium* (Fig. 13.4 a)
CLASS: Eocrinoida
GENUS: *Macrocystella* (Fig. 14.12 a)
CLASS: Crinoida
SUBCLASS: Somaseroidea
GENUS: *Villebrunaster* (Fig. 14.13 d)
SUBCLASS: Asteroidea
GENUS: (Starfish) (Fig. 14.13 e)
SUBCLASS: Ophiuroidea
GENUS: (Brittlestar) (Fig. 14.13 f)
PHYLUM: ANNELIDA (segmented worms)
GENUS: *Spriggina* (Fig. 13.7 c)
GENUS: *Worthanella* (Fig. 14.13 b)
GENUS: *Dicksonia* (Fig. 13.7 b)
PHYLUM: ARTHROPODA
CLASS: Trilobita (Fig. 12.4 a, b)
GENUS: *Olenellus* (Fig. 14.5 a)
CLASS: Onchyophora
GENUS: *Aysheaia* (Fig. 14.2 a)
GENUS: *Paripatus* (Fig. 14.2 b)
GENUS: *Opabinia* (Fig. 14.4)
CLASS: Crustacea:
SUBCLASS: Malacostraca
GENUS: *Canidaspis* (Fig. 14.3)
SUBPHYLUM: Chelicerata
GENUS: *Aglaspis* (Fig. 14.5 c)
GENUS: *Xephosura* (Fig. 14.5 e)
CLASS: Meristomata

CLASS: Meristomata (continued)

SUBCLASS: Eurypterida
GENUS: *Pterygotus* (Fig. 15.3 a)
CLASS: Arachnida — spiders and scorpions
CLASS: Insecta
SUBCLASS: Apterygota
ORDER: Collembal, oldest division
ORDER: Lepidoptera (butterflies)
GENUS: *Danais* (Fig. 12.2)
GENUS: *Papilio* (Fig. 12.3)
ORDER: Odonata (dragonflies)
GENUS: *Meganeura* (Fig. 18.7)
SUBKINGDOM: VERTEBRATA
PHYLUM: CHORDATA
SUBPHYLUM: Cephalochordata — Hemichordata
GENUS: *Rhabdopleura* (Fig. 14.14 a)
GENUS: *Cephalochordata* (Fig.14.14 b)
GENUS: *Amphioxus* (Fig. 14.15 a)
GENUS: *Pikaia* (Fig. 14.15 b)
PHYLUM: VERTEBRATA — The Vertebrates
SUPERCLASS: Pisces
CLASS: Agnatha — jawless fish
SUBCLASS: Monorhina — dorsal nostril between eyes
ORDER: Cephalaspida — armored ostracoderms, flattened head shield
GENUS: *Hemicyclaspis* (Fig. 15.5)
ORDER: Anapsida — small armored, deep bodied ostracoderms
SUBCLASS: Diplorrhina — no dorsal nostril
ORDER: Pteraspida — ostracoderm — head shield of large plates.
GENUS: *Pteraspis* (Fig. 16.1)
CLASS: Acanthodii — spiny fish
ORDER: Climatiformes — primitive acathodians
GENUS: *Climatius* (Fig. 16.2)
CLASS: Placodermi — early jawed fish — heavily armored
ORDER: Athrodira — armored fish with jointed necks
GENUS: *Dinichthys* (Fig. 17.1)
CLASS: Chondrichthyes — cartilaginous fish — sharks

CLASS: Chondrichthyes (continued)

- **SUBCLASS:** Elasmobranchii — sharks
  - **ORDER:** Cladoselacheformes — ancestral sharks
    - **GENUS:** *Cladoselache* (Fig. 17.2)

CLASS: Osteichthyes — bony fish

- **SUBCLASS:** Sarcopterygii — lobe-finned, air breathing fish
  - **ORDER:** Crossopterygii — progressive, air breathing fish
    - **SUBORDER:** Rhipidistia — ancestors of amphibians
      - **SUPERFAMILY:** Osteolepiformes leading to amphibians
        - **GENUS:** *Eusthenopteron* (Fig.17.4)
  - **ORDER:** Dipnoi — lungfish
    - **GENUS:** *Fleurantia* (Fig. 17.3)
- **SUBCLASS:** Actinoplerygii, ray-finned fish
  - **INFRACLASS:** Holostei, intermediate ray-finned fish
    - **ORDER:** Pycnodontiformes, deep-bodied holsteans
      - **GENUS:** *Proscinetes* (Fig. 21.1)

CLASS: uncertain

- **GENUS:** *Conodonta* (Fig. 15.4)

CLASS: Amphibia — amphibians, the earliest tetrapods or land living vertebrates

- **SUBCLASS:** Labyrinthodontia — late Paleozoic and Triassic, solid skulled amphibians with complex vertebrae
  - **ORDER:** Ichthyostegalia — ancestral labyrinthodonts
    - **GENUS:** *Ichthyostega* (Amphy, Fig. 17.5)
  - **ORDER:** Temnospondyli — late Paleozoic and Triassic labyrinthodonts
    - **SUBORDER:** Rhachitomi — the culmination of labyrinthodont evolution
      - **GENUS:** *Eryops* (Fig. 19.1)
- **SUBCLASS:** Lepospondyli — late Paleozoic amphibians with spool-shaped vertebrae
  - **ORDER:** Aistopoda — ancient limbless lepospondyls
    - **GENUS:** *Ophiderpeton* (Fig. 18.6)
  - **ORDER:** Microsauria — small early lepospondyls
    - **GENUS:** *Microbrachis* (Fig. 18.2)

CLASS: Reptilia — reptiles — scaled or armored tetrapods reproducing by amniotic egg

- **SUBCLASS:** Anapsida — reptiles with solid skull roof
  - **ORDER:** Cotylosauria — stem reptiles

ORDER: Cotylosauria (continued)

SUBORDER: Captohinomorpha — generally small carnivorous cotylosaurs
GENUS: *Hylonomus* (Fig. 18.4)
GENUS: *Seymouria* (Fig. 19.2)
GENUS: *Diadectes* (Fig. 19.3)
ORDER: Mesosauria — ancient aquatic reptiles
GENUS: *Mesosaurus* (Fig. 18.1)
GENUS: *Tylosaurus* (Fig. 22.5)
ORDER: Chelonia — turtles
SUBORDER: Proganochelydia — ancestral turtles
GENUS: *Triassochelys* (Fig. 20.7)
GENUS: *Archelon* (Fig. 22.3 c)
SUBORDER: Amphichelydia — primitive turtles
SUBCLASS: Synapsida — mammal-like reptiles with a skull opening below the post-orbital, squamosal bones
ORDER: Pelycosauria — early mammal-like reptiles
SUBORDER: Sphenacodontia — carnivorous pelycosaurs
GENUS: *Dimetrodon* (Fig. 18.5)
ORDER: Therapsida — varied advanced mammal-like reptiles
SUBORDER: Anomodontia — small to large hervivorous therapsids
INFRAORDER: Dromasauria — small generalized types
GENUS: *Galepus* (Fig. 19.9)
INFRAORDER: Dinocephalia — massive herbivores
GENUS: *Moschops* (Fig. 19.8)
INFRAORDER: Titanosuchia
GENUS: *Titanophoneus* (Fig. 19.7)
INFRAORDER: Dicynodontia — beaked therapsids often with tusks
GENUS: *Kannemeyeria* (Fig. 19.5)
GENUS: *Lystrosaurus* (Fig. 19.4)
SUBORDER: Theriodontia — advanced, carnivorous therapsids
INFRAORDER: Cynodontia — late advanced theriodonts
GENUS: *Thrinaxodon* (Fig. 19.6)
INFRAORDER: Tritylodontia — highly adapted theriodonts

INFRAORDER: Tritylodontia (continued)

GENUS: *Oligokyphus* (Fig. 19.10)

INFRAORDER: Ictidosauria — theriodonts close to mammals

SUBCLASS: Euryapsida — generally marine reptiles with a skull opening above the post-orbital, squamosal bones

ORDER: Protorosauria — ancestral land-living euryapsids

GENUS: *Tanystropheus* (Fig. 20.9)

ORDER: Ichthyosauria — ocean living reptiles of fishlike form

GENUS: *Ichthyosaur* (Fig. 18.3)

ORDER: Placodontia — mollusk eating eurapsids

GENUS: *Henodus* (Figure 20.6)

ORDER: Sauropterygia — marine euryapsids

SUBORDER: Nothosauria — small primitive sauropterygians

SUBORDER: Plesiosauria — large advanced sauropterygians

SUPERFAMILY: Pistosauroidea — ancestral plesiosaurs

SUPERFAMILY: Pliosauroidea — short-necked plesiosaurs

SUPERFAMILY: Plesiosauroidea — long-necked plesiosaurs

GENUS: *Morenosaurus* (Fig. 21.2)

SUBCLASS: Diapsida — the ruling reptiles, with two skull openings separated by the post-orbital, squamosal bones

INFRACLASS: Lepidosauria — primitive diapsids and their direct descendents

ORDER: Eosuchia — ancient lepidosaurs

ORDER: Rhynchocephalia — beaked lepidosaurs — tuataras

GENUS: *Scaphonyrx* (Fig 22.3 b)

GENUS: *Homoesaurus* (Fig. 21.8 a)

ORDER: Squamata - lizards and snakes

SUBORDER: Lacertilia — lizards

GENUS: *Kuehneosaurus* (Fig. 20.5)

SUBORDER: Ophidia — snakes

INFRACLASS: Archosauria — advanced diapsids

ORDER: Thecodontia — ancestral archosaurians

SUBORDER: Proterosuchia — early, primitive thecodonts

GENUS: *Longisquama* (Fig. 20.4)

ORDER: Thecodontia (continued)

SUBORDER: Pseudosuchia — varied thecodonts
GENUS: *Saltoposuchus* (Fig. 20.2)
SUBORDER: Phytosauria — aquatic, crocodile-like thecodonts
GENUS: *Mystriosuchus* (Fig. 20.8)
ORDER: Crocodilia
SUBORDER: Mesosuchia — Mesozoic crocodilians
GENUS: *Phobosuchus* (Fig. 22.1 a)
SUBORDER: Eusuchia — modern crocodilians, gavials, crocodiles, alligators
ORDER: Saurischia — saurischian dinosaurs
SUBORDER: Theropoda — carnivorous saurischians
INFRAORDER: Coelurosauria — small to medium sized carnivores
GENUS: *Coelophysis* (Fig. 20.3)
SUBORDER: Sauropodomorpha — herbivorous saurischians
INFRAORDER: Sauropoda — gigantic, swamp dwelling dinosaurs
GENUS: *Diplodocus* (Fig. 21.3)
ORDER: Ornithischia — ornithischian dinosaurs
GENUS: *Camptosaurus* (Fig. 21.4)
SUBORDER: Ornithopoda — duck-billed and related dinosaurs
GENUS: *Anatosaurus* (Fig. 22.2 a)
SUBORDER: Stegosauria — plated dinosaurs
GENUS: *Stegosaurus* (Fig. 21.5)
SUBORDER: Ceratopsia — horned dinosaurs
GENUS: *Protoceratops* (Fig. 22.3 a)
GENUS: *Monoclinus* (Fig. 22.4)
ORDER: Pterosauria — flying reptiles
SUBORDER: Rhamphorhynchoidea — primitive pterosaurs
GENUS: *Rhamphorhynchus* (Fig. 21.6)
CLASS: Aves — birds
SUBCLASS: Archaerorhithes — Jurassic toothed birds
ORDER: Archaeopterygiformes
GENUS: *Archaeopteryx* (Fig. 21.7)
ORDER: Diatrymiformes

CLASS: Aves (continued)

GENUS: *Diatryma* (Fig. 24.8)
GENUS: *Phororhacos* (Fig. 22.3 e)
SUBCLASS: Neornithes — true birds
ORDER: Hesperornithiformes
GENUS: *Hesperornis* (Fig. 22.7)
ORDER: Rheiformes — rheas
ORDER: Casuariiformes — cassowaries, emus
ORDER: Aepyornithiformes — elephant birds
ORDER: Dinornithiformes — moas
ORDER: Gaviiformes — loons
ORDER: Anseriformes — geese, swans, ducks (Fig. 22.2 c)
ORDER: Psittaciformes — lories, macaws, parrots (Fig. 22.3 f)
CLASS: Mammalia — mammals; tetrapods with hair, that suckle their young
SUBCLASS: Prototheria — egg laying mammals
ORDER: Monotremata — Recent platypus and echidna
GENUS: *Platypus* (Fig. 22.2 b)
SUBCLASS: Allotheria — a long line of early mammals
ORDER: Multituberculata — early mammals with multicuspid teeth
SUBORDER: Taeniolabidoidea — persisting multituberculates
GENUS: *Eozostrodon* (Fig. 20.10)
SUBCLASS: Theria — most of the mammals
INFRACLASS: Pantotheria — the first of the therians
ORDER: Eupantotheria — ancestors of marsupials and placentals
ORDER: Symmetrodonta — symmetrodonts
INFRACLASS: Metatheria — pouched mammals
SUPERORDER: Marsupialia — marsupials
ORDER: Marsupicarnivora — opossums, carnivorous marsupials
GENUS: *Didelphis* (Fig. 23.2, 23.3 a,c)
GENUS: *Borhyaena* (Fig. 26.1)
GENUS: *Thylacosmilus* (Fig. 27.1 a)
ORDER: Paucituberculata — opossum-rats and related fossil forms

SUPERORDER: Marsupialia (continued)

ORDER: Diprotodonta — phalangers, koala, wombat, diprotodonts, kangaroos, wallabies

INFRACLASS: Eutheria — placental mammals

ORDER: Insectivora — insectivores, the most primitive placentals

SUBORDER: Dilambdodonta — moles, hedgehogs, shrews

ORDER: Chiroptera — bats

SUBORDER: Microchiroptera — insectivorous bats

GENUS: *Palaeochiropteryx* (Fig. 24.7)

ORDER: Taeniodonta — taeniodonts

ORDER: Tillodontia — tillodonts

ORDER: Primates — lemurs, tarsiers, monkeys, apes, man

SUBORDER: Plesiadapoidea — ancestral primates

GENUS: *Plesiadapis* (Fig. 23.6)

SUBORDER: Lemuroidea — lemurs

GENUS: *Notharctus* (Fig. 24.9 a)

SUBORDER: Tarsiodea — tarsiers

GENUS: *Tetonius* (Fig. 24.9 b)

SUBORDER: Platyrrhini — New World monkeys

SUBORDER: Catarrhini — Old World monkeys, apes, man

SUPERFAMILY: Parapithecoidea — ancestral catarrhines

GENUS: *Aegyptopithecus* (Fig. 25.13 a)

GENUS: *Pliopithecus* (Fig. 25.13 b)

SUPERFAMILY: Ceropithecoidea — Old World monkeys

SUPERFAMILY: Hominoidea — apes and man

GENUS: *Dryopithecus* (Proconsul) (Fig. 26.12)

GENUS: *Australopithecus*

SPECIES: *afarensis* (Fig. 27.2)

SPECIES: *africanus* (Fig. 28.5)

GENUS: *Homo*

SPECIES: *habilis* (Fig. 28.6)

SPECIES: *erectus* (Fig. 28.7)

SPECIES: *sapiens neanderthalensis* (Fig. 28.8)

SUPERFAMILY: Hominoidea (continued)

SPECIES: *sapiens sapiens* (Fig. 21.8 c, 28.9)

ORDER: Xenarthra (Edentata) New World xenarthrans

SUBORDER: Cingulata — armadillos and glyptodonts

GENUS: *Dasypus* (Armadillo) (Fig. 23.7 a)

GENUS: (Glyptodont) (Fig. 23.7 b)

SUBORDER: Pilosa — ground sloths, anteaters

GENUS: *Nothrotherium* (Fig. 23.7 c)

ORDER: Rodentia — rodents

SUBORDER: Protogomorpha — primitive rodents

SUBORDER: Scuiromorpha — squirrels

SUBORDER: Myomorpha — mice, rats and relatives

SUBORDER: Caviomorpha — South American rodents

SUBORDER: Hystricida — Old World porcupines

SUBORDER: Castorida — beavers and their relatives

GENUS: *Castorides* (Fig. 28.3)

ORDER: Lagomorpha — hares, rabbits and pikas

ORDER: Cetacea — porpoises and whales

SUBORDER: Archaeoceti — ancestral whales

GENUS: Basilosaurus (Fig. 24.1)

SUBORDER: Odontoceti — porpoises, dolphins, toothed whales

SUBORDER: Mysticeti — whalebone whales

ORDER: Creodonta — ancient carnivorous placentals

SUBORDER: Oxyaenoidea — early creodonts

SUBORDER: Hyaenodontia — varied and persisting creodonts

GENUS: *Hyaenodon* (Fig. 25.2)

ORDER: Carnivora — modern carnivorous placentals

SUBORDER: Fissipedia — land living carnivores

SUPERFAMILY: Miacoidea — ancestral fissipeds

GENUS: *Daphoenodon* (bear-dog) (Fig. 25.3)

SUPERFAMILY: Canoidea — dogs, bears, pandas, raccoons, mustelids

GENUS: Canid (Fig. 12.4 c, d)

SUBORDER: Fissipedia (continued)

SUPERFAMILY: Feloidea — civets, hyenas, cats
GENUS: *Dinictes* (Fig. 25.1)
GENUS: *Smilodon* (sabertooth cat) (Fig. 27.1 b)
SUBORDER: Pinnipedia — marine carnivores, seals, sea lions, walruses
ORDER: Condylartha — ancestral hoofed mammals or ungulates
GENUS: *Phenacodus* (Fig. 23.4)
ORDER: Amblypoda — primitive ungulates
SUBORDER: Pantodontia — large early Tertiary ungulates
GENUS: *Barylambda* (Fig. 23.1)
SUBORDER: Dinocerata — gigantic horned amblypods
GENUS: *Uintatherium* (Fig. 23.5)
ORDER: Notoungulata — the most varied South American ungulates
SUBORDER: Typotheria — small rabbit-like
SUBORDER: Hegetotheria — small rabbit-like
SUBORDER: Toxodontia — specialized ungulates
GENUS: *Thomashuxleya* (Fig. 24.6)
GENUS: *Toxodon* (Fig. 28.4)
ORDER: Litopterna — camel-like and horse-like South American ungulates
GENUS: *Diadiaphorus* (Fig. 26.2)
GENUS: *Theosodon* (Fig. 26.3)
ORDER: Astrapotheria — large possibly amphibious South American ungulates
GENUS: *Astrapotherium* (Fig. 25.9)
ORDER: Pyrotheria — very large South American ungulates
GENUS: *Pyrotherium* (Fig. 25.10)
ORDER: Perissodactyla — odd-toed hoofed mammals
SUBORDER: Hippomorpha — horses, titanotheres
SUPERFAMILY: Equoidea — horses
GENUS: *Hyracotherium* (*Eohippus*) (Fig.24.3)
GENUS: *Paleotherium* (Fig. 24.4)
GENUS: *Miohippus* (Fig. 24.2)
GENUS: *Mesohippus* (Fig. 25.4)
GENUS: *Merychippus* (Fig. 24.2)

SUPERFAMILY: Equoidea (continued)

GENUS: *Pliohippus* (Fig. 24.2)
GENUS: *Equus* (Fig. 24.2)
GENUS: *Indricotherium* (Fig. 26.8)
SUPERFAMILY: Brontotherioidea — titanotheres
GENUS: *Eotitanops* (Fig. 24.5)
GENUS: *Brontops* (Fig. 25.5)
SUBORDER: Ancylopoda — chalicotheres
GENUS: *Moropus* (Fig. 26.9 a)
GENUS: *Chalicotherium* (Fig. 26.10)
SUBORDER: Ceratomorpha — rhinoceroses and tapirs
SUPERFAMILY: Tapiroidea — tapirs
SUPERFAMILY: Rhinocerotoidea — rhinoceroses
GENUS: *Hyracodon* (Fig. 25.6)
GENUS: *Metamynodon* (Fig. 25.7)
ORDER: Artiodactyla — even-toed hoofed mammals
SUBORDER: Paleodonta — early artiodactyls
SUPERFAMILY: Dichobunoidea — ancestral artiodactyls
SUPERFAMILY: Entelodontoidea — entelodonts
GENUS: *Dinohynus* (Fig. 26.7)
SUBORDER: Suina — pig-like artiodactyls
SUPERFAMILY: Suoidea — pigs and peccaries
SUPERFAMILY: Anthracothereoidea — anthracotheres
SUPERFAMILY: Hippopotamoidea — hippopotamuses
SUBORDER: Ancodonta — ancodonts
SUPERFAMILY: Merycoidodontoidea — oreodonts
GENUS: *Agriochoerus* (Fig. 25.8)
SUBORDER: Tylopoda — camels and llamas
GENUS: *Oxydactylus* (Fig. 26.5)
GENUS: *Alticamelus* (Fig. 26.6)
SUBORDER: Ruminantia — advanced artiodactyls
SUPERFAMILY: Cervoidea — deer
GENUS: *Megaloceros* (Fig. 10.1)
FAMILY: Palaeomerycidae
GENUS: *Blasomeryx* (Fig. 26.4)
SUPERFAMILY: Giraffoidea — okapis and giraffes
SUPERFAMILY: Bovidea — antelopes and cattle

SUPERFAMILY: Bovidea (continued)

GENUS: *Bison occidentalis* (Fig. 28.2)
ORDER: Proboscidea — elephants and their progenitors
SUBORDER: Moerithérioidea — ancestral proboscideans
GENUS: *Moeritherium* (Fig. 25.12 a)
SUBORDER: Dinotherioidea — dinotheres
GENUS: *Dinotherium* (Fig. 25.12 c)
SUBORDER: Euelephantoidea — mastodonts and elephants
SUPERFAMILY: Gomphotherioidea — long-jawed mastodonts and descendants
GENUS: *Phioma* (Fig. 25.12 b)
GENUS: *Mastodont* (Fig. 28.1 b)
SUPERFAMILY: Elephantoidea — mammoths and elephants
GENUS: *Mammuthus* (Fig. 28.1 a)
GENUS: *Platybeledon* (Fig. 28.1 c)
ORDER: Sirenia — sea cows
GENUS: *Protosiren* (Fig. 22.3 d)
ORDER: Desmostylia — desmostylids, large marine waders
GENUS: *Palaeoparadoxia* (Fig. 26.11)
ORDER: Hyracoidea — hyraxes
ORDER: Embrithopoda — gigantic rhinoceros-like mammal
GENUS: *Arsinotherium* (Fig. 25.11)

References (See Bibliography for full citation):

Brasier, 1980.
Ciochon and Fleagle, eds., 1987.
Clarkson, 1986.
Colbert, 1980.
Fraenkel-Conrat, 1976: Vol 8.
Kummel, 1961.
Kurtin, 1982.
Margulis, 1993.
Romer, 1966.
Simpson, 1980.
Stewart, 1983.
Whittaker, 1969.
Wilner, 1965.

Illustrations, Sources and Credits

The illustrations for this book were drawn under supervision of author by Bruce Cook and Dina Cohen except Figure 11.19, Tables 10.1, 10.2, and Plates 11.1, 11.2, 12.5, 12.6, and 12.7. Doris Walker, Clarion Pacific's composition director, orchestrated typography, with graphics assistance from artist Warren Schepp. Brice Beckham prepared computer graphics.

I thank the authors, publishers and copyright holders whose illustrations or data have served as bases for some of the illustrations and are credited here, in the text and/or in the bibliography. Many figures are from reconstructed and/or partial fossils, and these drawings involve speculation about final configuration.

Illustrations from Published Materials:

Denton, Michael 1986: 130. *Evolution: A Theory in Crisis.* NY: Adler and Adler, copyright © 1985. Plate 12.7.

Frakes, L. A. 1979: 261. *Climates Throughout Geologic Time.* NY: Elsevier Scientific Publishing Co. Table 10.1.

Haeckel, Ernst, 1866. *Generelle Morphologie der Organismen.* Berlin: Reimer. Plate 12.5.

Margulis, Lynn 1993. *Symbiosis in Cell Evolution; Microbial Communities in the Archean and Proterozoic Eons.* NY: W. H. Freeman and Company. Drawing by Rae Wallhauser after Laszlo Meszoly. Curtesy of Lynn Margulis. Figure 11.19.

Romer, Alfred S. 1966: 208. *Vertebrate Paleontology* (3rd ed.) (1st ed. 1933). IL: University of Chicago Press. Plate 12.6.

Computer graphics prepared under supervision of author, based on source as noted:

Einstein, Albert, 1961: 25. *Relativity; The Special and General Theory*. NY: Crown Publishers, Inc., Figure 5.1.

Illustrations by Bruce Cook based on multiple sources:

Figures 11.1 through 11.9, 11.12 through 11.16, 11.20 through 11.23, 11.25 through 11.27. Figures 12.2, 13.1a,b, 13.3, 20.10, 21.8b, 22.2b,c, 22.6, 23.2, 23.6, 23.7a,b, 25.10, 26.11, 28.5 through 28.9. Plates 12.1 through 12.3.

Illustrations by Bruce Cook based on sources as noted:

Abel, O. 1912. Grundzuge der Palaeobiologie der Wirbeltiere. Stuttgart. *Lehrbuch der Palaozoologie.* Jena. Figure 24.4

Andrews, C. W. 1906. *A Descriptive Catalogue of the Tertiary Vertebrata of the Fayum, Egypt.* London. Figure 25.13 a, b.

Bignot, Gerald 1985: 77, 92. *Elements of Micropalaeontology: Microfossils — Their Geological and Palaeobiological Applications.* London: Graham and Trotman Ltd. Figure 10.2.

Broom, Robert 1923: 661. On the structure of the skull in the carnivorous dinocephalian reptiles. London: *Proceedings Zoological Society.* Figure 19.9.

Brown, Barnum and E. M. Schlaikjer, 1940. The structure and relationship of Protoceratops. *Annals New York Academy Science* 40: 133-266. Figure 22.3 a, 22.4.

Briggs, D. E. G., 1977. The morphology, mode of life and affinities of *Canadaspis perfecta* (Crustacea Phyllocarida) Middle Cambrian, Burgess Shale, British Columbia.

Philosophical Transactions Royal Society, London B, 281: 439-487. Figure 14.3.

Callomon, J. H., 1963. Sexual dimorphism in Jurassic ammonites. *Transactions Leicester Literary and Philosophical Society*. 57: 21-56. Figure 17.6.

Carroll, Robert L., 1988. *Vertebrate Paleontology and Evolution.* NY: W. H. Freeman and Co. Figure 23.6.

Clark, David L. 1987: 636-662. Phylum Conodonta (Chapter 20) *Fossil Invertebrates*. Ed: Richard S. Boardman, Alan H. Cheetham and Albert J. Rowell. CA: Blackwell Scientific Publications. Reprinted by permission of Blackwell Scientific Publications, Inc. Figure 15.4.

Clark, W. E. Le Gros, 1965. *History of the Primates; An Introduction to the Study of Fossil Man.* IL: University of Chicago Press. Figure 26.12, 28.9.

Clarkson, E. N. K. 1986: 133, 329. *Invertebrate Palaeontology and Evolution.* (1st ed., 1979) London: Allen & Unwin. Figures 12.4 a, b, 13.9 b, 14.6 c, 15.1 c.

Cowen, Richard, 1976: 15. *History of Life.* NY: Mcgraw-Hill Book Co., Inc. Figure 13.1 c.

Cox, L. R, 1960. In R. C. Moore, editor, *Treatise on Invertebrate Paleontology.* Part I. Mollusca I. NY: Geological Society of America and University of Kansas Press. Figure 14.10 b,c.

Denton, Michael, 1986: 218. *Evolution: A Theory in Crisis.* NY: Adler & Adler. Copyright © 1985. Figure 20.1.

Gaudry, A. 1867. *Animaux Fossiles et Geologie de l'Attique.* Paris. Figure 25.12 c.

George C. Page Museum, Los Angeles County. Skeleton. Figures 21.8 c, 27.1 b.

Giedion, S. 1962: 469-473. *Eternal Present: The Beginnings of Art.* Trustees, National Gallery of Art, Washington, DC. Figure 29.1.

Gould, Stephen Jay, 1989: 322. *Wonderful Life, The Burgess Shale and the Nature of History.* NY: W. W. Norton and Co., Inc. Figure 14.15 b.

Graessner, M. F. and M. Wade, 1966. The late Precambrian

fossils from Ediacara, South Australia. *Palaeontology* 9: 599-628. Figure 13.4.

Graham-Smith,W. and T. S. Westoll, 1937. On a new long-headed dipnoan fish from the Upper Devonian of Scaumenac Bay, P. Q. Canada. *Transactions Royal Society Edinburgh* 59: 241-266. Figure 17.3.

Gregory, William K., 1920. On the structure and relations of *Notharctus,* an American Eocene primate. *Memoirs American Museum Natural History*. N.S. 3: 49-243. Figure 24.9 a.

Gregory, William K., 1951: 73, 395, 387, 510, 834. *Evolution Emerging* Vol 2. NY: Macmillan Co. Courtesy Department of Library Services, American Museum of Natural History. Figures 14.15 a, 19.2, 19.3.

Gregory, Wiliam K., 1963: 48-49. (1st. edition, 1929). *Our Face from Fish to Man, A Portrait Gallery of our Ancient Ancestors and Kinsfolk together with a Concise History of Our Best Features*. NY: Hafner Publishing Co. Figure 23.3.

Gross, W., 1967. Usber das gebiss der Acanthodier und Placodermen in Fossil Vertebrates. Eds. C. Patterson and P. H. Greenwood. *Journal Linnaean Society (Zoology)*. Vol 47: 121-130. Figure 16.3 a, b.

Hay, O P., 1908. The fossil turtles of North America. *Publication Carnegie Institution,* Washington. No. 75: 1-568. Figure 22.3 c.

Hinde, G. J., 1887. A monograph of the British fossil sponges. *Palaeontogr. Society Monographs*. 1-254. Figure 14.1 a, b.

Holland, W. J. and O. A. Peterson, 1913. The Osteology of the Chalicotheroidea, with Special Reference to a Mounted Skeleton of *Moropus elatus* Marsh, installed in Carnegie Museum. *Memoirs Carnegie Museum.* 3: 189-406. Figure 26.9.

Huene, F. von, 1939. Die Verwandtschaftsgeschichte der Rhynchosauriden. *Physis* 14: 499-522. Figure 22.3 b.

Hutt, Frederick B. 1979: 179. *Genetics for Dog Breeders*. Curtesy of Lena Tamboer and American Kennel Club. S.F.: W. H. Freeman. Figure 12.4 b, c.

Hyman, L. H., 1940: 523. *The Invertebrates.* NY: McGraw-Hill Book Co. Figure 13.8 b, c.

Jarvik, Erik, 1980: 345. *Basic Structure and Evolution of Vertebrates.* Vol. 2. NY: Academic Press. Figure 16.3 a, b.

Jefferies, R. P. S., 1979: 452. On origin of the Chordates — A methodological essay. *The Origin of Major Invertebrate Groups,* (ed. M. R. House) NY: Academic Press. pp 443-477. The pterobranch hemichordate Cephalodiscus after van der Horst, Schepotieff and Andersson. Figure 14.14 b.

Jell, P. A. and J. S. Jell, 1976. Early Middle Cambrian Corals from Western New South Wales. *Alcheringa* 1: 181-195. Figure 13.9 a

Johanson, Donald and Maitland Edey, 1981: 182, 354-357. *Lucy: The Beginnings of Humankind.* NY: Warner Books. Figure 27.2 a, b.

Johnson, Leland G., 1983: 91, 114, 297, 423, 539, 940, 970. *Biology.* IA: Wm. C. Brown Co., Publishers. Figures 11.17, 11.18, 13.1 a, 13.2, 13.5 a, 13.6 a, 14.13 e.

Lemche illustration in: *Treatise on Invertebrate Paleontology,* Part I, Mollusca 1., 1960: 89, 91, 103. NY: Geological Society of America and University of Kansas Press. Figure 14.7 b,c.

LeRoy, L. W. 1953: 61-65. *Biostratigraphy of the Maqfi Section, Egypt.* Geological Society of America. Figure 12.1 a.

Marshall, N. B., 1978. *Aspects of Deep Sea Biology.* London: Cassell PLC. Figure 14.1 c

Matthews, S. C. and V. Massarzhevsky, 1975. Small shelly fossils of late Precambrian and early Cambrian age: a review of recent work. London: *Quarterly Journal Geological Society,* 131: 289-304. Figure 13.7 a, b, c.

McGregor, J. H., 1908: 302-336. On *Mesosaurus brasiliensis,* Nov. Sp. from the Permian of Brazil. *Commisao de Estudos das Minas de Carvao de Pedra do Brazil.*

Figure 18.1.
Moore, Raymond E., Cecil G. Lalicker, & Alfred G. Fisher, 1952: 157, 158, 212, 228, 229, 272, 335, 336, 347, 355, 401, 415, 457, 463, 471, 487, 554, 555, 557, 558, 559, 568, 579, 660, 666, 717. *Invertebrate Fossils.* By permission of McGraw-Hill Book Co., Inc. Copyright © 1952. Figures 13.6 b, 14.2 b, 14.5 a-f, 14.6 a, b, d, 14.8 a-c, 14.9 b, c, 14.11 a, c, 14.12 a-d, 14.13 b, c, d, f, 14.14 a, 15.1 a, b, 15.2 a, b, 15.3 a, 18.7.
Natural History Museum of Los Angeles County. Figure 24.8.
Nester, Eugene W, C. Evens Roberts, Brian McCarthy and Nancy Pearsal, 1978. *Microbiology, Molecules, Microbes and Man.* Copyright © 1978 by Holt, Rinehart and Winston. Figure 11.24.
Orvig, T., 1973. Acanthodian dentition and its bearing on the relationships of the group. *Palaeontographica (A)* 143: 119-150. Figure 16.3 c.
Osborn, H. F. 1896. The titanotheres of ancient Wyoming, Dakota and Nebraska. Vols. I, II, *U. S. Geological Survey Monograph 55.* Figure 24.5.
Osborn, Henry Fairfield, 1910: 60, 126, 155, 162, 164, 214, 220, 231, 236, 364, 483. *The Age of Mammals in Europe, Asia and North America.* NY: The Macmillan Co. Figures 23.4, 23.5, 23.7 a, b, 24.3, 24.9 b, 25.6, 25.7, 25.8, 27.1 a, 28.1 a, b, c.
Pessagno, Emile A. Jr., Charles D. Blome & José F. Longoria 1984: 1-52. *Bulletins of American Paleontology.* NY: Paleontological Research Institution. Vol. 87, # 320. Figure 12.1 b.
Pojeta, John Jr., Bruce Runnegar, Jiri Kriz, 1973. *Fordilla troyensis* Barande: The oldest known pelecypod. Published 5/25/74. *Science.* NY: 180: 866-868. Copyright © 1974 by the American Association for the Advancement of Science. Figure 14.9 a.
Revilloid, P. 1917-1922. Contribution a l'etude des Chiropteres des terrains Tertiaires. *Mem. Soc. Pal. Suisse* 43; 1-57;

44: 63-128; 45: 133-195. Figure 24.7.

Runnegar, Bruce and & John Pojeta Jr., 1974, Molluscan phylogeny: the paleontological viewpoint. NY: *Science* 186: 313, 314. Copyright © 1974 by the American Association for the Advancement of Science. Figures 14.7 a, 14.10 a.

Scott, W. B., 1928. Astrapotheria of the Santa Cruz beds. *Rept. Princeton Univ. Exped. Patagonia* 6: 301-342. Figure 25.9.

Scott, W. B., 1937. *A History of Land Mammals in the Western Hemisphere.* NY: Macmillan. Figures 25.2, 26.2.

Sickenberg, O., 1934. Beitrage zur Kenntnis tertiarer Sirenen. *Mem. Mus. Hist. Nat. Belgique* 63: 1-352. Figure 22.3 d.

Simpson, G. G., J. L. Minoprio and B. Patterson, 1962. The mammalian fauna of the Divisadero Largo Formation, Mendoza, Argentina. *Bulletin Museum Comparative Zoology* 127: 239-293. Figure 24.6.

Sinclair, W. J. 1906. Mammalia of the Santa Cruz Beds: Marsupialia. *Report Princeton University Expedition Patagonia* 4: 333-460. Figure 26.1.

Snodgrass, R. E., 1952. *A Textbook of Arthropod Anatomy.* Ithaca, NY: Cornell University Press. Copyright © 1952 by Cornell University. Adapted from R. E. Snodgrass. Used by permission of the publisher, Cornell University Press. Figure 15.3 b.

Storer, Usinger, Mybakken and Stebbens illustration. In: Denton, Michael, 1986. *Evolution: A Theory in Crisis.* NY: Adler & Adler. Figure 24.2.

Trauth illustration. In: R. C. Moore, editor, *Treatise on Invertebrate Paleontology, 1960.* Part l. Mollusca l. NY: Geological Society of America and University of Kansas Press. Figure 15.1 b.

University of Nebraska State Museum, skeleton. Figures 10.1, 25.12 a, b, 28.3.

Wade, M., 1972, Hydrozoa and scyphozoa and other medusoids from the Precambrian Ediacara fauna, South Australia. *Palaeontology* 15: 197-225. Cambridge

University Press. Figure 13.8 a.
Watson, D. M. S., 1914: 749. An order of mammal-like reptiles. London: *Proceedings Zoological Society.* Figure 19.4.
Watson, James D., 1968: 117, 135. *The Double Helix.* NY: Macmillan Publishing Co. Adapted with the permission of Atheneum Publishers, an imprint of Macmillan Publishing Co. Copyright © 1968 by James D. Watson. Figures 11.10, 11.11.
Whittington, H. B., 1975. The enigmatic animal *Opabinia regalis. Philosophical Transactions Royal Society,* B. 271: 1-43. Figure 14.4.
Whittington, H. B., 1978. The lobopod animal *Aysheaia peduculata,* Middle Cambrian, Burgess Shale, British Columbia. *Philosophical Transactions Royal Society,* London B, 284: 165-197. Figure 14.2 a.
Wolcott, Charles E., 1896: 615. Fossil jelly fishes from the Middle Cambrian terrane. *Proceedings of the U. S. National Museum,* Vol xviii. No. 1086. Figure 14.13 a.
Yochelson, E. L., R. H. Flower and G. F. Webers, 1973. The bearing of the new Late Cambrian monoplacophoran genus *Knightoconus* upon the origin of the Cephalopoda. *Lethaia* 6: 275-309. Figure 14.11 b.
Zapfe, H., 1979. Chalicotherium grande (Blainv) aus der miozanen Spallenfullung von Neudord an der March. Tschechoslowakei-Neue Denkscher. *Naturhis. Mus. Wien.* 2: 1-282. Figure 26.10.

Illustrations by Dina Cohen based on multiple sources:

Figures 13.1 c, 22.2 a, 23.7 c.

Illustrations by Dina Cohen based on sources as noted:

American Museum of Natural History, skeleton. Figure 19.8.
Andrews, C. W., 1906. *A Descriptive Catalogue of the Tertiary Vertebrata of the Fayum, Egypt.* London. Figure 25.11.

Brink, A. S., 1957. Speculations on some advanced mammalian characteristics in the higher mammal-like reptiles. *Palaeontology Africa* 4: 77-96. Figure 19.6.

Broom, R., 1932. *The Mammal-like Reptiles of South Africa and the Origin of Mammals.* London. Figure 19.5.

Carroll, R. L., 1964. The earliest reptiles. *Journal Linnaean Society,* London (Zool.) 45: 61-83. Figure 18.4.

Colbert, E. H., 1964. Relationships of the saurischian dinosaurs. *Amer. Mus. Novit.* No. 2181: 1-24. Figure 20.3.

Dean, B., 1909. Studies on fossil fishes (sharks, chimaeroids and arthrodires). *Memoirs American Museum Natural History.* 9: 211-287. Figure 17.2.

Fritsch, A., 1879-1901. *Fauna der Gaskohle und der Kalksteine der Permformation Bohmens.* 4 vols. Prague. P. 492. Figure 18.6.

Gregory, W. K. 1951: 371, 379, 387. *Evolution Emerging.* Vol 2. NY: Macmillan Publishing Company. Figures 17.4, 19.1.

Gidley, J. W., 1913. A recently mounted *Zeuglodon* skeleton in the United States National Museum. *Proceedings U. S. National Museum,* 649. Figure 24.1.

Gilmore, C. W., 1909. Osteology of the Jurassic reptile *Camptosaurus. Proceedings U. S. National Museum,* 36: 197-332. Figure 21.4.

Gilmore, C. W., 1914. Osteology of the armored Dinosauria in the United States National Museum, with special reference to the genus *Stegosaurus. Bulletin U. S. Natural Museum,* 89: 1-143. Figure 21.5.

Hatcher, J. B., 1901. *Diplodocus* (Marsh): Its osteology, taxonomy, and probable habits, with a restoration of the skeleton. *Memoirs Carnegie Museum,* Vol 1: 1-1903. Figure 21.3.

Heilmann, Gerhard, 1927. *The Origin of Birds.* Copyright 1927 by D. Appleton and Co., renewed © 1955 by Gerhard Heilmann. Used by permission of the publisher, Dutton, an imprint of New American Library, a division of Penguin Books USA Inc. Figure 21.7.

Heintz, A., 1932: 115-224. *The Structure of Dinichthys; A Contribution to our Knowledge of the Arthrodira.* Bashford Dean Memorial Volume. New York. Figure 17.1.

Huene, F. von, 1936. The constitution of the Thecodontia. *American Journal Science* Ser. 5, 32: 207-217. Permission of *American Journal of Science.* Figure 20.2.

Huene, Friedrich R. Freiherr von, 1956. *Palaeontologie und Phylogenie der Niederen Tetrapoden.* Jena: Veb Gustav Fischer Verlag. Figure 20.6.

Jackel, O., 1914-1916. Die Wirbeltierfunde aus dem Keuper von Halberstadt. *Paleont. Zeitschr.* 1: 155-215; 2: 88-214. Figure 20.7.

Jarvik, Erik, 1980: 235. *Basic Structure and Evolution of Vertebrates.* Vol. 2. NY: Academic Press. Figure 17.5.

Kuhne, W. G., 1956. The Liassic Therapsid *Oligokyphus. British Museum of Natural History.* London. Figure 19.10.

Lull, R. S. and N. E. Wright, 1942. Hadrosaurian dinosaurs of North America. *Geological Society America, Special Papers* No. 40. Figure 22.2 a.

Marsh, O. C., 1880. Odontornithes; a monograph on the extinct toothed birds of North America. *Report Geological Exploration 40th Parallel.* Figure 22.7.

Matthew, W. D., 1909. The Carnivora and Insectivora of the Bridger Basin, Middle Eocene. *Memoirs American Museum Natural History,* 9: 466-484. Figure 25.1.

Matthew W. D., 1929 Reclassification of the Artiodactyla. *Bulletin Geological Society America,* 40: 403-408. Figure 26.4.

McClung, C. E., 1908. Restoration of the Skeleton of *Bison occidentalis. Kansas University Science Bulletin.* Vol IV, No.10. Figure 28.2.

McGregor, J. H., 1906: 27. The Phytosauria, with especial reference to *Mystriosuchus* and *Rhytidodon. Memoirs American Museum Natural History.* Figure 20.8.

Merriam, J. C. 1908. Triassic *Ichthyosauria* with special

reference to the American forms. *Memoirs University California* 1: 1-196. Figure 18.3.

Natural History Museum of Los Angeles County. Figures 18.5, 21.2.

Orlov, J. A., 1960. Les dinocephales rapaces de la faune d'Isheevo (*Titanosuchia*). *Trea., Bur. Rechr. Grp; Min.,* Paris. No. 2244. Figure 19.7.

Osborn, Henry Fairfield 1910: 209, 364, 483. *The Age of Mammals in Europe, Asia and North America.* NY: The Macmillan Co. Figures 22.5, 25.5, 26.6, 26.8.

Patterson, B. 1939. New Pantodonta and Dinocerata from the Upper Paleocene of western Colorado. *Field Museum Natural History, Geology Series,* 6: 351-384. Figure 23.1.

Peterson, O. A., 1904. Osteology of *Oxydactylus. Annals Carnegie Museum,* 2: 434-476. Figure 26.5.

Peterson, O. A., 1910. Description of new carnivores from the Miocene of western Nebraska. *Memoirs Carnegie Museum,* 4: 205-278. Figures 25.3, 26.7.

Peyer, B., 1931-1937. Die Triasfauna der Tessiner Kalkalpen. 1-12. Abhandl. Schweiz. *Palaont. Ges.* Vols. 50-59. Figure 20.9.

Robinson, P. L., 1962. Gliding lizards from the Upper Keuper of Great Britain. *Proceedings Geological Society,* London. No.1601: 137-146. Figure 20.5.

Scott, W. B., 1910. Litopterna of the Santa Cruz Beds. *Report Princeton University Expedition Patagonia:* Figure 26.3.

Scott, W. B., 1912, 6: 287-300. Mammalia of the Santa Cruz Beds: Order Toxodontia. *Report Princeton University Expedition Patagonia.* Figure 28.4.

Scott, W. B., 1937. *Land Mammals of the Western Hemisphere,* by permission of Macmillan Co., publishers. Figures 25.4.

Seeley, H. G., 1901. *Dragons of the Air; An Account of Extinct Flying Reptiles.* NY. Figure 21.6.

Sharov, A, USSR. Figure 20.4.

Smith-Woodard, A., 1889-1901. *Catalogue of the Fossil Fishes in the British Museum,* 4 Vols. London. Pp 2493 + 70 plates. Figure 21.1.

Steen, M., 1938. On the fossil Amphibia from the Gas Coal of Nyrany and other deposits in Czechoslovakia. *Proceedings Zoological Society,* London B 108: 205-283. Figure 18.2.

Stensio, E. A., 1927. The Downtonian and Devonian vertebrates of Spitzbergen. I. Family Cephalaspidae. *Skr. om Svalbard og Nordishavet,* No. 12. 2 Vols. pp. 391 + 112 plates. Figure 15.5.

University of Nebraska State Museum, reconstruction and skeleton. Figures 26.8, 28.3.

Watson, D. M. S., 1937, The acanthodian fishes. *Philosophical Transactions Royal Society,* London, B 228: 49-146. Figure 16.2.

White, E. I., 1935. The ostracoderm *Pteraspis* Kner and the relationships of the agnathous vertebrates. *Philosophical Transactions Royal Society,* London, B 225: 381-457. Figure 16.1.

Williston, S. W., 1898. *The Osteology of the Reptiles.* Figure 21.8 a.

Acknowledgments

Evolution and Reason would not have been written without Lyn Harris Hicks' faith in me, her editing expertise and philosophical collaboration during the manuscript's seven years of development. How can such dedication be fully credited?

I especially value the enthusiasm of the late Norman Cousins in his comments after he read an early draft of my manuscript.

My husband John Elliott Boberg gave enduring patience and support during the long period of the book's gestation. My son Mark Boberg and daughter-in-law Rebecca Boberg gave constructive suggestions for the project and my grandchildren Kurt and Cory stimulated my incentive. Brothers Herman Kurth, Dr. Robert Kurth and nieces Jan Morton and Jo Hanna Kurth buoyed my spirits.

Friends who commented on chapters of the manuscript and illustrations, and also cheered me on are Ellen Prior, Meri Beckham, Roberta O'Brien, Gloria Van Gieson, Francis Little, K. C. Hicks, Mark Lit, Dr. Estelle Lit, Edith Roth, Ione Buie, Doris Bryant, Evelyn and Peter Wolenty, Betty Jo McCone and Susanne McCann.

It was a pleasure to work with Doris Walker and Warren Schepp in the composition of the book.

Artists Bruce Cook and Dina Cohen capably brought the illustrations from my mind and from numerous resources into conjunction in finished form.

I value the contributions of those whose works are cited in Illustrations, Sources and Credits; Bibliography — and of other academics who encouraged my adventuresome speculation: Dr. Carol Smallenburg, Dr. Thomas Love, Dr. William Emboden, Dawn Dressler, Dr. Gilbert Albelo and Dr. Jack Lessinger.

I benefitted from editing and stimulating challenges of numerous other scientists and educators, especially Dr. Lynn Margulis, Dr. Darwin Smith, Dr. Christine Brewer, Dr. M. L. Rosensweig, Dr. Bonnie Paller, Dr. Ti Grace Atkinson, Dr. Lorraine Heisler, Dr. Jane Gray, Dr. Jane Aloi, Dr. Ross Pohlo, Dr. James Brune, Dr. David Tracer, Dr. Richard Petersen, Dr. Lester Newman, Dr. John Minch, Dr. Lawrence G. Barnes, Dr. J. D. Stewart, Dr. Steven Oppenheimer, Dr. James Dale Smith, the Reverend David Richardson, and the Reverend Loren Bush.

Technical assistance was provided by Dr. Stephen Jay Gould, Dr. and Mrs. Edwin Colbert, Dr. J. H. Callomon, Dr. June Stephenson, and F. Bentley Mooney, Esq.

I appreciate and respect the individuals, their philosophies and their generous sharing of information and perspectives in our common search for knowledge of our intriguing and often-times magnificent evolution heritage.

About the Author

Occasionally a scholar climbs to new philosophical heights and in a sense blazes a trail for adventurous others. Author Dorothy Kurth Boberg is such a trail-blazer in evolution philosophy.

In her concept **the reciprocal moment** she discovered the secret of adaptation and other feedback loops in nature. This led to her **principle of responsive salmutations** — the processes of origination of new kinds of organisms in evolution.

This pioneer interdisciplinarian of evolution sciences was born and educated in Lincoln, Nebraska, USA. An honored Phi Beta Kappa scholar of the University of Nebraska, Boberg earned a Bachelor of Arts Degree with an interdisciplinary major and was elected to honorary fraternities Psi Chi and Alpha Kappa Delta.

Boberg's study of evolution, genetics, and the evolution of scientific thought from Newton to Einstein stimulated a lifetime of reading and pondering of philosophical and scientific works. She participated in graduate studies, University of Southern California and California State University Northridge, and has been a guest lecturer at California State University Northridge and the California State University Los Angeles. *Evolution and Reason* is a synthesis of Boberg's lifetime scholastic endeavor.

In her first professional position, as librarian of the Nebraska Legislative Counsel, Boberg expanded her enjoyment of books and libraries, and initiated her experience in civic involvement, which has continued through her life.

Her role as a pioneer in interdisciplinary evolution theory is not a singular pioneering role. She is esteemed for her thorough research and her effective advocacy in futurist environmental issues and has been honored for her visionary action in civic improvement, environmental preservation and global concerns. In recognition of these contributions, the librarians of California State University Northridge have established the Dorothy Kurth Boberg archive collection.

In 1983, Boberg presented a paper on her concept **complementarity dynamics** to a Universal Peace Conference in India. That favorable reception of her intriguing and innovative ideas inspired her to write *Evolution and Reason — Beyond Darwin*.

Her perception of the **complementarity dynamics** of competitive and cooperative processes in nature and society is a foundation of this text.

Boberg has given presentations of her evolution theories to groups of science professors of universities in Oregon (Portland State, Hood and University of Oregon) and University of Washington, Seattle. She is a member of the California Academy of Sciences, the American Association for the Advancement of Science, the American Association of University Women and the Society for the Study of Evolution.